ANIMAL CELL TECHNOLOGY: BASIC & APPLIED ASPECTS

JAACT'97 NAGOYA

JAACT'97 Organizing Committee

Animal Cell Technology: Basic & Applied Aspects

Volume 10

Proceedings of the Tenth Annual Meeting of the
Japanese Association for Animal Cell Technology,
Nagoya, Japan, November 5–8, 1997

Edited by

Y. KITAGAWA
Bioscience Center, Nagoya University,
Nagoya, Japan

T. MATSUDA
Department of Applied Biological Sciences,
Graduate School of Agricultural Sciences,
Nagoya University, Nagoya, Japan

and

S IIJIMA
Department of Biotechnology,
Graduate School of Engineering,
Nagoya University, Nagoya, Japan

SPRINGER-SCIENCE+BUSINESS MEDIA, B.V.

Library of Congress Cataloging-in-Publication Data

Japanese Association for Animal Cell Technology. Meeting (10th : 1997 : Nagoya-shi, Japan)
Animal cell technology : proceedings of the Tenth Annual Meeting of the Japanese Association for Animal Cell Technology, Nagoya, Japan, November 5-8, 1997 / edited by Y. Kitagawa, T. Matsuda, and S. Iijima.
p. cm.
"Volume 10"
Includes index.
ISBN 978-94-017-4332-7 ISBN 978-0-306-46865-0 (eBook)
DOI 10.1007/978-0-306-46865-0
1. Animal cell biotechnology--Congresses. I. Kitagawa, Yasuo. II. Matsuda, Tatsurō. III. Iijima, S. (Shinji), 1952- . IV. Title.
TP248.27.A53J36 1997
660.6--dc21 98-45569

ISBN 978-94-017-4332-7

Printed on acid-free paper

Contents

Symposium 4

Immunological Manipulation

Symposium 5

DT40 Cells as a System of Homologous Gene Recombination

Symposium 6

Transgenic Livestock

Symposium 7

Transgenic Fish: Standardization of Technology in the Model System and Aquaculture

Symposium 8

Growth, Immortality and Death of Cells in Culture: Present and Future Impact for the Development of Health Care Products

Symposium 9

Gene Transfer to Cells and Tissues of Living Animals: Application to Production and Therapeutics

Oral Session

Cell Culture for Assessment of Physiologically Functional Substances

Expression of Recombinant Animal Genes in Cultured Animal Cells

Cell Culture Engineering

Cost Performance for Producing Biochemicals

Establishment of Functional Cell Lines for Production of Biologicals

Application of Human Monoclonal Antibody

Cell Characterization and Validation

Poster Session

Cell Culture Engineering

Expression of Recombinant Animal Genes in Cultured Animal Cells

Cell Characterization and Validation

Transplantation and Artificial Organs

Transgenic Technology

Antibody Technology

Cytokines and Signal Transduction

Cell Culture for Assessment of Physiologically Functional Substances

Preface

Animal cell technology is a newly growing discipline of cell biology which aims not only to understand structure, function and behavior of differentiated animal cells but also to uncover their ability useful for industrial and medical purpose. The goal of animal cell technology includes clonal expansion of differentiated cells with useful ability, optimization of their culturing in industrial scale, modulation of their ability for production of pharmaceutical proteins and monoclonal antibodies. When we regard the genetic information expressed during differentiation of fertilized eggs into over 200 different cell types of adult animal bodies as our natural resources, there is vast region to be explored by animal cell technologist. Most of the genetic information is expressed during limited period of animal development, and we might have many opportunity of utilizing such genetic information by cell technology. The subjects of animal cell technology are physical, physiological and genetical modulation of animal cells for full expression of their activity of our interest.

The last nine Annual Meetings of the Japanese Association for Animal Cell Technology (JAACT) had attracted increasing number of participants. At the Tenth Meeting (JAACT'97) held in Nagoya from November 5 through 8, 1997, we had more than 60 participants from outside of Japan and 140 from Japan. We have here completed the Meeting Proceedings which contains about 60 articles. We hope that this book will help worldwide researchers to envisage the present status and future trends in animal cell technology.

The editors express their sincere gratitude to all researchers who joined the meeting, to the organizers of the Symposium Sessions, to members of the organizing committee who dedicated themselves in assuring the Meeting's success, and the graduate students of Nagoya University who supported management of the Meeting. We also thank to Nagoya Convention & Visitors Bureau and Nagoya University Foundation for the financial support.

The Editors

THE FUNCTION OF DALLY, A DROSOPHILA GLYPICAN, IN DPP SIGNALING

HIROSHI NAKATO[1], SUSUMU IZUMI[1], SCOTT B. SELLECK[2]
[1]*Department of Biology, Tokyo Metropolitan University, Tokyo 192-03, Japan*
[2]*Department of Molecular and Cellular Biology, University of Arizona, Tucson AZ 85721, U.S.A.*

1. Introduction

division abnormally delayed (*dally*) was identified in a screen for genes affecting cell division patterning in the developing visual system of the fruitfly, *Drosophila melanogaster* [1]. *dally* mutants show defects in G2-M progression for specific sets of dividing cells in the larval central nervous system and eye disc. Mutations in *dally* also affect the morphogenesis of several adult tissues including the antenna, wing, and genitalia. Molecular cloning and sequencing of a *dally* cDNA revealed extensive homology to glypicans, a family of integral-membrane proteoglycans. Glypicans bear heparan sulfate chains and are anchored to the cell membrane via a glycosylphosphatidylinositol linkage [2].

The heparan sulfate modified integral membrane proteoglycans have been implicated as co-receptors required for growth factor signaling [3]. Studies using cell culture systems have shown that these molecules can affect signaling mediated by FGF [4, 5], TGF-β [6], HGF [7], heparin-binding EGF [8], and Wingless [9]. We show here that *dally* is required for normal function of Decapentaplegic (Dpp), a *Drosophila* homologue of the TGF-β/Bone Morphogenetic Proteins (BMPs), during post-embryonic development.

2. Genetic Interactions Between the *dally* and *dpp* Genes

To determine if *dally* affects events directed by Dpp, we examined genetic interactions between *dally* and *dpp*. Reducing *dpp* function increases the severity and penetrance of *dally* phenotypes. For example, flies heterozygous for *dpp* and *dally* (Fig. 1B) show phenotypes never observed in animals heterozygous for either *dpp* or *dally* alone. In *dally* homozygotes, heterozygosity for a *dpp* mutation enhances the eye phenotype (Fig. 1D). In all combinations of *dally* and *dpp* mutations we have tested (two *dally* alleles and five *dpp* alleles), reductions in *dpp* gene function increase both expressivity and penetrance of *dally* eye, antenna, and genitalia phenotypes. Consistent with these results, *dally* phenotypes in these tissues were rescued by increasing the normal gene dosage of *dpp* (data not shown, see ref [10]). The genetic interaction

Y. Kitagawa et al. (eds.), Animal Cell Technology: Basic & Applied Aspects, Vol. 10, 1–5.

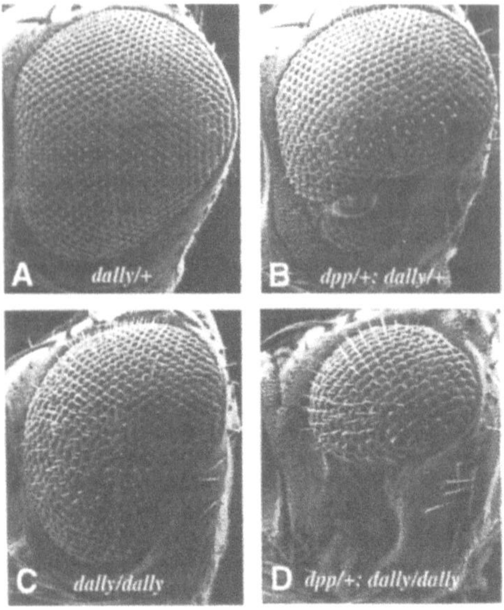

Figure 1. *dpp* mutation enhances *dally* eye phenotypes. Scanning electron micrographs of adult eyes are shown for $dally^{P2}/+$ (A), $dpp^{blk}/+$; $dally^{P2}/+$ (B), $dally^{P2}/dally^{P2}$ (C), and $dpp^{blk}/+$; $dally^{P2}/dally^{P2}$ (D).

between *dally* and *dpp* genes strongly suggests that these two genes function in the same signaling pathway during development.

3. *dally* Mutations Can Suppress Phenotypes Produced by Ectopic Expression of Dpp

If *dally* affects cellular responses to Dpp, reducing *dally* function should alter phenotypes produced by ectopic expression of Dpp. We therefore tested if *dally* mutations can suppress the defects produced by *Moonrat* (*Mrt*), a dominant allele of *hedgehog* (*hh*), that shows ectopic expression of *hh* and subsequently *dpp*, in the anterior compartment of wing disc (Fig. 2H, see ref [11]). Ectopic Dpp expression in *Mrt* wing discs causes consequent overgrowth along the wing margin of adult wing blades (Fig. 2B). Heterozygosity of a *dally* hypomorphic allele, $dally^{P2}$, partially rescued *Mrt* wing phenotype (Fig. 2C). Complete suppression was observed when flies were made homozygous for $dally^{P2}$ (Fig. 2D). Similar suppression was observed for another severe *dally* allele, $dally^{\Delta P\text{-}527}$ (Fig. 2E), while $dally^{P1}$, a mild *dally* allele, failed to rescue the *Mrt* phenotype (Fig. 2F). Ectopic expression of *dpp* still remains in *Mrt*, *dally/dally* wing discs (Fig. 2I), suggesting that *dally* mutation can affect Dpp signaling without affecting expression of *dpp* itself.

The yeast GAL4/UAS system is a useful tool to express a specific gene of interest in a particular set of cells [12]. In combination with a GAL4 enhancer trap line, *A9*, *UAS-dpp*$^+$ transgene ectopically expresses Dpp in the wing pouch region of wing discs, resulting in extra wing vein structures in the adult wing (Fig. 2J). This

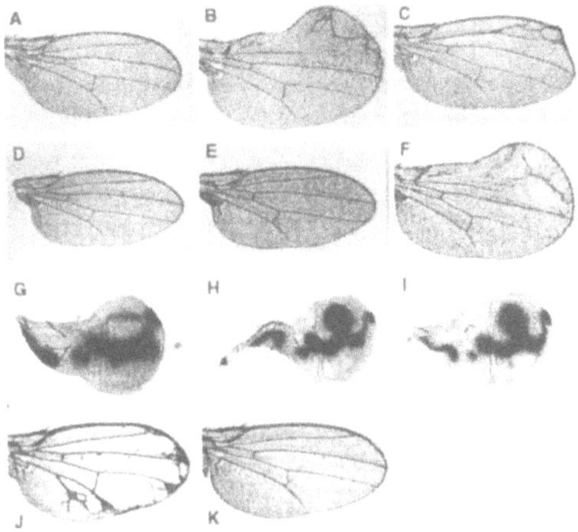

Figure 2. Effects of *dally* mutations on defects produced by ectopic *dpp* expression. Adult wings were shown for *wild-type* (A), *Mrt/+* (B), *Mrt/dally*^{P2} (C), *Mrt, dally*^{P2}*/dally*^{P2} (D), *Mrt, dally*$^{\Delta P\text{-}527}$*/dally*$^{\Delta P\text{-}527}$ (E), *Mrt, dally*^{P1}*/dally*^{P1} (F), *GAL4 (A9); UAS-dpp/+* (J), and *GAL4 (A9); UAS-dpp/dally*$^{\Delta P\text{-}527}$ (K). *dpp* expression was followed by a *dpp-lacZ* reporter in wing discs from *wild-type* (G), *Mrt/+* (H), and *Mrt, dally*$^{\Delta P\text{-}527}$*/dally*$^{\Delta P\text{-}527}$ (I).

abnormality was rescued by *dally* mutation (Fig. 2K). The experiments show that *dally* can affect cellular responses to Dpp.

4. Expression of Dpp Target Genes Were Reduced in *dally* Mutants

If *dally* influences *dpp* signaling, the activation of *dpp* target genes should be reduced in *dally* mutants. Hence, we examined the expression of *optomotor-blind* (*omb*) and *spalt* (*sal*), two downstream targets of *dpp* signaling [13, 14]. *omb* expression is significantly reduced in *dally* eye-antennal discs (Fig. 3B), while expression of *dpp* in *dally* mutants is only modestly affected (Fig. 3D). We have also observed decreases in *sal* expression in imaginal discs of *dally* homozygotes (data not shown). These findings support the conclusion that *dally* is a component of the Dpp signaling system.

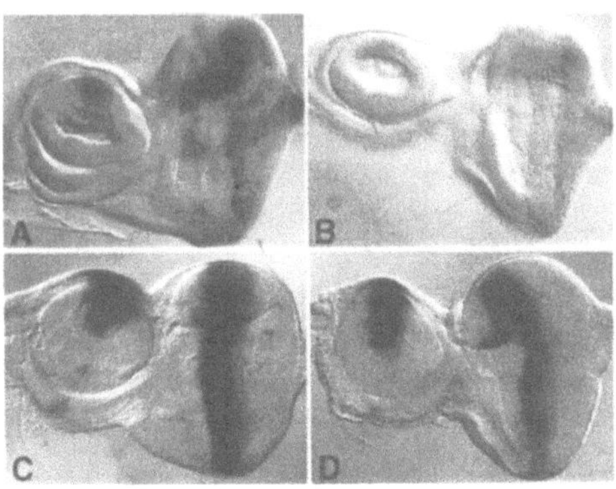

Figure 3. Expression of *optomoter-blind* (*omb*) in eye-antennal discs of *dally* mutants. Expression of *omb* was detected using an enhancer-trap line, *omb*^{P1}, in eye-antennal discs of *wild-type* (A) and *dally*$^{\Delta P\text{-}527}$ homozygous backgrounds (B). *dpp* expression is shown for *wild-type* (C) and *dally*$^{\Delta P\text{-}527}$*/dally*$^{\Delta P\text{-}527}$ (D).

5. Dally Can Modulate the Patterning Activity of Dpp

Dpp acts as morphogen, directing different cellular responses at different extracellular concentrations. To determine if *dally* can modulate the patterning activity of Dpp, we have expressed *dally*$^+$ ectopically in the presence of ectopic *dpp*$^+$ in wing discs. In *Mrt* mutants, where *dpp* expression is induced at the future wing margin region in the anterior compartment, ectopic *dally*$^+$ function converted wing overgrowth to duplications of wing structures (Fig. 4B and C). The result indicates that *dally* can alter the patterning activity of Dpp.

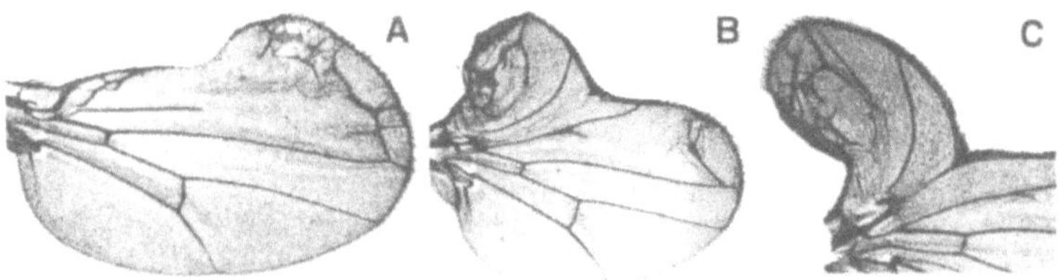

Figure 4. Increased expression of *dally* enhances *Mrt* and *Pka*^{DN} wing phenotypes. Adult wings of Mrt in the absence (A) or presence (B and C) of *dally* misexpression. *hs-dally* transgene converts the *Mrt* wing overgrowth to wing duplications. The high magnification view shows the mirror-image symmetry of the duplicated structures (C).

6. Discussion

The availability of *dally* mutants provide a unique opportunity to study the role of cell surface proteoglycans in growth factor signaling in vivo. In this study, we demonstrated that, 1) reductions in *dpp* function enhance *dally* eye, antenna, and genitalia phenotypes, and increasing the dosage of *dpp* rescues these phenotypes, 2) *dally* mutations can suppress phenotypes produced by ectopic expression of *dpp*$^+$, 3) *dally* mutations reduce expression levels of Dpp-target genes, and 4) ectopic *dally*$^+$ expression can alter the patterning activity of Dpp. These findings are consistent with Dally serving as a co-receptor of Dpp, modulating its activity. A link between these two genes is further supported by the observation that *dpp* and *dally* affect the cell cycle progression of the same set of dividing cells in developing eye [15].

Recently, mutations in a human glypican gene, GPC3, were shown to be responsible for the human overgrowth syndrome, Simpson-Golabi-Behmel Syndrome (SGBS) [16]. SGBS patients show pre- and postnatal overgrowth of somatic tissues, congenital malformations, and a predisposition to tumors. The genetic interactions we observe between a *Drosophila* glypican and a member of TGF-β superfamily raise the possibility that SGBS patients have defects in TGF-β signaling. Although the molecular mechanisms by which glypicans affects growth factor signaling remains to be elucidated, it is clear that this class of integral-membrane proteoglycans plays a critical role in cell division and differentiation in a variety of animal species.

7. References

[1] Nakato, H., Futch, T.A., and Selleck, S.B. (1995) Development 121, 3687-3702.
[2] David, G., Lories, V., Decock, B., Marynen, P., Cassiman, J.-J., and Van den Berghe, H. (1990) J. Cell Biol. 111, 3165-3176.
[3] David, G. (1993) FASEB J. 7, 1023-1030.
[4] Rapraeger, A.C., Krufka, A., and Olwin, B.B. (1991) Science 252, 1705-1708.
[5] Olwin, B.B., and Rapraeger, A. (1992) J. Cell Biol. 118, 631-639.
[6] Lopez-Casillas, F., Wrana, J. L., and Massague, J. (1993) Cell 73, 1435-1444.
[7] Zioncheck, T.F., Richardson, L., Liu, J., Chang, L., King, K.L., Bennett, G.L., Fugedi, P., Chamow, S.M., Schwall, R.H., and Stack, R.J. (1995) J. Biol. Chem. 270, 16871-16878.
[8] Aviezer, D., and Yayon, A. (1994) Proc. Natl. Acad. Sci. (USA) 91, 12173-12177.
[9] Reichsman, F., Smith, L., and Cumberledge, S. (1996) J. Cell Biol. 135, 819-827.
[10] Jackson, S.M., Nakato, H., Sugiura, M., Jannuzi, A., Oakes, R., Kaluza, V., Golden, C. and Selleck, S.B. (1997) Development 124, 4113-4120.
[11] Tabata, T. and Kornberg, T.B. (1994) Cell 76, 89-102.
[12] Brand, A. H., and Perrimon, N. (1993) Development 118, 401-415.
[13] Grimm, S., and Pflugfelder, G.O. (1996) Science 271, 1601-1604.
[14] Kuhnlein, R.P., Frommer, G., Friedrich, M., Gonzalez-Gaitan, M., Weber, A., Wagner-Bernholz, J.F., Gehring, W.J., Jackle, H., and Schuh, R. (1994) EMBO J. 13, 168-179.
[15] Penton, A., Selleck, S.B., and Hoffmann, F.M. (1997) Science 275, 203-206.
[16] Pilia, G., Hughes-Benzie, R.M., MacKenzie, A., Baybayan, P., Chen, E.Y., Huber, R., Neri, G., Cao, A., Forabosco, A., and Schlessinger, D. (1996) Nature Genet. 12, 241-247.

NOVEL FORMS OF PROTEIN O-GLYCOSYLATION

R. S. HALTIWANGER

State University of New York at Stony Brook Stony Brook, N Y

1 1794-5215 USA

1. Introduction

The past decade has seen rapid progress in the field of Glycobiology. Increasingly, the biological implications of modifying proteins with specific carbohydrate moieties is being unraveled. In prior years our understanding of the function of the oligosaccharide chains on proteins was severely hampered by the tremendous diversity of carbohydrate structures and the difficulty in obtaining sufficient quantities for analysis. Much of the recent progress in this field has been fueled by the development of highly sensitive techniques capable of determining the precise structure of the oligosaccharide modifications found on extremely small amounts of biologically interesting proteins. In combination with the power of molecular biology, the next decade promises to bring tremendous growth in our knowledge of the structure and function of glycoproteins.

2. Classical and novel forms of protein glycosylation

While the oligosaccharide modifications found on proteins in mammalian systems are amazingly diverse in structure, they can be grouped into a few simple classes. Grossly,

Y. Kitagawa et al. (eds.), Animal Cell Technology: Basic & Applied Aspects, Vol. 10, 7–14.

glycoproteins are divided into two groups based on the atom involved in the linkage between the sugar and the protein. Thus, N-linked oligosaccharides (N-glycans) are linked to protein through the amide nitrogen of asparagine side chains, whereas O-linked oligosaccharides (O-glycans) are linked through the hydroxyl oxygen of either serine or threonine. N-glycans all share a common core structure but can be further subdivided into high mannose, complex, and hybrid-type structures depending on the modifications found on the outer portions of the oligosaccharide chains (1). O-glycans fall into two major classes. Classical, or mucin-type, O-glycans consist of an N-acetylgalactosamine O-linked to serine/threonine. The GalNAc can be elongated to form a large variety of structures typically terminating with sialic acids (2). The other major type of O-glycans are the glycosaminoglycans which are O-linked to proteins through xylose residues (3).

Recently several novel forms of O-glycosylation have been demonstrated to exist on proteins in mammals (Fig. 1). Each of these modifications occurs on serine and/or threonine residues and has been demonstrated to exist on a distinct subset of proteins. For instance, O-linked N-acetylglucosamine (O-GlcNAc) is found exclusively on nuclear and cytoplasmic proteins (4-6). The other two modifications (O-linked fucose and O-linked glucose) both occur within epidermal growth factor (EGF)-like modules of secreted or cell surface proteins (7). While O-GlcNAc appears only as a monosaccharide, both O-fucose and O-glucose exist as larger structures (Fig. 1). In addition, both O-fucose and O-glucose modifications appear to occur at strict consensus sequences (7). Several serum proteins have been identified with these modifications, and recently the peptide: O-fucosyltransferase responsible for the addition of the fucose to the proteins has been purified and characterized (8,9). We have recently identified the β1,3 glucosyltransferase responsible for elongation of O-fucose (10). Nothing is known about the enzyme responsible for the addition of the O-glucose proteins.

3. Modification of nuclear and cytoplasmic proteins with O-GlcNAc

A great deal of work has been done on the O-GlcNAc modification over the past decade (4-6). O-GlcNAc modifies numerous proteins in eukaryotic cells and appears to be a major form of post-translational modification. Proteins modified with O-GlcNAc include RNA polymerase II, several RNA polymerase II transcription factors, and many others (6). Sites of glycosylation have been identified on several of these proteins (6). While there is no absolute consensus sequence which can be discerned, most of the sites have a proline residue within a few amino acids of the modified serine or threonine. Enzymes capable of addition and removal of O-GlcNAc have been purified and characterized. A UDP-GlcNAc: polypeptide N-acetylglucosaminyltransferase (O-GlcNAc transferase) appears to be responsible for the addition of O-GlcNAc to most proteins (11-14). It is present and active in most eukaryotic cells which have been examined (13,14). Similarly, a recently purified O-GlcNAc-specific β-N-acetylglucosaminidase (O-GlcNAc'ase) is believed to be the enzyme responsible for removal of O-GlcNAc from proteins (15). Both of these enzymes exist as large multisubunit soluble proteins in the cytoplasm of cells.

Recent work has suggested that O-GlcNAc may serve as a regulated modification much like phosphorylation. The O-GlcNAc transferase and O-GlcNAc'ase are capable of competitive addition and removal of the sugar from proteins, forming a cycle analogous to a kinase/phosphatase cycle (Fig. 2) (5,6). Alterations in the activity of either of these enzymes, or in the intracellular levels of UDP-GlcNAc, could result in changes in O-GlcNAc levels on proteins. In addition, several recent reports have provided evidence that O-GlcNAc competes with phosphate for modification of certain proteins (4-6), indicating that O-GlcNAc modification may add an extra layer of control to signal transduction pathways.

4. Inhibition of O-GlcNAc removal with the O-GlcNAc'ase inhibitor PUGNAc

Recently we have begun to examine inhibitors of the O-GlcNAc'ase and O-GlcNAc transferase as potential tools for analyzing O-GlcNAc function within cells. Analogous studies using inhibitors of kinases or phosphatases have been quite successful over the years in the study of phosphorylation. All of the known inhibitors of the O-GlcNAc transferase are highly charged and unlikely to easily enter cells (11, 12). In contrast, several of the known inhibitors of the O-GlcNAc'ase are uncharged and have hydrophobic groups which could allow the inhibitor to cross the plasma membrane (15). The best of these in vitro O-GlcNAc'ase inhibitors is PUGNAc (Fig. 3). We have examined whether this inhibitor can be used to alter O-GlcNAc levels on proteins within tissue culture cells. Using the human colon cancer cell line, HT 29, we have demonstrated that PUGNAc causes a time and concentration dependent increase in O-GlcNAc levels on numerous proteins (16). The effect is reversible by simply removing the inhibitor from the media. The extent of change was somewhat protein dependent, with some proteins showing as much as a 4-fold increase in O-GlcNAc levels.

We have used PUGNAc to test the hypothesis that O-GlcNAc and phosphate compete for individual sites on proteins (16). The transcription factor, Sp1, is known to be modified with both O-GlcNAc and phosphate (17,18). Treatment of HT 29 cells with PUGNAc causes approximately a four-fold increase in the level of glycosylation of Sp1. Interestingly, PUGNAc also causes approximately a 50% decrease in the level of Sp1 phosphorylation, indicating that there is at least partial competition between glycosylation and phosphorylation sites on Sp1. Surprisingly, in spite of the effects seen on both the glycosylation and phosphorylation states of numerous proteins, there was essentially no growth effect of PUGNAc on any of the cells tested.

5. Conclusions

In summary, O-GlcNAc is a major form of post-translational modification on nuclear and cytoplasmic proteins of eukaryotic cells. Several lines of evidence suggest that O-GlcNAc is a regulated modification, much like phosphorylation. We have tested a potent in vitro inhibitor of the O-GlcNAc'ase on cells in culture and demonstrated that we can significantly increase the level of glycosylation on numerous proteins. Interestingly, the PUGNAc-induced increase in the O-GlcNAc level on the transcription factor Sp1 is accompanied by a complementary decrease in phosphorylation. These data support the hypothesis that O-GlcNAc and phosphorylation may compete with each other for sites on proteins. Future studies using inhibitors such as PUGNAc should help us to understand the functional implications of O-GlcNAc modification on proteins.

6. Acknowledgments

Original work was supported by NIH GM 48666. RSH is the recipient of an American Cancer Society Junior Faculty Research Award.

7. References

1. Cummings, R.D.: Synthesis of asparagine-linked oligosaccharides: pathways, genetics, and metabolic regulation. In *Glycoconjugates: Composition, Structure, and Function.* Allen, H.J. and Kisailus, E.C. editors. Marcel Dekker, Inc., New York 333-360, 1992
2. Schachter, H. and Brockhausen, I.: The biosynthesis of serine (threonine)-N-acetylgalactosamine-linked carbohydrate moieties. In *Glycoconjugates. Composition, Structure, and Function.* Allen, H.J. and Kisailus, E.C. editors. Marcel Dekker, Inc., New York. 263-332, 1992
3. Bhavanandan, V.P. and Davidson, E.A.: Proteoglycans: Structure, synthesis, function. In *Glycoconjugates: Composition, Structure, and Function.* H.J. Allen and E.C. Kisailus, editors. Marcel Dekker, Inc., New York. 167-202, 1992
4. Hart, G W.: Dynamic O-linked glycosylation of nuclear and cytoskeletal proteins. *Annu.Rev. Biochem.* **66** (1997)315-335.
5. Hart, G. W., Kreppel, L. K., Comer, F. I., Arnold, C. S., Snow, D. M., Ye, Z., Cheng, X., DellaManna, D., Caine, D. S., Earles, B. J., Akimoto, Y., Cole, R. N., and Hayes, B. K.: O-GlcNAcylation of key

nuclear and cytoskeletal proteins: reciprocity with O-phosphorylation and putative roles in protein multimerization. *Glycobiol.* **6** (1996) 711-716.

6. Haltiwanger, R. S., Busby, S., Grove K., S. Li, Mason, D., Medina, L., Moloney, D. J., Philipsberg, G. A. and Scartozzi, R.: O-Glycosylation of nuclear and cytoplasmic proteins: Regulation analogous to phosphorylation? *Biochem. Biophys. Res. Commun.* **231** (1997) 237-242.
7. Harris, R. J. and Spellman, M. W.: O-linked fucose and other post-translational modifications unique to EGF modules. *Glycobiol.* **3** (1993) 219-224.
8. Wang, Y., Lee, G. F., Kelley, R. F. and Spellman, M. W.: Identification of a GDP-L-fucose: polypeptide fucosyltransferase and enzymatic addition of O-linked fucose to EGF domains. *Glycobiol.* **6** (1996) 837-842.
9. Wang, Y. and Spellman, M. W. Purification and characterization of a GDP-fucose: polypeptide fucosyltransferase from Chinese hamster ovary cells. *J. Biol. Chem.* **273** (1998) 8112-8118.
10. Moloney, D. J. and Haltiwanger, R. S.: The O-linked fucose glycosylation pathway: Identification and characterization of a UDP-glucose: O-fucose β 1 ,3-glucosyltransferase. *submitted*, 1998
11. Haltiwanger, R. S., Holt, G. D. and Hart, G. W.: Enzymatic addition of O-GlcNAc to nuclear and cytoplasmic proteins. Identification of a uridine diphospho-N-acetylglucosamine:peptide β-N-acetylglucosaminyltransferase. *J. Biol. Chem.* **265** (1990) 2563-2568.
12. Haltiwanger, R. S., Blomberg, M. A. and Hart, G. W.: Glycosylation of nuclear and cytoplasmic proteins. Purification and characterization of a uridine diphospho-N-acetylglucosamine: polypeptide β-N-acetylglucosaminyltransferase. *J. Biol. Chem.* **267** (1992) 9005-9013.
13. Kreppel, L. K., Blomberg, M. A. and Hart, G. W.: Dynamic glycosylation of nuclear and cytosolic proteins- Cloning and characterization of a unique O-GlcNAc transferase with multiple tetratricopeptide repeats. *J. Biol. Chem.* **272** (1997) 9308-9315.
14. Lubas, W. A., Frank, D. W., Krause, M. and Hanover, J. A.: O-linked GlcNAc transferase is a conserved nucleocytoplasmic protein containing tetratricopeptide repeats. *J. Biol. Chem.* **272** (1997) 9316-9324.
15. Dong, L. -Y. D. and Hart, G. W.: Purification and characterization of an O-GlcNAc selective N-acetyl-β-D-glucosaminidase from rat spleen cytosol. *J. Biol. Chem.* **269** (1994) 19321-19330.
16. Haltiwanger, R. S., Grove, K. and Philipsberg, G. A.: Modulation of O-GlcNAc levels on nuclear and cytoplasmic proteins in vivo using the peptide O-GlcNAc-β-N-acetylglucosaminidase inhibitor O-(-acetamido-2-deoxy-D-glucopyranosylidene) amino-N-phenylcarbamate. *J. Biol. Chem.* **273** (1998) 3611-3617.
17. Jackson, S. P. and Tjian, R.: O-Glycosylation of eukaryotic transcription factors: Implications for mechanisms of transcriptional regulation. *Cell* **55** (1998) 125-133.
18. Jackson, S. P., MacDonald, J. J., Lees-Miller, S. and Tjian, R.: GC box binding induces phosphorylation of Spl by a DNA-dependent protein kinase. *Cell* **63** (1990) 155-165.

<u>O-linked N-acetylglucosamine (nuclear and cytoplasmic proteins)</u>

GlcNAcβ1->O-Ser/Thr

<u>O-linked fucose (EGF-like modules)</u>

Fucα1->O-Ser/Thr

Glcβ1,3Fucα1->O-Ser/Thr

Siaα2,6Galβ1,4GlcNAcβ1,3Fucα1->O-Ser/Thr

<u>O-linked glucose (EGF-like modules)</u>

Xylα1,3Glcβ1->O-Ser

Xylα1,3Xylα1,3Glcβ1->O-Ser

Fig. 1: Novel forms of protein O-glycosylation. Known structures derived from three novel forms of O-glycosylation. For reviews on each of these modifications: O-linked GlcNAc (4-6), O-linked glucose and fucose (7).

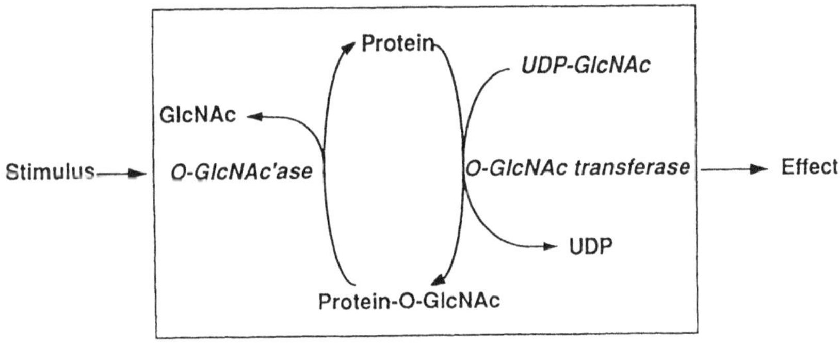

Fig. 2: The cyclic nature of the O-linked GlcNAc modification. O-GlcNAc levels on proteins can be changed by alterations in the activity of the O-GlcNAc transferase, O-GlcNAc'ase, or in the intracellular levels of UDP-GlcNAc.

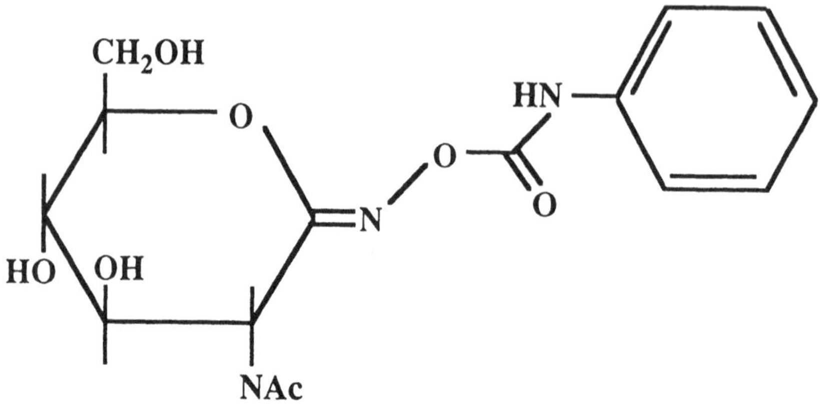

Fig. 3: PUGNAc (O-(2-acetamid O-2-deoxy-D-glucopyranosylidene) amino-N-phenylcarbamate). PUGNAc has been demonstrated to be a potent in vitro inhibitor of the O-GlcNAc'ase (15,16).

SYNTHESIS OF GLYCOCONJUGATES : THEIR USE TO STUDY THE INTRACELLULAR TRAFFIC OF ENDOGENOUS LECTINS AND TO DELIVER OLIGONUCLEOTIDES AND GENES

M. MONSIGNY, C. QUETARD, S. BOURGERIE, V. CARRIERE, E. DUVERGER, V. ALTEMAYER, C. PICHON, R. MAYER, P. MIDOUX, and A.C. ROCHE
Glycobiologie, Centre de Biophysique Moléculaire Centre National de la Recherche Scientifique, Rue Charles-Sadron, 45071 - Orléans cedex 02 - France

1. Introduction

The use of glycoconjugates - to characterize endogenous lectins, to study their intracellular localization and motion and to help drugs, oligonucleotides and genes to be selectively taken up by animal cells - is based on the discovery by G. Ashwell, A. Morell and their coworkers at the end of the seventies that animal cells express sugar-binding proteins (lectins) on their surface and that these lectins very efficiently induce the endocytosis of their ligands (see Ashwell and Harford, 1982 for a review).

Since, lectins have been found to be expressed on a very large number of normal as well as transformed animal cells. The present paper describes the preparation and some properties of synthetic glycoconjugates (neoglycoproteins, glycosylated poly- lysines and new types of derivatized oligosaccharides), their use to discover new intracellular lectins and to study their role, and finally their use to increase the activity of antisense oligonucleotides and the efficiency of animal cell transfection.

2. Preparation of Glycoconjugates

2.1. NEOGLYCOPROTEINS

Neoglycoproteins were developed independently by Lee and coworkers (1976) and ourselves (Privat *et al.*, 1974) as tools to study the properties of lectins. In both cases, neoglycoproteins were made by adding activated sugars on serum albumin. About half the lysine residues of bovine serum albumin are easily substituted upon adding glycosylphenylisothiocyanate (Roche *et al.*, 1983, Monsigny *et al.*, 1984) : neoglycoproteins are then made fluorescent by adding fluoresceinylisothiocyanate.

Fluorescein labeled neoglycoproteins contain usually about 23 ± 3 sugars and 2.5 ± 0.5 fluorescein residues : for instance, if the sugar is mannose : Os_{23}-, $Flu_{2.5}$-BSA ; where Os stands for α-mannopyranosylphenyl-thiocarbamyle and Flu for fluoresceinyl-

Y. Kitagawa et al. (eds.), Animal Cell Technology: Basic & Applied Aspects, Vol. 10, 15–21.

thiocarbamyle residues, respectively. The apparent affinity of neoglycoproteins for lectins is much higher (more than 1000) than the corresponding free sugar, for instance 500 nM instead of 2 mM.

2.2. GLYCOSYLATED POLYLYSINE AND DERIVATIVES

Poly-L-lysine (about 190 residues) was made free of halogenure by adsorption on an anion exchange resin and transformed into a *para*-toluene sulfonate salt (Derrien *et al.*, 1989).

This poly-L-lysine, *para*- toluene sulfonate is soluble in organic solvent (such as *N,N*-dimethyformamide) allowing a very efficient substitution by either glycosylphenyl-isothiocyanate or fluoresceinylisothiocyanate or both. For the purpose of an efficient gene transfer, a glycosylated polylysine (dp : 190) should content about 60 sugar residues (Midoux *etal.*, 1993) and for that carrying oligonucleotides the poly-lysine (dp : 455) should content about 100 sugar residues (Stewart *et al.*, 1996).

Recently, another modified polylysine was prepared for the purpose of gene transfection ; it (dp: 190) contains about 50 gluconoyl residues, and 15 glycosyl residues ; the gluconoyl residues are used to increase the solubility of the polymer/plasmid complexes and to decrease the strength of these complexes (Erbacher *et al.*, 1997).

2.3. OLIGOSACCHARYL PYROGLUTAMYL DERIVATIVES

Neoglycoproteins and glycosylated polymers must contain several tens of simple sugars to be efficiently recognized by lectins, while glycoconjugates containing a single complex oligosaccharide tightly bind on lectins. Indeed, as shown by Y. C. Lee in the case of the asialoglycoprotein receptor of parenchymal cells, the affinity of an oligosaccharide containing a single galactose residue is about 10^3 L $\times$ mole^{-1}, with 2 galactose residues, about 10^6 L $\times$ mole^{-1} and with 3 or 4 galactose residues about 10^9 L $\times$ mole^{-1} (Lee *etal.*, 1983). Therefore, a single triantennary oligosaccharide will allow a protein or a polymer to be efficiently recognized by this lectin. In addition, such glycoconjugates containing a complex oligosaccharide are more specific than the glycoconjugates containing tens of simple sugars. Oligosaccharides isolated from living organisms, or released by hydrolysis from glycoconjugates may be easily transformed into glycosynthons which can be used to prepare fluorescent oligosaccharides or glycoconjugates with an excellent yield (Sdiqui *et al.*, 1995 ; Quétard *et al.*, 1997, 1998 ; see Monsigny *et al.*, 1998 for a review).

2.3.1. *Glycosynthons*

Reducing oligosaccharides readily react with the amino group of a glutamyl derivative (glutamyl-*p*-nitroanilide, or glutamyl-amido-ethyl-dithio-2-pyridine) leading to a β-glycosylamine. Then, by activating the carboxylic group of the glutamyl side chain with BOP (benzotriazole-1-yl-oxy-tris-(dimethylamino)-phosphonium hexafluorophosphate), the conjugate is quantitatively transformed in a β-glycosyl-pyroglutamyl derivative. These glycosynthons are quite stable and the sugar bound to the pyroglutamyl nitrogen is a pyranoside in a β anomeric configuration.

2.3.2. *Fluorescent oligosaccharide derivatives*
β-oligosaccharyl-pyroglutamyl-amido-ethyl-dithio-2-pyridine was quantitatively reduced by one equivalent of tris(carboxyethyl)phosphine, and then transformed into a fluorescent conjugate by reacting with a slight excess of iodoacetamido-fluorescein or another iodoacetamide derivative of a fluorophore. The conjugate β-oligosaccharyl-pyroglutamyl-amido-ethyl-thio-acetamido-fluorescein, for instance, is readily purified by gel filtration and characterized by nuclear magnetic resonance and by mass spectrometry (Quétard *et al.*, 1997, 1998).

3. Characterization and Putative Function of New Intracellular Lectins

3.1. USE OF FLUORESCENT NEOGLYCOPROTEINS AND FLUORESCENT OLIGO-SACCHARIDES TO LOCALIZE INTRACELLULAR BINDING PROTEINS

Precursors of macrophages do not express on their surface the mannose specific lectin present in macrophages (Shepherd *et al.*, 1982). However, upon permeabilization with saponine, the cytoplasm of these cells is labeled by fluorescent neoglycoproteins bearing α-mannopyranosyl residues (Pimpaneau *et al.*, 1991). The labeling occurs through out the cytoplasm at low level and at a high level in area close to the nucleus, which may correspond to the intermediate compartment (see 3.2). Conversely, with fluoresceinylated oligomannosides, the cytoplasm of permeabilized cells is not labeled but, depending on the number of mannose present in the oligosaccharides (from 5 to 9), the number and the size of the labeled area is different (unpublished data). Therefore, fluorescent oligosaccharides are clearly more specific than the neoglycoproteins made with simple sugars.

3.2. MR 60 : A MANNOSIDE-SPECIFIC LECTIN SHUTTLING BETWEEN ENDOPLASMIC RETICULUM AND GOLGI-APPARATUS

On the basis of the labeling obtained with the neoglycoprotein containing mannose, a mannose-binding protein was isolated from intracellular membranes of the human line : HL60 cells ; this protein called MR 60 was partially sequenced (Arar *et al.*, 1995) and its sequence was then deduced from the corresponding cDNA. It appeared that MR 60 is identical with a protein called ERGIC 53 which was isolated from the intermediate compartment and shown to recycle between endoplasmic reticulum and *cis*-Golgi apparatus by H.P. Hauri and coworkers (see also Roche and Monsigny, 1996). This intracellular membrane lectin was found to be present in a variety of cells from mice, rabbits and rats. On the basis of similarities with various plant lectins two amino acids (one aspartate and one asparagine) - which could be involved in the binding of its sugar ligand -, were separately exchanged for alanine. The mutated proteins were not any more able to bind the neoglycoprotein containing mannose (Itin *etal.*, 1996) showing that MR 60 is a lectin and that its similarities with plant lectins are real.

3.3. SUGAR DEPENDENT NUCLEAR IMPORT OF GLYCOCONJUGATES FROM CYTOSOL

Since the publication of our report (Sève *et al.*, 1986), showing for the first time that nuclei express sugar binding proteins, several nuclear lectins have been characterized and isolated. We recently succeeded to demonstrate that glycoconjugates introduced in the cytosol, upon plasma membrane permeabilization for instance, were able to reach the nucleus in a sugar dependent manner and without using the classical nuclear localization signal (NLS) mechanism (for a review see Nigg, 1997). The nuclear import of glycoconjugates requires energy, is inhibited by wheat germ agglutinin known to block the nuclear pore as well as by an excess of the related neoglycoprotein but not by NLS. Neoglycoproteins bearing GlcNAc residues correctly presented (GlcNAcβ 4GlcNAcβ-linked to BSA) are imported into the nucleus (Duverger *et al.*, 1996). Knowing that numerous cytosolic (and nuclear proteins) are glycosylated (*N*-acetyl-β-glucosaminides linked to the β-hydroxy amino acids) (see for a review, Hart *et al.*, 1989), this new mechanism could be involved in the importation of glycosylated proteins (with a Mr above 40000) as well as complexes between glycosylated proteins and other macromolecules.

4. Glycosylated Polylysines as Carriers of Nucleic Acids

On the basis of the efficient internalization of ligands by various membrane embedded lectins, we studied the efficiency of glycosylated polylysine to increase the efficiency of antisense oligonucleotides and genes by helping these nucleic acids to enter specifically and efficiently inside the cells.

4.1. ENHANCEMENT OF THE ACTIVITY OF ANTISENSE OLIGONUCLEOTIDES

Glycosylated polylysines of relatively large size (dp $\geqq$ 455) were shown to give stable complexes with oligonucleotides containing about 20 bases. In addition, when the ratio of charges born by the glycosylated polymers (cation) and of those born by the oligonucleotides is close to 1, the efficiency is optimal. This was clearly shown by using a phosphorothioate oligonucleotide, complementary of a sequence in an intron close to the 3' end of the ICAM-1 mRNA. This antisense oligonucleotide is inefficient even at concentration as high as 5 µM when it is used alone, it becomes efficient when it is added as a complex with lipocations (Bennett *et al.*, 1992) or when it is complexed on partially fucosylated polylysine (Stewart *et al.*, 1996). In the last case, the IC_{50} is about 0.5 µM .

4.2. ENHANCEMENT OF THE EFFICIENCY OF GENE TRANSFECTION

Polylysine as a polycation interacts with nucleic acid as a polyanion. This interaction is quite strong, the complex is stable even at high ionic strength up to 1.2 M NaCl, for instance. The formation of the complex induces an efficient compaction of this plasmid leading toroid particles having about 50 nm in diameter for a 5 kb plasmid and a polylysine containing about 200 lysine residues. Such complexes are relatively inefficient to transfer the plasmid in cultured cells ; conversely polylysine transferrin conjugates, polylysine asialo-orosomucoid conjugates or glycosylated polylysine are quite efficient (Wagner *et al.,* 1994 ; Frese *et al.,* 1994 ; Monsigny *et al.*, 1994a, 1994b, respectively for reviews). However, to be efficient, the complexes obtained with these substituted polylysines, must be used in conjunction with either chloroquine or fusogenic peptides. Glycosylated polylysines as well as glycosylated polylysines partially substituted with gluconoyl residues (Erbacher *et al.*, 1997) specifically transfect cultured cells on a sugar dependent manner. Lactosylated polylysine is suitable to transfect the hepatocyte line $HepG_2$, rabbit smooth cells, human air way epithelial cells ; mannosylated polylysine is suitable for human, mouse, rat macrophages, but not for $HepG_2$ cells ; fucosylated polylysine suitable for macrophages. The high efficiency of glycosylated (and gluconoylated) polylysine on suitable cells is mainly due to two reasons : one is related to the sugar itself acting as a recognition signal, the other is due to a decrease in the strength of the interaction between the polymer and DNA, allowing a more efficient release of the plasmid in the cell (Erbacher *et al.*, 1997).

5. Future Developments

Oligonucleotides acting as specific modulators of the expression of a gene may be made cell-specific by their substitution with an oligosaccharide, and the activity of the glyco-oligonucleotides may be still enhanced by the substitution with a peptide containing a KDEL stretch in a C terminal position. Such a peptide, indeed, enhances the activity of an oligonucleotide by helping it to reach the endoplasmic reticulum (Pichon *et al.*, 1997). Gene transfection based on the use of glycosylated (and gluconoylated) polylysine will be improved by using complex oligosaccharides instead of simple sugars and by adding on the polylysine an agent which selectively destabilizes the endosomal membrane. Such a novel polylysine derivative was prepared and found to be more active than any other previously described derivatives and, in addition, it did not require any helper (chloroquine or fusogenic peptides) ; furthermore, in contrast with the great majority of plasmid complexes containing lipocations or cationic polymers, the new polylysine derivative is fully efficient when it is added to cells in the presence of serum (Midoux and Monsigny, unpublished data).

6. Acknowledgements

This work was partly supported by grants from ANRS *(Agence Nationale de Recherches sur le SIDA* ; AIDS) and ARC (*Association pour la Recherche sur le Cancer*). MM is Professor of Biochemistry at the University of Orléans. CQ received a fellowship from the « *Région Centre* », SB a fellowship from ANRS, VC a fellowship from the Ministry of Research ; ED, VA and CP are *Maître de Conférences* at the University of Orléans, RM is Research Director at the *Centre National de la Recherche Scientifique*, PM and ACR are Research Director at the *Institut National de la Santé et de la Recherche Médicale.*

7. **References**

Arar, C., Carpentier, V., Le Caer, J. P., Monsigny, M., Legrand, A., and Roche, A. C. (1995) ERGIC-53, a membrane protein of the ER-Golgi Intermediate compartment is identical to MR60, an intracellular mannose specific lectin of myelomonocytic cells. *J. Biol. Chem.* **270**, 3551-3553.

Ashwell, G., and Harford, J. (1982) Carbohydrate-specific receptor of the liver. *Annu. Rev. Biochem.* **51**, 531-554.

Bennett, C. F., Chiang, M. Y., Chan, H. J., Shoemaker, E. S., and Mirabelli, C. K. (1992) Cationic lipids enhance cellular uptake and activity of phosphorothioate antisense oligonucleotides. *Mol. Pharmacol.* **41**, 1023-1033.

Derrien, D., Midoux, P., Petit, C., Nègre, E., Mayer, R., Monsigny, M., and Roche, A. C. (1989) Muramyl dipeptide bound to poly-L-lysine substituted with mannose and gluconoyl residues as macrophage activators. *Glycoconjugate J.* **6**, 241-255.

Duverger, E., Roche, A. C., and Monsigny, M. (1996) *N*-acetylglucosamine-dependent nuclear import of neoglycoproteins. *Glycobiology* **6**, 381-386.

Erbacher, P., Roche, A. C., Monsigny, M., and Midoux, P. (1997) The reduction of the positive charges of polylysine by partial gluconoylation increases the transfection efficiency of DNA/polylysine complexes. *Biochim. Biophys. Acta* **1324**, 27-36.

Frese, J., Wu, C. H., and Wu, G. Y. (1994) Targeting of genes to the liver with glycoprotein carrier. *Adv. Drug Delivery Rev.* **14**, 137-152.

Hart, G. W., Haltwinger, R. S., Holt, G. D., and Kelly, W. G.(1989) Glycosylation in the nucleus and cytoplasm. *Annu. Rev. Biochem.* **58**, 841-874.

Itin, C., A. C. Roche, M. Monsigny, and H. P. Hauri. (1996) ERGIC-53 is a functional mannose-selective and calcium-dependent human homologue of leguminous lectins. *Mol. Biol. Cell* **7**, 483-493.

Lee, Y. C., R. R. Townsend, M. R. Hardy, J. Lonngren, J. Arnap, M. Haraldsson, and H. Lonn. (1983) Binding of synthetic oligosaccharides to the hepatic Gal/GalNAc lectin. Dependence on fine structural features. *J. Biol. Chem.* **258**, 199-202.

Lee, Y. C., C.P. Stowell and M.J. Krantz (1976) 2-imino 2- methoxyethyl -1-thio-glycoside : new reagents for attaching sugars to proteins. *Biochemistry*, **15**, 3956-3963.

Midoux, P., C. Mendes, A. Legrand, J. Raimond, R. Mayer, M. Monsigny, and A. C. Roche. (1993) Specific gene transfer mediated by lactosylated poly-L-lysine into hepatoma cells. *Nucleic Acids Res.* **21**, 871-878.

Monsigny, M., C. Quétard, S. Bourgerie, D. Delay, C. Pichon, P. Midoux, R. Mayer, and A. C. Roche. (1998) Glycotargeting : the preparation of glyco-amino acids and derivatives from unprotected reducing sugars. *Biochimie* in press .

Monsigny, M., A. C. Roche, and P. Midoux. (1984). Uptake of neoglycoproteins via membrane lectin(s) of L 1210 cells evidenced by quantitative flow cytofluorometry and drug targeting. *Biol. Cell* **51**, 187-196.

Monsigny, M., A. C. Roche, P. Midoux, and R. Mayer. (1994a). Glycoconjugates as carriers for specific delivery of therapeutic drugs and genes. *Adv. Drug Delivery Rev.* **14**, 1-24.

Monsigny, M., A. C. Roche, P. Midoux, and R. Mayer. (1994b) Sugar specific delivery of drugs, oligonucleotides and genes. In *Targeting of drugs, 4 : Advances in system constructs. NATO ASI Series A : Life Sciences, Vol. 273.* G. Gregoriadis, B. McCormack, and G. Poste, editors. Plenum Press, New-York. 31-50.

Nigg, E. R. (1997) Nucleocytoplasmic transport : signals, mechanisms and regulation. *Nature* **386**, 779-787.

Pichon, C., K. Arar, A. J. Stewart, M. Duc Dodon, L. Gazzolo, P. J. Courtoy, R. Mayer, M. Monsigny, and A. C. Roche. (1997) Intracellular routing and inhibitory activity of oligonucleopeptides containing a KDEL motif. *Mol. Pharmacol.* **51**, 434-438.

Pimpaneau, V., P. Midoux, M. Monsigny, and A. C. Roche. (1991) Characterization and isolation of an intracellular mannose specific receptor from the human promyelocytic HL60 cells. *Carbohydr. Res.* **213**, 95-108.

Privat, J. P., F. Delmotte, and M. Monsigny. (1974) Protein-sugar interaction. Association of (β-4) linked *N*-acetyl-D-glucosamine oligomer derivatives with wheat germ agglutinin (lectin). *FEBS Lett.* **46**, 224-228.

Quétard, C., S. Bourgerie, N. Normand-Sdiqui, R. Mayer, G. Strecker, A. C. Roche, and M. Monsigny. (1998) Novel glycosynthons for glycoconjugate preparation : oligosaccharyl-pyroglutamyl-anilide derivatives. *Bioconjugate Chem.* in press :.

Quétard, C., N. Normand-Sdiqui, R. Mayer, G. Strecker, A. C. Roche, and M. Monsigny. 1997. Simple synthesis of novel glycosynthons : oligosylpyroglutamyl derivatives. *Carbohydr. Lett.* **2**, 415-422.

Roche, A. C., M. Barzilay, P. Midoux, S. Junqua, N. Sharon, and M. Monsigny. (1983) Endocytosis of glycoconjugates by Lewis lung carcinoma cells. *J. Cell Biochem.* **22**, 131-140.

Roche, A. C., and M. Monsigny. (1996) Trafficking of endogenous glycoproteins mediated by intracellular lectins : facts and hypotheses. *Chemtracts Biochem. Mol. Biol.* **6**, 188-201.

Sdiqui, N., A. C. Roche, R. Mayer, and M. Monsigny.(1995) New synthesis of Glycopeptides. *Carbohydr. Lett.* **1**, 269-275.

Sève, A. P., J. Hubert, D. Bouvier, C. Bourgeois, M. Bouteille, P. Midoux, A. C. Roche, and M. Monsigny. (1986) Quantitative flow cytofluorometry analysis of sugar-binding sites in mammalian cell nuclei. *Proc. Natl Acad. Sci.* (USA) **83**, 5997-6001.

Shepherd, V. L., E. J. Campbell RM Senior, and P. D. Stahl. (1982) Characterization of the mannose/fucose receptor on human mononuclear phagocytes. *J. Reticuloendothelial Soc.* **32**, 423-431.

Stewart, A. J., C. Pichon, L. Meunier, P. Midoux, M. Monsigny, and A. C. Roche. (1996) Enhanced biological activity of antisense oligonucleotides complexed with glycosylated poly-L-lysine. *Mol. Pharmacol.* **50**, 1487-1494.

Wagner, E., D. Curiel, and M. Cotten. (1994) Delivery of drugs, proteins and genes into cells using transferrin as a ligand for receptor-mediated endocytosis. *Adv. Drug Delivery Rev.* **14**, 113-135.

OLIGO/POLYSIALIC ACIDS IN SEA URCHIN: A DEVELOPMENTALLY REGULATED SEA URCHIN POLYSIALYLTRANSFERASE

J. W. CHO
Department of Biology, Yonsei University
134 Shinchon-Dong, Seodaemun-Ku, Seoul 120-749, Korea

1. Polysialic Acid-containing Glycoproteins in the Jelly Coat of Sea Urchin Eggs and the Sea Urchin Egg Receptor for Sperm

A gelatinous jelly coat sea urchin eggs consists of a mixture of Sia-rich glycoproteins and fucose-rich glycoproteins [1]. Chemical and physical studies recently revealed the presence of a novel polysialic acid (polySia) structure in the sea urchin glycoproteins isolated from *Hemicentrotous pulcherrimus*, designated polySia-gp(*H*) and *Stronylocentrotus purpuratus* designated polySia-gp(*S*) [2]. The structure of the polySia chains was characterized as ($\rightarrow$5-$O_{glycolyl}$-Neu5Gcα2$\rightarrow$)$_n$, where n ranges from 4 to more than 40 Neu5Gc residues. In the case of polySia-gp(*H*), the polyNeu5Gc chains were attached to core oligosaccharides that were O-glycosidically linked to threonine residues. Each polySia-gp(*H*) contained about 17 O-linked polysialylated sugar chains and its apparent molecular weight was 180,000. In the case of polySia-gp(*S*), its apparent molecular weight 250,000 and about 25 polySia chains were O-glycosidically linked to both threonins (2/3) and serine (1/3) residues [2]. The ($\rightarrow$5-$O_{glycolyl}$-Neu5Gc2$\rightarrow$)$_n$ chains were not sensitive to the treatment of endo-*N*-acylneuraminidase (Endo-N) and exosialidase from bacteriophasge K1F and *Arthrobacter ureafaciens*, respectively. Discovery of these ($\rightarrow$5-$O_{glycolyl}$-Neu5Gc2$\rightarrow$)$_n$ chains adds a new class of naturally occurring polySia to the structurally diverse family of polysialylated glycoproteins.

From the *S. purpuratus* sea urchin egg receptor for sperm and *H. pulcherrimus* egg cell surface complex, the same type of polyNeuGc chains were identified [3]. But their non-reducing terminal Neu5Gc were sulfated at their carbon 9 positions. The structure of the polySia chains was characterized as (SO_4-)-9Neu5Gcα2($\rightarrow$5-$O_{glycolyl}$-Neu5Gcα2$\rightarrow$)$_n$, where n averages 3 Neu5Gc residues. The polyNeu5Gc chain was a component of a GalNAc-containing chain and was O-glycosidically linked. The competitive inhibition of fertilization bioassay showed the sulfated oligosialic acid.

2. Polysialyltransferase Activity in Gastrula Stage Embryos

The recent chemical identification of polysialylated glycoproteins in the sea urchin egg receptor for sperm and the jelly coat of the sea urchin egg raises important question

Y. Kitagawa et al. (eds.), Animal Cell Technology: Basic & Applied Aspects, Vol. 10, 23–26.

about their biosynthesis and possible function. Using CMP-[^{14}C]Neu5Ac and a membrane fraction, the existence of a CMP-Sia:α2,8-Polysialyltransferase (polyST) in eggs and embryos of the sea urchin *Lytechinus pictus*. The polyST had maximal enzyme activity at 20°C and pH 7.0. No obligatory metal ion requirement could be demonstrated, although 10 mM Mg^{+2} stimulated activity. The specific activity of the polyST in the membrane fraction prepared from gastrula stage embryos was 45 pmol[^{14}C]Neu5Ac incorporated/mg protein/hour. The fact that over 90% of the total polyST activity in the low speed cell-free supernatant fraction was recovered in the high speed pellet indicated that the polyST was membrane associated. The polyST activity dependent on the concentration of CMP-Neu5Ac and the apparent Km of the sugar nucleotide was determined to be 0.176 mM.

3. Characterization of the Polysialylated Endogenous Acceptors

The sialylated or polysialylated endogenous acceptor products synthesized by early gastrula stage embryo membranes were examined by SDS-PAGE before and after treatment with Endo-N, Exo-N, Pronase, PNGase F or ceramide glycanase (Figure 1). Two major [14C]-sialylated bands were observed that corresponded to apparent molecular masses of 164 kDa and 38 kDa (Figure 1A and 1B; lane 1). Endo-N digestion showed that the 38 kDa band (arrow in Figure 1) was polysialylated because

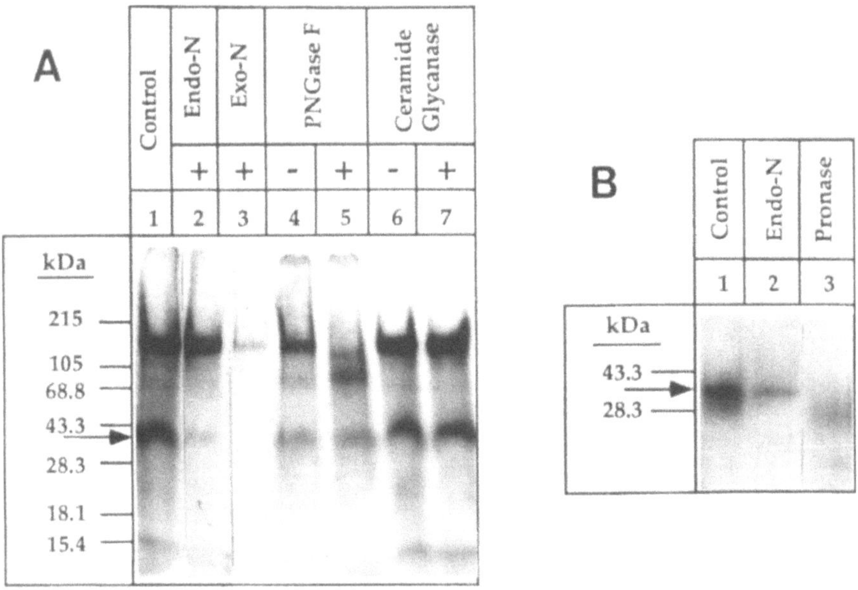

Figure 1. Partial characterization of [^{14}C]sialic acid labeled polysialylated and sialylated proteins. The membrane fraction prepared from gastrula stage embryos was labeled by incubation with CMP-[^{14}C]Neu5Ac and treated with the enzymes indicated. In panel B, the 38 kDa polysialylated endogenous acceptor was isolated by electroelution from SDS-PAGE before being treated the enzymes indicated. The arrow indicates the 38 kDa polysialylated endogenous acceptor molecule.

radioactivity in this band disappeared after treatment with Endo-N (Figure 1 A and 1B; lane 2). The 38 kDa polysialylated endogenous acceptor exhibited the properties of a protein because the band was shifted and greatly reduced in intensity after Pronase treatment (Figure 1B lane 3). Finally, this polysialylated product was not sensitive to treatment with either ceramide glycanase (Figure 1A; lane 7) or PNGase F (Figure 1A; lane 5), indicating that the polySia chains were not linked to a ganglioside or to a protein through an N-glycosidic linkage.

4. Gangliosides as Exogenous Acceptor Substrates for the gastrula Stage Polysialyltransferase

To determine if the level of polysialylation depended on the level of the polyST, or on the level of the endogenous acceptor, I searched for an exogenous acceptor to establish assay conditions in which acceptor was not limiting. Based on the recent finding that certain gangliosides function as exogenous acceptors for the polyST in neuroinvasive *Escherichia coli* K1 [4] a number of gangliosides were tested for their activity as exogenous acceptors for the sea urchin polyST (Table 1). Mammalian gangliosides were added to the polyST reaction mixture to determine if the gangliosides could stimulate the polyST activity. Stimulation of the polyST was observed for all gangliosides tested except GM3. For example, GD3 stimulated the polyST activity 5.5-fold. The α2,8-disialyl glycotope in GD3, GT1b and GQ1b was a preferred glycotope for polysialylation by the gastrula stage polyST. The decrease in radioactivity after Endo-N treatment of the sialylated gangliosides established that the disialyl moiety in the oligosaccharide chain had been polysialylated.

TABLE 1. Summary of the efficacy of gangliosides as exogenous acceptors of $[^{14}C]$Neu5Ac residues in gastrula stage membrane

Gangliosides	PolyST Activity[a]/nmol of Gangliosides Added	Relative Activity (%)
GQ1b	194.6	100
GT1b	145.5	75
GD3	125.7	65
GD1a	90.4	46
GM1a	18.4	9

[a] polyST Activity, pmol$[^{14}C]$Neu5Ac incorporated/mg protein/hour

5. Developmental Expression of the Polysialyltransferase in the Presence or Absence of GD3

Having established optimum conditions for assay of when the acceptor substrate was in excess, membrane fractions were prepared at various stages of development and assayed for polyST activity. Maximum polyST activity was found to be expressed at gastrula stage in the presence or absence of GD3 (Figure 2). In both cases, there is an attenuation in formation of the polysialylated product in the later developmental stages (prism stage). Because this decrease in polyST activity occurred even in the presence

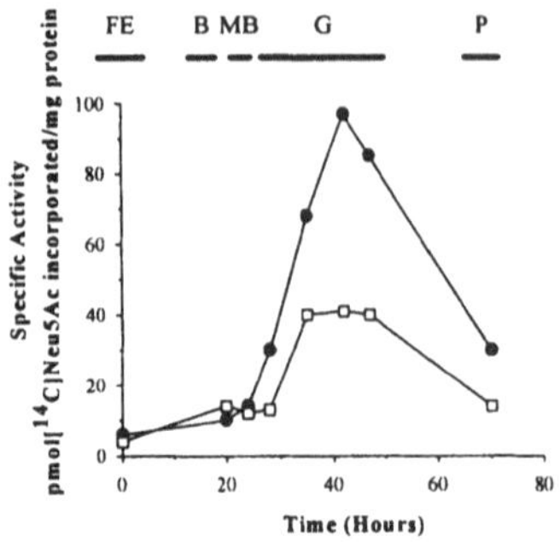

Figure 2. Developmental expression of the polyST activity in embryos extracts. The polyST reaction was carried out using these extracts in the presence (●) and absence (□) of GD3.

of an excess of acceptor (GD3) molecules, we conclude that it is the activity of the polyST which changes with development, rather than the level of the endogenous acceptor.

At early mesenchyme blastula stage where the polyST activity is first detected, the primary mesenchyme cells separate from the vegetal plate, migrate into the blastocoel, and begin to move along the blastula wall toward the animal pole. At gastrula stage where the maximal polyST activity is detected, invagination of the endodermal plate into the blastocoel occurs and an archenteron is formed. From the tip of the archenteron, secondary mesenchyme cells are formed and move toward the animal pole. It has been postulated that the increase level of polySia on N-CAM is associated with more plasticity in cell-cell interactions during cell migration in vertebrate nerve [5] and muscle developmental system [6], possibly the sea urchin embryonic polySia is also involved in synthesis of glycosylated proteins that mediate the cell migrations that occur during early mesenchyme blastula and gastrula stages.

This work was supported by grant from HRC-0101.

1. SeGall, G.K. and Lennarz, W.J.: Chemical characterization of the component of the jelly coat from sea urchin eggs responsible for induction of the acrosome reaction, *Dev Biol* **184** (1972), 33-48.
2. Kitazume, S., Kitajima, K., Inoue, S., Troy, F.A., Cho, J.-W., Lennarz, W.J., and Inoue, Y.: Identification of polysialic acid-containing glycoprotein in the jelly coat of sea urchin eggs. Occurrence of a novel type of polysialic acid structure, *J Biol Chem* **269** (1994), 22712-22718.
3. Kitazume, S., Kitajima, K., Inoue, S., Haslam, S.M., Morris, H.R., Dell, A., Lennarz, W.J., Inoue, Y.: The occurrence of novel 9-*O*-sulafated *N*-glycolylneuraminic acid-capped $\alpha 2 \rightarrow 5\text{-}O_{glycolyl}$-linked oligo/polyNeu5Gc chains in sea urchin egg cell surface glycoprotein. Identification of a new chain termination signal for polysialyltransferase, *J Biol Chem* **271** (1996), 6694-6701.
4. Cho, J.-W. and Troy, F.A.: Polysialic acid engineering: Synthesis of polysialylated neoglycosphingolipid by using the polysialyltransferase from neuroinvasive *Escherichia coli* K1, *Proc Natl Acad Sci USA* **91** (1994), 11427-11431.
5. Rutishauser, U., Acheson, A., Hall, A.K., Mann, D.M., and Sunshine, J.: The neural cell adhesion molecule (NCAM) as a regulator of cell-cell interaction, *Science* **240** (1988), 53-57.
6. Fredette, B., Rutishauser, U., and Landmesser, L.: Regulation and activity-dependence of N-cadherin, NCAM isoforms, and polysialic acid on chick myotubes during development, *J Cell Biol* **123** (1993), 1867-1888.

CLONING AND APPLICATION OF GLYCOSYLTRANSFERASES INVOLVED IN THE BIOSYNTHESIS OF SELECTIN LIGANDS

K. SASAKI, K. KURATA-MIURA, M. HASEGAWA AND T. NISHI
Tokyo Research Laboratories, Kyowa Hakko Kogyo Co. Ltd.
3-6-6, Asahi-machi, Machida-shi, Tokyo 194-8533, Japan

1. Introduction

The sialyl Lewis x (sLe^x; NeuAcα2-3Galβ1-4(Fucα1-3)GlcNAc) oligosaccharide structure and related structures have evoked considerable interest because they serve as ligands for the three known selectins (E-, P-, and L-selectins), which are involved in the recruitment of leukocytes to lymphoid tissues and the sites of inflammation [1, 2]. The compounds which function as selectin antagonists or inhibitors for expression of these adhesion molecules have therapeutic potential as anti-inflammatory drugs.

The sLe^x oligosaccharides are synthesized by sequential action of α2,3-sialyltransferases and α1,3-fucosyltransferases. However, there was no information about the molecular mechanisms regulating their biosynthesis. We tried to isolate and identify glycosyltransferases participating in the biosynthesis of the sLe^x oligosaccharides especially in leukocytes. To date, two types of α2,3-sialyltransferases (ST3Gal III and ST3Gal IV) and five types of α1,3-fucosyltransferases (Fuc-TIII, Fuc-TIV, Fuc-TV, Fuc-TVI, and Fuc-TVII) were isolated [2]. In this paper, we describe the cloning and application of ST3Gal IV and Fuc-TVII that are involved in the biosynthesis of the sLe^x oligosaccharides in leukocytes [3, 4].

2. Results and Discussion

2. 1. ISOLATION AND CHARACTERIZATION OF TWO TYPES OF α2,3-SIALYLTRANSFERASES (ST3Gal III AND ST3Gal IV) [3]

2. 1. 1. *Expression Cloning of Two Types of α2,3-sialyltransferase cDNAs*

We constructed an expression cloning system using both an Epstein-Barr virus-based cloning vector pAMo and Burkitt lymphoma cell line Namalwa KJM-1 as a host. To isolate glycosyltransferases involved in the biosynthesis of the sLe^x determinants, we developed an expression cloning system using lectin-resistance selection. *Ricinus communis* agglutinin 120 (RCA_{120}) is a cytotoxic lectin specific to galactose residues, especially Galβ1-4GlcNAc structure, which is a backbone of the sLe^x structure.

Y. Kitagawa et al. (eds.), Animal Cell Technology: Basic & Applied Aspects, Vol. 10, 27–31.

Therefore, it is predicted that the modification of this structure by glycosyltransferases may increase the levels of resistance of transfected cells to this lectin.

cDNA libraries of human melanoma cell line WM266-4 were constructed in vector pAMo and stably transfected into Namalwa KJM-1 cells. Those cells were selected in the presence of RCA_{120} to obtain RCA_{120}-resistant clones. From these clones we rescued transfected plasmids and identified two plasmids pAMo-ST3 and pAMo-ST4, which conferred the RCA_{120}-resistance phenotype on the transfected cells. One cDNA encoded a human Galβ1-3(4)GlcNAc α2,3-sialyltransferase, which was named ST3Gal III, and the other encoded a novel α2,3-sialyltransferase, which was designated ST3Gal IV.

2. 1. 2. *Enzymatic Properties of ST3Gal III and ST3Gal IV*

To examine *in vitro* enzymatic properties, the putative catalytic domain of ST3Gal III or ST3Gal IV was expressed as secretable forms fused with IgG-binding domain of *Staphylococcus aureus* protein A in Namalwa KJM-1 cells. These fusion proteins were adsorbed to IgG Sepharose and the sialyltransferase activities were assayed using various pyridylaminated oligosaccharides as acceptor substrates.

ST3Gal IV transferred sialic acid via α2,3-linkage toward type 2 chains more preferentially than type 1 chains. In contrast, ST3Gal III showed more preferential substrate specificity for type 1 chains than type 2 chains. Interestingly, ST3Gal IV, but not ST3Gal III, sialylated a Lewis x oligosaccharide (Galβ1-4(Fucα1-3)GlcNAcβ1-3Galβ1-4Glc) to form a sLe^x structure, albeit with low efficiency.

To test whether ST3Gal IV or ST3Gal III is involved in the biosynthesis of the sLe^x oligosaccharides, we compared the levels of the sLe^x antigens on transfected Namalwa KJM-1 cells by fluorescence activated cell sorter (FACS) using anti-sLe^x monoclonal antibodies (mAbs), CSLEX-1 and KM93. The levels of the sLe^x antigens were considerably increased by expression of ST3Gal IV, but not by that of ST3Gal III.

2. 1. 3. *Expression Levels of ST3Gal III and ST3Gal IV mRNAs in Various Cells*

Using competitive reverse transcriptase-polymerase chain reaction (RT-PCR) analyses, we examined the expression levels of ST3Gal IV and ST3Gal III mRNAs in a variety of cells including polymorphonuclear cells (PMN), monocytes, their immortalized cell lines (HL-60, U-937, and THP-1), and colon carcinoma cell lines (Colo205, SW1116, and LS180), which were known to express the sLe^x antigens. ST3Gal IV transcripts were significantly expressed in all the cells examined, whereas ST3Gal III transcripts were hardly expressed in PMN, monocytes, and two colon carcinoma cell lines, SW1116 and LS180. These results together with enzymatic properties suggest that ST3Gal IV is most likely to participate in the biosynthesis of the sLe^x antigens on leukocytes and colon cancer cells.

2. 2. ISOLATION AND CHARACTERIZATION OF A NOVEL α1,3-FUCOSYLTRANSFERASE (Fuc-TVII) [4]

2. 2. 1. *Expression Cloning of a novel α1,3-Fucosyltransferase (Fuc-TVII) cDNA*

We isolated a novel α1,3-fucosyltransferase (Fuc-TVII) cDNA by expression cloning using FACS selection. A cDNA library of human monocytic leukemia cell line THP-1 was constructed in vector pAMo and stably transfected into Namalwa KJM-1 cells. The cells expressing a large amount of sLe^x determinants were enriched by three rounds of sorting with anti-sLe^x mAb KM93, resulting in the recovery of a plasmid pAMo-FT7 carrying Fuc-TVII cDNA. The primary sequence of Fuc-TVII showed 42 -47% identity with those of other α1,3-fucosyltransferases.

2. 2. 2. *Enzymatic Properties of Fuc-TVII*

We constructed a secreatable Fuc-TVII fused with protein A and examined its acceptor substrate specificity. Fuc-TVII showed an α1,3-fucosyltransferase activity toward only an α2,3-sialylated type 2 oligosaccharide, but not toward either non-sialylated type 2 oligosaccharides or type 1 oligosaccharides.

2. 2. 3. *Expression Levels of Five Types of α1,3-Fucosyltransferases in Various Cells*

Using competitive RT-PCR analyses, we examined the expression levels of five cloned α1,3-fucosyltransferases in various cells. Both Fuc-TVII and Fuc-TIV mRNAs were significantly detected in PMN, monocytes, and myeloid-lineage cell lines that expressed the selectin ligands. Fuc-TIV was considerably expressed in other cells, whereas the expression of Fuc-TVII was restricted in leukocytes. Both Fuc-TIII and Fuc-TVI were highly expressed in colon carcinoma cell lines (Colo205, SW1116, and LS180) which expressed the sLe^x and sialyl Lewis a determinants. These results indicate that both Fuc-TIV and Fuc-TVII are candidates for α1,3-fucosyltransferases participating in the biosynthesis of the sLe^x determinants in leukocytes.

2. 2. 4. *Fuc-TVII is Involved in the Biosynthesis of Selectin Ligands in Leukocytes*

To test whether Fuc-TVII or Fuc-TIV is participating in the biosynthesis of the sLe^x antigens, we compared the levels of the sLe^x antigens on transfected Namalwa KJM-1 cells by FACS using three anti-sLe^x mAbs, CSLEX-1, KM93 and FH6. Transfection of Fuc-TVII efficiently increased the levels of sLe^x epitopes recognized by any of those mAbs. On the other hand, Fuc-TIV transfection led to a slight increase of the sLe^x level recognized by CSLEX-1, but by neither KM93 nor FH6.

In addition, we examined the activity of transfected Namalwa KJM-1 cells to bind to soluble E-selectin coated plates. Fuc-TVII-transfected Namalwa cells exhibited significant binding to E-selectin, although the binding capacity was lower than that of Fuc-TIII transfectants, Fuc-TVI transfectants, or U-937 cells. Their adhesion was inhibited completely by mAb KM93 but not by a control mAb. In contrast, Fuc-TIV transfectants exhibited virtually no binding to soluble E-selectin.

These results together with expression patterns of five α1,3-fucosyltransferases strongly suggest that Fuc-TVII may be involved in the biosynthesis of the selectin ligands in leukocytes. Recently, Maly *et al.* directly demonstrated the involvement of Fuc-TVII in the biosynthesis of selectin ligands by their characterization of Fuc-TVII-deficient mice [5].

2. 3. APPLICATION OF ST3Gal IV AND Fuc-TVII

2. 3. 1. *Possible Application of ST3Gal IV and Fuc-TVII*

Both ST3Gal IV and Fuc-TVII are most likely to be involved in the biosynthesis of the selectin ligands in leukocytes. Therefore, inhibitors for Fuc-TVII or ST3GalIV have therapeutic potential as anti-inflammatory drugs. In addition, Fuc-TVII and ST3Gal IV may be useful for enzymatic synthesis of the sLe^x oligosaccharides and their derivatives *in vitro* and *in vivo*, which should have therapeutic value as selectin antagonists.

2. 3. 2. *Production of ST3Gal IV and Fuc-TVII in Namalwa KJM-1 Cells [6]*

Namalwa KJM-1 cells producing a large amount of soluble ST3Gal IV or soluble Fuc-TVII were obtained using the dihydrofolate reductase gene co-amplification method. Clones S4C9 and F7C4 produced about 11 μg/ml of the soluble ST3Gal IV and about 0.6 μg/ml of the soluble Fuc-TVII, respectively. These enzymes were purified from the conditioned media using an IgG Sepharose column chromatography. Purity was about 90% for ST3Gal IV and 60-70% for Fuc-TVII.

2. 3. 3. *Inhibitory Effects of Various Nucleotides on Fuc-TVII [6]*

Inhibitory effects of various nucleotides (GTP, GDP, GMP, guanosine, inosine diphosphate, ADP, CDP, and UDP) on Fuc-TVII were compared with those on Fuc-TV [7]. The K_i values of GDP and guanosine for Fuc-TVII were 0.035 and >20 mM, respectively, *i. e.* nearly consistent with their inhibitory potencies as to Fuc-TV. On the other hand, GTP was a weaker inhibitor for Fuc-TVII (K_i = 0.17 mM) than Fuc-TV (0.03 mM), whereas GMP was a stronger one for Fuc-TVII (K_i = 0.094 mM) than Fuc-TV (K_i = 0.7 mM). Furthermore, the K_i values of inosine diphosphate, ADP, and CDP for Fuc-TVII (0.77, 7.3, and 11 mM, respectively) were different from those for Fuc-TV (0.069, 0.58, and 3.5 mM, respectively). These differences may be due to subtle differences in the structures of GDP-binding domains of these enzymes. These results present the rationale for the generation of Fuc-TVII-selective inhibitors.

2. 3. 4. *One-pot Enzymatic Synthesis of a sLe^x Oligosaccharide Using ST3Gal IV and Fuc-TVII [6]*

Fuc-TVII showed an acceptor substrate specificity restricted to an α2,3-sialylated type 2 oligasaccharide, indicating that sLe^x oligosaccharides could be synthesized from type 2 oligosaccharides in one-pot reaction using both ST3Gal IV and Fuc-TVII. Simultaneous addition of ST3Gal IV and Fuc-TVII converted lacto-*N*-*neo*tetoraose (LNnT; Galβ1-4GlcNAcβ1-3Galβ1-4Glc) into a sLe^x hexasaccharide efficiently without any detectable by-product. In contrast, when LNnT was incubated with both ST3Gal IV and Fuc-TVI, a Lewis x pentasaccharide, an undesirable by-product, was formed in a yield of ~20%, leading to a decreased yield (78%) of sLe^x hexasaccharide. One-pot reaction system using ST3Gal IV and Fuc-TVII may be useful for a large-scale production of sLe^x oligosaccharides.

2. 3. 5. *Production of Fuc-TVII in Sf9 Insect Cells [8]*

To obtain a large amount of soluble Fuc-TVII fused with protein A, it was expressed in Sf9 insect cells under the control of polyhedrin promoter. As a result, about 10 μg/ml of the soluble Fuc-TVII was produced. The soluble Fuc-TVII enzyme was purified using an IgG Sepharose column chromatography and its enzymatic properties were compared with those of Namalwa KJM-1-produced enzyme. Although the Sf9-produced Fuc-TVII had small-sized *N*-linked sugar chains, its enzymatic properties were almost the same as those of Namalwa KJM-1-produced enzyme.

3. Summary

We constructed an efficient expression cloning system and isolated a novel α2,3-sialyltransferase (ST3Gal IV) and a novel α1,3-fucosyltransferase (Fuc-TVII). Enzymatic properties and expression patterns of ST3Gal IV and Fuc-TVII indicated that these enzymes were most likely to be involved in the biosynthesis of selectin ligands. ST3Gal IV and Fuc-TVII were produced as secretable forms fused with protein A in a large quantity in Namalwa KJM-1 cells or in Sf9 insect cells. These enzymes are useful for synthesis of sLe^x oligosaccharides and their derivatives, which should have therapeutic value as selectin antagonists. In addition, compounds inhibiting enzymatic activity or expression of these enzymes are expected to have therapeutic potential as anti-inflammatory drugs. These enzymes may be also useful for modifying sugar chains of therapeutic glycoproteins.

4. References

1. Vestweber, D. : Ligand-specificity of the selectins, *J. Cell. Biochem.* **61** (1996), 585-591.
2. Lowe, J. B. : Selectin ligands, leukocyte trafficking, and fucosyltransferase genes, *Kidney Int.* **51** (1997), 1418-1426.
3. Sasaki, K., Watanabe, E., Kawashima, K., Sekine, S., Dohi, T., Oshima, M., Hanai, N., Nishi, T., and Hasegawa, M. : Expression cloning of a novel Galβ(1-3/1-4)GlcNAc α2,3-sialyltransferase using lectin resistance selection, *J. Biol. Chem.* **268** (1993), 22782-22787.
4. Sasaki, K., Kurata, K., Funayama, K., Nagata, M., Watanabe, E., Ohta, S., Hanai, N., and Nishi, T. : Expression cloning of a novel α1,3-fucosyltransferase that is involved in biosynthesis of the sialyl Lewis x carbohydrate determinants in leukocytes, *J. Biol. Chem.* **269** (1994), 14730-14737.
5. Maly, P., Thall, A., Petryniak, B., Rogers, C. E., Smith, P. L., Marks, R. M., Kelly, R. J., Gersten, K. M., Cheng, G., Saunders, T. L., Camper, S. A., Camphausen, R. T., Sullivan, F. X., Isogai, Y., Hindsgaul, O., von Andrian, U. H., and Lowe, J. B. : The α(1,3)fucosyltransferase Fuc-TVII controls leukocyte trafficking through an essential role in L-, E-, and P-selectin ligand biosynthesis, *Cell* **86** (1996), 643-653.
6. Shinoda, K., Morishita, Y., Sasaki, K., Matsuda, Y., Takahashi, I., and Nishi, T. : Enzymatic characterization of human α-1,3-fucosyltransferase Fuc-TVII synthesized in a B cell lymphoma cell lines, *J. Biol. Chem.* **272** (1997), 31992-31997.
7. Murray, B. W., Takayama, S., Schultz, J., and Wong, C. H.: Mechanism of human α-1,3-fucosyltransferase V: glycosidic cleavage occurs prior to nucleophilic attack, *Biochemistry* **35** (1996) , 11183-11195.
8. Shinkai, A., Shinoda, K., Sasaki, K., Morishita, Y., Nishi, T., Matsuda, Y., Takahashi, I., and Anazawa, H. : High-level expression and purification of a recombinant human α-1,3-fucosyltransferase in baculovirus-infected insect cells, *Protein Expression and Purification* **10** (1997), 379-385.

INDUCTION OF GANGLIOSIDE BIOSYNTHESIS, NEURITE OUTGROWTH AND FUNCTIONAL SYNAPSE FORMATION BY A SYNTHETIC CERAMIDE ANALOG

JIN-ICHI INOKUCHI
Seikagaku Corporation, Tokyo Research Institute, 1253 Tateno 3-chome, Higashiyamato-shi, Tokyo 207, JAPAN

Abstract

To address the role of brain gangliosides in synaptic activity, the ceramide analogs, D-*threo*-1-phenyl-2-decanoylamino-3-morpholino-1-propanol (D-PDMP) and its enantiomer, L-PDMP, were used to inhibit and stimulate ganglioside biosynthesis in cultured cortical neurons. Prolonged treatment with both PDMP isomers exhibited opposite effects on functional synapse formation measured by spontaneous synchronized oscillatory activity of intracellular Ca^{2+} between the neurons: suppression by D-PDMP and facilitation by L-PDMP. Up-regulation of synaptic activity by L-PDMP could be correlated with the slow but robust activation of p42 mitogen-activated protein kinase. Treatment with L-PDMP after transient forebrain ischemia in rats ameliorated the deficit of a well-learned spatial memory by an 8-arm maze task, suggesting a new potential therapeutic approach for neurodegenerative disorders.

Results

The effects of PDMP isomers on spontaneous synchronized oscillatory activity of intracellular Ca^{2+} between the neurons, which represents the functional synapse formation, have been investigated. The glucosylceramide synthase inhibitor, D-PDMP was used to deplete glycosphingolipids (GSLs) of cultured rat cerebral cortical neurons and to identify the role of endogenous ganglioside(s) in the functional synapse formation (1). Treatment with D-PDMP resulted in dose- and time-dependent decrease in the frequency of synchronous oscillations and the decreased frequency by D-PDMP was specifically normalized by the supplementation of ganglioside GQ1b to the ganglioside depleted neurons, suggesting that *de novo* synthesis of a particular molecular species of the gangliosides, GQ1b, might be essential for synapse formation and synaptic activity (1). In contrast to the effect of the D-isomer, L-PDMP was proved its stimulatory action on GSL biosynthesis in several cell lines (2-5). Recently, we demonstrated that treatment of primary cultured rat embryonic cortical explants with L-PDMP resulted in the stimulation of both neurite outgrowth and ganglioside biosynthesis (5).

Therefore, we asked whether or not L-PDMP is able to facilitate the functional synapse formation in the long term primary culture of cortical neurons (6). Treatment of the cells with continuous presence of 20 and 40 μM L-PDMP for 8 days resulted in the significant increase of the frequency of the spontaneous synchronous oscillation between the

Y. Kitagawa et al. (eds.), Animal Cell Technology: Basic & Applied Aspects, Vol. 10, 33–36.

neurons. The prolonged treatment by L-PDMP was required to reveal the significant enhancement of the functional synapse formation.

In order to examine the effect of L-PDMP on *de novo* synthesis of gangliosides, the cortical cells were cultured in the presence of absence 20 μM L-PDMP for 8 days and the cells were metabolically labeled with [^{14}C]galactose for 6 h at the last period of each culture. The radioactivities of total GSL fractions of the treated cells were significantly increased, and the selective acceleration of the biosynthesis of GM3, GD3 and GQ1b was observed. Next, we performed the direct measurement of the activities of these ganglioside-specific sialyltransferases using cellular lyzates from the cells cultured for 8 days with or without 20μM L-PDMP. The significant and selective stimulatory effect of L-PDMP on GM3 synthase (200 %), GD3 synthase (303 %) and GQ1b synthase (440 %) was demonstrated. The stimulatory effect of L-PDMP could be demonstrated only when the cells were precultured with this drug before making cell lyzates. These findings confirmed our previous observation that L-PDMP may not interact with these GSL synthesizing enzymes directly and influence the translational or post-translational modifications of these enzymes in intact cells (3,5). In addition, there was a positive correlationship between the chronological changes of GQ1b synthase activities after L-PDMP treatment and that the facilitation of functional synapse formation (6).

Several lines of evidence have indicated that glycosyltransferase activities can be altered by inducers of differentiation. For example, In PC 12 cells, nerve growth factor (NGF) induces not only cellular differentiation but also enhance the expression of some ganglioside moieties and their synthase activities. Neurotrophin activation of Ntk/Trk tyrosine kinase receptors elicits signal transduction cascade leading to activation of mitogen-activated protein kinases (MAPKs), which are excellent candidates for regulators of synaptic plasticity in neurons. It has been demonstrated that p42 MAPK, but not p44 MAPK, is activated in response to NMDA receptor stimulation to elicit long-term potentiation in hippocampal CA1 neurons. Brain derived neurotrophic factor (BDNF) could potentiate synaptic activity in cultured cortical neurons and in hippocampal neurons probably through MAPK activation. From these observations, we measured the expression and activity of MAPK in the cortical cells treated chronologically with L-PDMP (6). p42 MAPK was selectively and sustainably activated by L-PDMP in the presence of serum. Time course of MAPK activation was similar to that of GQ1b synthase activity.

In the course of trials to evaluate the efficacy of L-PDMP on memory in vivo, we have tested the effect of both PDMP isomers on the deficit of previously acquired spatial memory after transient forebrain ischemia in rats (6). Rats treated with 40 mg/kg of L-PDMP showed significant recall of the acquired spatial memory. By contrast, D-PDMP did not help the recall and, possibly, even produced more errors than the controls. Both PDMP isomers had no effect on heart rates, blood pressure, or body temperature.

Concluding Remarks and Future Aspects

1) We could successfully demonstrate a new approach for treating memory deficit by up-regulating *de novo* synthesis of gangliosides and synaptic function using the synthetic ceramide analog, L-PDMP.

2) Functional synapse formation, measured by spontaneous synchronized oscillatory activity of intracellular Ca^{2+} in cultured rat cerebral cortical neurons, was inhibited by depletion of the endogenous gangliosides with the addition of a glucosylceramide

synthase inhibitor, D-PDMP, into the culture medium. This suppressed synaptic activity was normalized by supplementation with ganglioside GQ1b, but not with other gangliosides.

3) The importance of polysialylated gangliosides, including GQ1b, in the central nervous system was further indicated by the observations that mice deficient in β1,4-N-acetylgalactosaminyltranseferase (GM2/GD2 synthase), exhibited reduction in both conduction velocity of somatosensory nerves and performance in a passive avoidance passive task test (7). Thus, it seemed possible that L-PDMP could augment normal levels of gangliosides in neurons and thereby produce beneficial effects.

4) Transfection of the GD3 synthase gene into neuroblastoma cells induced the expression of b-series gangliosides including GD3 and GQ1b, and also caused differentiation into cholinergic neuron-like cells (8). Our results presented here demonstrate the primary importance of *de novo* biosynthesis of gangliosides, especially those of b-series gangliosides, on neuronal differentiation and synaptogenesis. Further studies are in progress to clarify the mechanism of activation of especially GQ1b synthases elicited by L-PDMP.

5) Since GPI-anchored proteins are believed to aggregate with GSLs in the trans-Golgi network of polarized cells and are consequently cotransported to the apical plasma membrane, it is interesting to examine whether PDMP can modulate cotransport of the complex of GSLs and GPI-anchored protein. It has been reported that b-series gangliosides were preferentially transported from the trans-Golgi network of the neuronal soma to the perikaryon plasma membrane and further delivered to neurites.

6) At present the regulatory mechanisms of glycosyltransferase activities involved in the biosynthesis and expression of gangliosides in various tissues are still poorly understood. Therefore, the further investigation on relationship between the activation of both ganglioside synthases and p42 MAPK by L-PDMP is a matter of interest.

7) We are currently asking; i) whether or not the phosphorylation of extracellular domains of MAP1B or related proteins is correlated with the expression of GQ1b on synaptic membranes, ii) the presence of a novel GQ1b-dependent ectokinase, and iii) the following signal transduction system with possible involvement of p42 MAPK activation. Studies along this line will be aided by the use of inhibitor and stimulator of ganglioside biosynthesis, D-PDMP and L-PDMP.

References

1. Mizutani, A., Kuroda, Y., Muramoto, K., Kobayashi, K., Yamagishi, K., and Inokuchi, J. (1996) Effects of glucosylceramide synthase inhibitor and ganglioside GQ1b on synchronous oscillations of Intracellular Ca^{2+} in cultured cortical neurons, *Biochim. Biophys. Res. Commun.* **222,** 494-498.

2. Inokuchi, J., K. Momosaki, K., H. Shimeno, H., A. Nagamatsu, A., and Radin, N.S. (1989) Effect of D-*threo*-PDMP, an inhibitor of glucosylceramide synthetase, on expression of cell surface glycolipid antigen and binding to adhesive proteins by B16 melanoma cells, *J.Cell. Physiol.* **141,**573-583.

3. Inokuchi, J., Usuki, S, and Jimbo, M. (1995) Stimulation of glycosphingolipid biosynthesis by L-*threo*-1-phenyl-2-decanoylamino-1-propanol and its homologs in B16 melanoma cells, *J. Biochem.* (Tokyo) **117,** 766-773.

4. Inokuchi, J., Jimbo, M., Momosaki, K., Shimeno, H., and Nagamatsu, A. (1990) Inhibition of experimental metastasis of murine Lewis lung carcinoma by an inhibitor of glucosylceramide synthase and its possible mechanism of action, *Cancer Res.* **50,** 6731-6737.

5. Usuki, S., Hamanoue, M., Kohsaka, S., and Inokuchi, J. (1996) Induction of Ganglioside Biosynthesis and neurite outgrowth of primary cultured neurons by L-*threo*-1-phenyl-2-decanoylamino-3-morpholino-1-propanol, *J. Neurochem.* **67**, 1821-1830.

6. Inokuchi, J, Mizutani, A, Usuki, S, Jimbo, M, Motizuki, H., Muramoto, K., Kobayashi, K., Kuroda, Y., Iwasaki, K., Ohgami, Y., and Fujiwara, M. (1997) Up-regulation of Ganglioside Synthesis, Functional Synapse Formation and Memory Retention by Synthetic Ceramide Analog (L-PDMP), *Biochem. Biophys. Res. Commun.* **238**, 595-600.

7. Takamiya, K., Yamamoto, A., Furukawa, K., Yamashiro, S., Shin, M. Okada, M., Fukumoto, S., Haraguchi, M., Takeda, N., Fujimura, K., Sakae, M., Nishikawa, M., Shiku, H., Furukawa, K., and Aizawa, S. (1996) Mice with disrupted GM2/GD2 synthase gene lack complex gangliosides but exhibit only subtle defects in their nervous system, *Proc. Natl. Acad. Sci. USA* **93**, 10662-10667.

8. Kojima, K., N. Kurosawa, N., T. Nishi, T., N. Hanai, N., and & S. Tsuji, S. (1994) Induction of cholinergic differentiation with neurite sprouting by *de novo* biosynthesis and expression of GD3 and b-series gangliosides in Neuro2a cells, *J. Biol. Chem.* **269,** 30541-30456.

HUMAN CYTOTOXIC T LYMPHOCYTES FOR TUMOR THERAPY
Induction on formalin-fixed tumor tissues and expansion on immobilized lectins

T. OHNO
RIKEN Cell Bank, The Institute of Physical and Chemical Research (RIKEN), Koyadai 3-1-1, Tsukuba Science City, 305, Japan

Introduction

Cytotoxic T lymphocytes (CTL) exhibit strong killing activity against tumor cells and, therefore, are expected to be employed in adoptive immunotherapy for human tumors. Generation of CTL requires autologous/syngeneic target tumor cells or tumor-derived antigenic peptides presented on antigen-presenting cells. However, successful generation of CTL has been limited to cases where sufficient tumor cells (Crowley *et al.*, 1991; Kawakami *et al.*, 1994; Wolfel *et al.*, 1994) or tumor-derived antigenic peptides (Slingluff *et al.*, 1994; Tjoa *et al.*, 1994) were available for repeated stimulation of CTL growth for a prolonged culture period. The supply of target tumor cells is a key problem due to the difficulty of establishing a tumor cell line from every patient (Hay *et al.*, 1994). Therefore, an alternative and stable source of tumor-specific antigens is desirable.

We have reported that, when cultured on formalin-fixed paraffin-embedded tumor sections as a substitute for live tumor cells, human autologous CTL with high specificity and strong killing activity against the target tumor cells were induced (Liu *et al.*, 1996a), since formalin fixation preserved the specific antigenicity of the target (Pollard and Holgate, 1987). The supply of tumor sections, however, was insufficient when the CTL grew to a considerably large number, and therefore, we needed to restimulate the CTL with anti-CD3 monoclonal antibody to maintain the culture (Liu *et al.*, 1995). Repeated stimulation with this antibody sometimes results in loss of cytotoxic activity in the CTL population (Muul *et al.*, 1987). In clinical practice, conditioning of CTL by timely restimulation in prolonged culture is essential for repeated administration to a patient.

To solve this timing problem, two approaches were examined in the present study. One approach is to employ formalin-fixed primary target cells, such as primary cultured (but not yet completely established) tumor cells, tumor cells dissociated from tumor tissues, or tumor cells in ascites, as a stable source of tumor antigens, available any time, for further stimulation of CTL. Another approach is to activate the CTL by means of non-specific stimulatory molecules immobilized on an insoluble matrix.

Restimulation by formalin-fixed tumor cells

TKB-1p-CTL are known to kill the autologous target TKB-1p cells, but not allogeneic renal carcinoma cells or gastric carcinoma cells (Tsurushima *et al.*, submitted for publication).

We attempted to stimulate the growth of these CTL using live TKB-1p cells previously irradiated with X-rays or formalin-fixed TKB-1p cells as target cells at E/T

Y. Kitagawa et al. (eds.), Animal Cell Technology: Basic & Applied Aspects, Vol. 10, 37–41.

ratios of 1 and 10. As shown in Fig. 1, the number of control CTL gradually decreased after culturing for 2 days. This control culture consisted of all the components except the target cells. With the formalin-fixed TKB-1p cells as target cells, the CTL grew continuously at E/T ratios of 1 and 10 with mean doubling times of 95 hr and 119 hr, respectively. With the X-ray-irradiated live TKB-1p cells as target cells, the growth rates of the CTL were nearly the same at both E/T ratios of 1 and 10; mean doubling times were 138 hr and 106 hr, respectively. These values were similar to the mean doubling time observed for CTL stimulated using formalin-fixed tumor cells as target cells at the E/T ratio of 10 and were greater than that observed using these fixed target cells at the E/T ratio of 1.

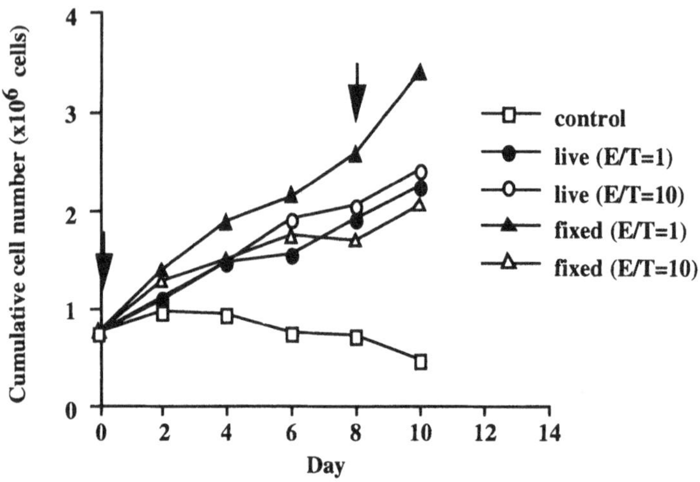

Fig. 1 Continuous stimulation of CTL growth with formalin-fixed tumor cells

Arrows indicate restimulation with X-ray-irradiated live or formalin-fixed TKB-1p cells at the E/T ratio indicated. Each point is the mean value of duplicate observations. The control culture consisted of all the components except target cells.

The results suggest that CTL are continuously expandable when cultured on formalin-fixed autologous tumor cells as well as when cultured on live cells of an autologous established tumor cell line. Even though we have failed to establish an autologous tumor cell line, many investigators may have target tumor cells available found in small pieces of dissected tumor tissue, in the fluid of the thoracic cavity and ascites, or in quiescent primary cultures. If fixed and supplied properly, these cells could also serve as a source of tumor antigens available any time for further stimulation of CTL .

Restimulation by immobilized Con A

The plant-derived lectin, concanavalin A (Con A) (Baker *et al.*, 1979; Quentmeier *et al.*, 1992; Shinomura *et al.*, 1986) is able to induce cytokine production in peripheral blood mononuclear cells. Considering the cost of long term culture for expansion of CTL, Con A is one of the candidates for use as a non-specific stimulator of CTL growth. However, since Con A itself has a toxic effect on lymphocytes at high concentrations, one should be careful when adding Con A directly to a culture of CTL (Baker *et al.*; 1979, Mukherji *et al.*, 1989; Quentmeier *et al.*, 1992; Shinomura *et al.*, 1986). In the present study, we investigated the effect of Con A immobilized on carrier beads, expecting to reduce its toxicity and stimulate continuous growth of CTL in the absence of tumor-derived antigens.

Crosslinked alginate beads were prepared by incubating 0.5 g of water-soluble sodium alginate beads (50-100 μm in diameter) in 40 ml of BDDE-containing basic dimethylformamide aqueous solution (prepared by mixing 1 part of 0.1 N NaOH with 3 parts of DMF, and adding 2% (w/v) BDDE) with constant shaking for 12 hr. The beads were washed, dried, then activated with tresyl chloride in pre-dried acetone (Liu *et al.*, 1996b). Con A was coupled onto the activated beads by overnight incubation of the beads (2 g) in 10 ml of a phosphate-buffered solution (PBS, pH 8.0) of Con A (1.0 mg/ml) at room temperature. The unreacted sulfonyl groups on the beads were quenched by replacing the reaction solution with glycine ethyl ester hydrochloride-PBS solution (100 mM, in 0.5 M PBS, pH 8.0) followed by incubation at room temperature for another 1 hr, then the beads were washed with a large volume of acetate buffer (0.02 M, pH 3.5), or phosphate buffer (0.01 M, pH 5.6 or pH 8.0) until free Con A became undetectable in the washing solution.

Allogeneic CTL were generated as described (Liu *et al.*, 1995) by co-culture with a duodenum papilloma cell line, TGBC18TKB, of which the human leukocyte antigen (HLA)-type was A2, A11, matching that of peripheral blood mononuclear cells (PBMC) derived from a healthy donor. The CTL were $CD4^+$ $CD8^-$ and their HLA-type was A2. The CTL killed half of the seeded TGBC18TKB cells at an effector/target (E/T) ratio of 10. However, they did not affect three other cell lines tested at this E/T ratio, i.e., HLA-type matched brain tumor TKB-1 (HLA A2), lung squamous carcinoma SQ-5 (HLA A2), and HLA-type mismatched gastric adenocarcinoma GT3TKB (HLA A2402).

Fig. 2 shows the growth of $CD4^+$ CTL in the presence of soluble and immobilized Con A. Soluble Con A inhibited CTL growth at 0.1 μg/ml or more, while immobilized Con A increased CTL growth slightly in the range of 0.025-2 μg/ml. The total amount of Con A in the beads was used to calculate the concentration of immobilized Con A. It seems likely that the Con A effective in stimulation of live cells was limited only to those molecules on the surface of the beads. This must be one of the reasons why the CTL retained their growth potential at the higher concentration of immobilized Con A.

To confirm whether the CTL can keep growing without antigen stimulation in the presence of immobilized Con A, long-term culture of CTL was investigated. When the $CD4^+$ CTL were cultured without tumor antigen stimulation in the presence of Con A-immobilized beads (Con A concentration, 2 μg/ml) or naked beads as the control under serum-free conditions (RHAMα medium supplemented with 0.5% human serum albumin and 67 U/ml IL-2) for 45 days, the CTL grew continuously under the former conditions, but, under the latter conditions, the growth rate decreased gradually after 3 weeks and growth ceased after 34 days (Fig. 3).

Specific cytotoxicity of the CTL was retained against the target TGBC18TKB cells after long-term culture with immobilized Con A. These CTL did not kill cells of the other 3 malignant tumor cell lines, TKB-1, SQ-5, and GT3TKB, or additionally tested renal carcinoma cell lines, OS-RC-2 (HLA-type A9), Hpt.10 (A2,2402), and TUHR4TKB (A2402,25).

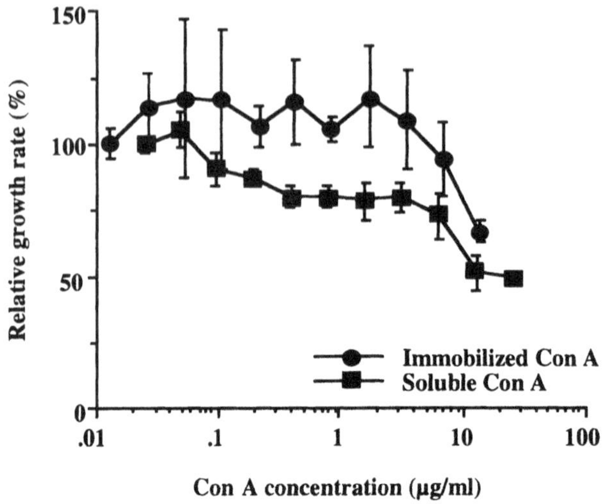

Fig. 2 Effects of immobilized and soluble Con A on the growth of CTL. Dose-dependence of CTL growth in the presence of immobilized or soluble Con A upon 48-h incubation.

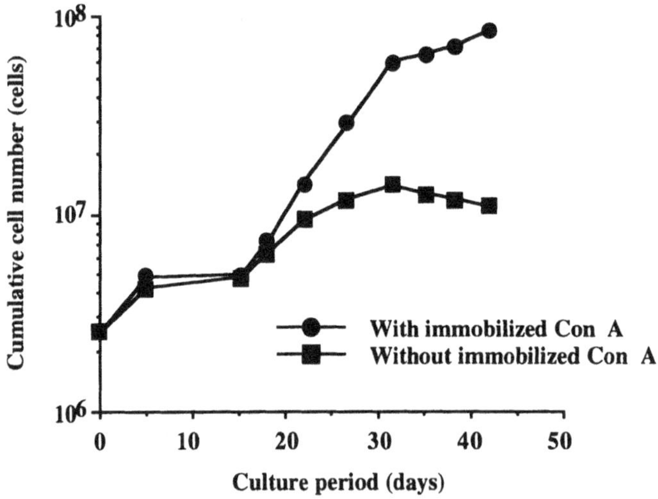

Fig. 3 Growth of tumor-specific $CD4^+$ CTL in the presence or absence of immobilized Con A, without antigen stimulation.

References

Baker, P.E., Gillis, S., and Smith, K.A. (1979) Monoclonal cytolytic T-cell lines. *J. Exp. Med.* **149**, 273-278.

Crowley, N.J., Darrow, T.L., Quinn-Allen, M.A., and Seigler, H.F. (1991) MHC-restricted recognition of autologous melanoma by tumor-specific cytotoxic T cells - evidence for restriction by a dominant HLA-A allele. *J. Immunol.* **146**, 1692-1699.

Hay, R.J., Park, J.G., and Gazdar, A. (ed.) (1994) *Atlas of human tumor cell lines*, Academic Press, Inc., C.A., USA.

Kawakami, Y., Eliyahu, S., Delgado, C.H., Robbins, P.F., Sakaguchi, K., Appella, E., Yannelli, J.R., Adema, G.J., Miki, T., and Rosenberg, S.A. (1994) Identification of a human melanoma antigen recognized by tumor-infiltrating lymphocytes associated with *in vivo* tumor rejection. *Proc. Natl. Acad. Sci. USA.* **91**, 6458-6462.

Liu, S.Q., Saijo, K, Todoroki, T., and Ohno, T. (1995) Induction of human autologous cytotoxic T lymphocytes on formalin-fixed and paraffin-embedded tumour sections. *Nature Med.* **1**, 267-271.

Liu, L.S., Liu, S.Q., Ng, S.Y., Froix, M., Ohno, T., and Heller, J. (1996b) Controlled release of interleukin-2 for tumour immunotherapy using alginate/chitosan porous microspheres. *J. Control. Release* **43**, 65-74.

Liu, S.Q., Shiraiwa, H., Kawai, K., Hayashi, H., Akaza, H., Kim, B.S., Oki, A., Nishida, M., Kubo, T., Hashizaki, K., Saijo, K., and Ohno, T. (1996a) Tumor-specific autologous cytotoxic T lymphocytes from tissue sections. *Nature Med.* **2**, 1283-1283.

Mukherji, B., Guha, A., Chakraborty, N.G., Sivanandham, M., Nashed, A.L., Sporn, J.R., and Ergin, M.T. (1989) Clonal analysis of cytotoxic and regulatory T cell responses against human melanoma. *J. Exp. Med.* **169**, 1961-1976.

Muul, L.M., Speiess, P.J., Drector, E.P., and Rosenberg, S.A. (1987) Identification of specific cytolytic immuno response against autologous tumor in humans bearing malignant melanoma. *J. Immunol.* **138**, 989-995.

Pollard, K. and Holgate, C.S. (1987) Surface antigen preservation by PLP and formalin fixation and the effects of tissue processing. *J. Immunol. Methods* **96**, 145-145.

Quentmeier, H., Klaucke, J., Muhlradt, P.F., and Drexler, H.G. (1992) Role of IL-6, IL-2 and IL-4 in the in vitro induction of cytotoxic T cells. *J. Immunol.* **149**, 3316-3320.

Shinomura, N., Bluestone, J.A., and Sachs, D.H. (1986) Cloned cytotoxic T lymphocytes that recognized an I-A region product in the context of a class I antigen. *J. Exp. Med.* **163**, 972-980.

Slingluff, C.L., Cox, A.L., Stover, J.M.J., Moore, M.M., Hunt, D.F., and Engelhard, A.H. (1994) Cytotoxic T-lymphocyte response to autologous human squamous cell cancer of the lung: epitope reconstitution with peptides extracted from HLA-Aw68. *Cancer Res.* **54**, 2731-2737.

Tjoa, B.A. and Kranz, D.M. (1994) Generation of cytotoxic T-lymphocytes to a self-peptide/class I complex: a model for peptide-mediated tumor rejection. *Cancer Res.* **54**, 204-208.

Wolfel, T., Van Pel, A., Brichard, V., Schneider, J., Seliger. B,, Meyer zum Buschenfelde, K.H. and Boon, T. (1994) Two tyrosinase nonapeptides recognized on HLA-A2 melanomas by autologous cytolytic T lymphocytes. *Eur. J. Immunol.* **24**, 759-764.

ORGANOID FORMATION WITH BRANCHED CAPILLARY-LIKE NETWORKS

Toshiaki TAKEZAWA*, and Katsutoshi YOSHIZATO
*Department of Biological Science, Faculty of Science, Hiroshima University, 1-3-1 Kagamiyama, Higashihiroshima 739-8526, Japan. *Present Address: Tokyo Research Laboratories, KYOWA HAKKO KOGYO CO., LTD., 3-6-6 Asahi-machi, Machida, Tokyo 194-8533, Japan.*

Abstract

A body organ can be simply grasped as a self-assembled multicellular mass containing both mesenchymal cells and epithelial cells which is well vascularized by blood capillaries as a nutrient supplier and a remover of metabolites. To artificially reconstruct such a body organ, we have developed the two key technologies of tissue engineering for the reconstitution of an ideal organoid from animal cells cultured *in vitro*. One is a preparation method for multicellular hetero-spheroids composed of mesenchymal cells and epithelial cells utilizing a thermo-responsive culture substratum. The other is a preparation method for a three-dimensionally reconstituted multicellular mass (3-DRMM) with a medium circulating system utilizing cotton-gauze. In this paper we summarize their characteristics and the application strategy of such a reconstituted organoid in near future.

1. Hetero-spheroids composed of mesenchymal cells and epithelial cells

The thermo-responsive culture substratum was prepared by coating the culture dish with the mixture of type-I collagen and poly-N-isopropyl acrylamide that is a typical thermo-responsive polymer and has a lower critical solution temperature (LCST) of about 30°C[1). We firstly seeded human dermal fibroblasts as typical mesenchymal cells on the substratum and cultured them up to a confluent state. Then, epithelial cells such as human epidermal keratinocytes, rat primary hepatocytes, or a human cholangioadeno-carcinoma cell line (MEC) were seeded onto the fibroblast monolayer. The epithelial cells-attached mesenchymal cell monolayer was detached from the substratum as a cell sheet by decreasing the temperature below LCST. The hetero-spheroid was prepared by culturing it on the non-adhesive substratum such as an agarose-coated culture dish. Figure 1 shows the schematic protocol for the preparation of a hetero-spheroid composed of mesenchymal cells and epithelial cells. The hetero-

Y. Kitagawa et al. (eds.), Animal Cell Technology: Basic & Applied Aspects, Vol. 10, 43–47.

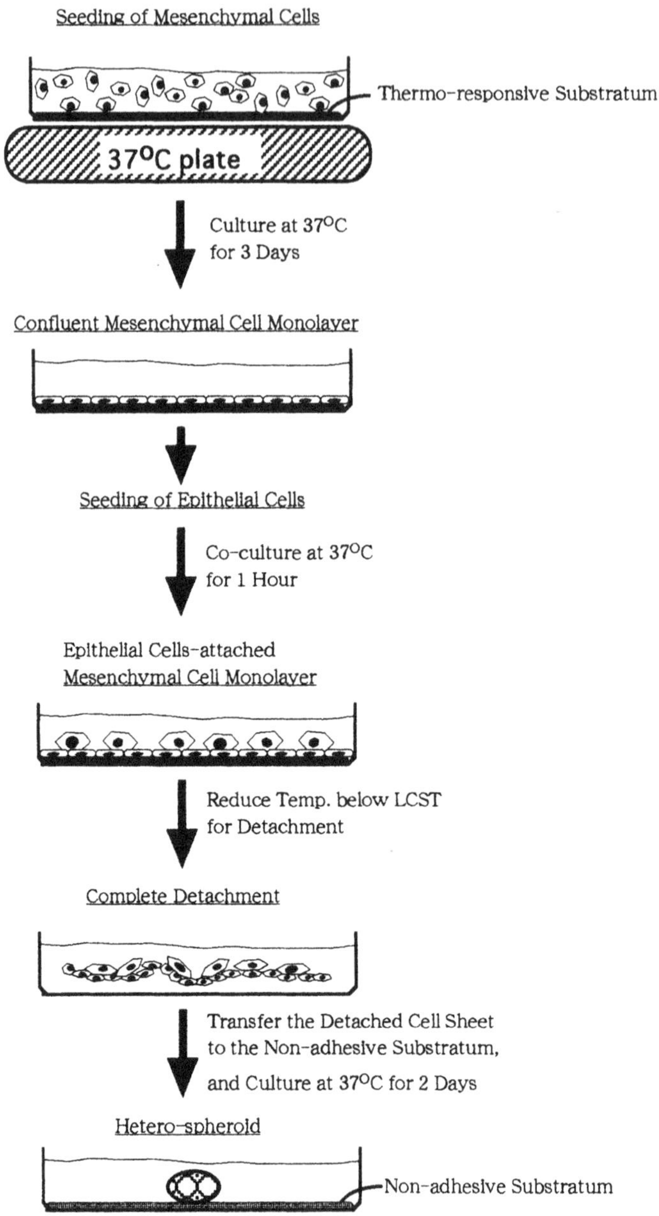

Figure 1 Schematic protocol for the preparation of a hetero-spheroid composed of mesenchymal cells and epithelial cells.

spheroids were immuno-histochemically characterized by using antibodies against keratin, rat albumin, and CA 19-9, depending on the type of epithelial cells incorporated. In the hetero-spheroid composed of fibroblasts and keratinocytes, the keratinocytes migrated to the surface and expressed differentiated phenotypes. In the hetero-spheroid composed of fibroblasts and hepatocytes, both types of cells were randomly distributed in the spheroid which was fully covered with a few layers of squamous fibroblasts and the hepatocytes expressed albumin for up to at least 3 weeks. TEM study of the hetero-spheroid showed the presence of structures morphologically similar to the Disse's space and the bile canaliculus, which are features characteristic of liver. And in the hetero-spheroid composed of fibroblasts and cholangiocarcinoma cells, the carcinoma cells formed duct structures which are similar to the original cancer tissue *in vivo*. These hetero-spheroids were considered as organoids because they morphologically and functionally resembled tissues or organs *in vivo*[2, 3, 5]. However, the viability of cells within the spheroid was solely dependent upon the diffusion efficiency of medium because there were no capillary networks within it[4].

2. A 3-DRMM with a medium circulating system

To improve the viability of cells within the three-dimensional reconstituted multicellular mass (3-DRMM) such as a hetero-spheroid mentioned above, we have tried to incorporate the capillary networks into the 3-DRMM. The cotton gauze-framed collagen substratum was prepared by immersing a cut of cotton gauze in type-I collagen solution and drying it in a culture dish. Human dermal fibroblasts cultured on the substratum adhered well to the collagen-coated fibers of cotton gauze, proliferated well, and formed multilayers of cells in the presence of L-ascorbic acid 2-phosphate. Fibroblast sheets multilayered on the cotton gauze were physically detached from a dish and cultured on the non-adhesive substratum. The detached fibroblast sheet gradually self-assembled and formed a 3-DRMM involving the cotton gauze. The cotton fibers were able to circulate culture medium into the multicellular mass when the end of the gauze was connected with a medium flow system. Pyknotic and necrotic cells were histologically observed in a 21 day-cultured multicellular spheroid with a diameter of 0.6 mm which had been prepared by the above mentioned method without capillary-like networks and medium circulation. In contrast, no such necrotic cells were seen in the same-day-cultured 3-DRMMs with capillary-like networks, in spite of the fact that their sizes were much larger than the spheroid. Fibroblasts in the 3-DRMM elongated well and showed histologically healthy appearance (Figure 2). TEM study revealed the presence of assemblies of collagen fibrils between cells in the 3-DRMM involving the cotton gauze-incorporated collagen fibrils, but not in the conventional multicellular spheroids. Thus, we established a novel culture system of organoids with branched capillary-like networks[6]. Recently, we also found that this new culture method of 3-DRMM is applicable to cells embedded in collagen gels.

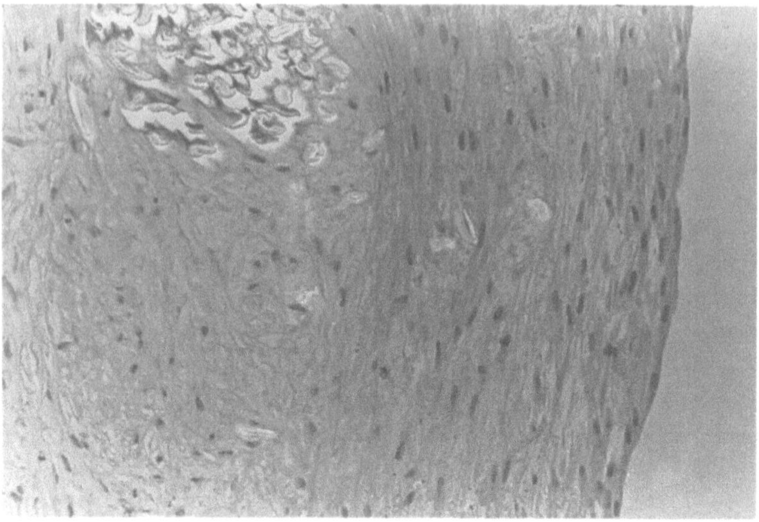

Figure 2 Hematoxylin and Coomassie Brilliant Blue-stained histological section of a 3-DRMM. The 3-DRMM involving the cotton gauze-incorporated collagen fibrils in the medium circulatory system was cultured for 21 days. (x190)

3. Future Perspectives

What is "an ideal organoid *in vitro*"? It should 1) be a three-dimensional reconstituted multicellular mass (3-DRMM) which is formed by self-assembling of epithelial cells and mesenchymal cells, 2) possess a capillary network system which can provide fresh culture medium toward cells and remove cellular metabolites, 3) involve stem and/or progenitor cells to replace the terminally differentiated cells. In the near future, we hope the ideal organoid reconstituted *in vitro* could be applied to the practical stages such as an *in vitro* alternative model of experimental animals to assess drug effects and/or toxicity, an *ex vivo* extracorporeal device to assist a defective organ *in vivo*, and an *in vivo* graft (Figure 3).

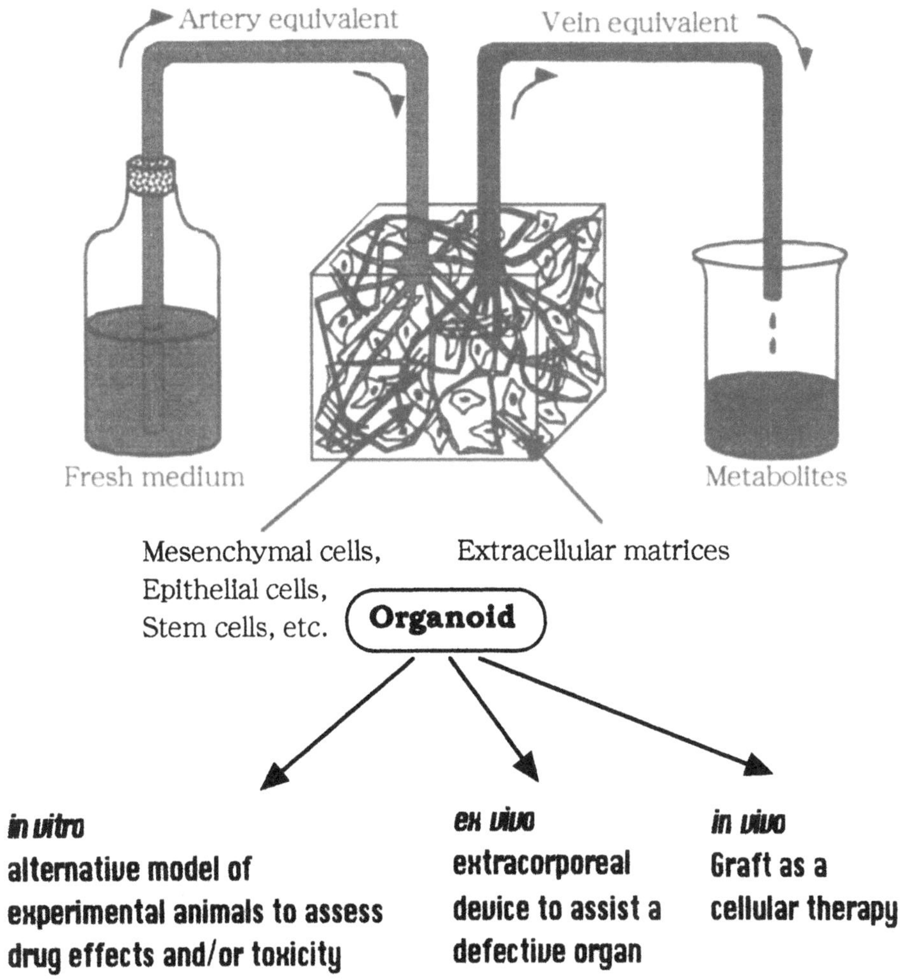

Figure 3 Application strategy of reconstituted organoid.

References

1) Takezawa, T., Mori, Y., and Yoshizato, K. (1990) Cell culture on a thermo-responsive polymer surface, Bio/Technol. **8**, 854-856.

2) Takezawa, T., Yamazaki, M., Mori, Y., Yonaha, T., and Yoshizato, K. (1992) Morphological and immuno-cytochemical characterization of a hetero-spheroid composed of fibroblasts and hepatocytes, J. Cell Sci. **101**, 495-501.

3) Kubota, S., Takezawa, T., Mori, Y., Tomizawa, H., and Takakuwa, T. (1992) A new technique of forming a multicellular spheroid of a human cholangioadenocarcinoma cell line cocultured with fibroblasts on a thermo-responsive polymer, J. Clin. Exper. Med. (Igaku no Ayumi, *In Japanese*) **161**, 539-540.

4) Takezawa, T., Mori, Y., Yonaha, T., and Yoshizato, T. (1993) Characterization of morphology and cellular metabolism during the spheroid formation by fibroblasts, Exp. Cell Res. **208**, 430-441.

5) Takezawa, T., Ito, A., and Mori, Y. (1994) Histological observations on hetero-spheroid consisting of human dermal fibroblasts and human epidermal keratinocytes, Connective Tissue, **26**, 111-115.

6) Takezawa, T., and Yoshizato, K. (1997) Mass transport via naturally branched scaffolds maintains viability of a reconstituted model of connective tissue, Tissue Eng. **3**, 329-343.

CYTOTOXIC ANTIBODIES TO ADIPOCYTES:AGRICULTURAL AND CLINICAL POTENTIAL

S. KESTIN[1], C.E. FUTTER, E. TONNER, R. KENNEDY, S. BUTTERWITH[2], S.M. RHIND[3], H. GRIFFIN[2], A. CRYER[4] and D.J. FLINT
Hannah Research Institute, Ayr KA6 5HL, UK; [1]University of Bristol, Department of Meat Animal Science, Langford, Bristol BS18 7DY, UK; [2]Roslin Institute, Roslin, Midlothian EH25 9PS, UK; [3]Macaulay Land Use Research Institute, Craigiebuckler, Aberdeen AB0 2QJ, UK; [4]University of Wales Cardiff, Cardiff CF1 3XP, UK

Abstract

Hormonal growth promoters (growth hormone (GH), ß-adrenergic agonists, steroids) which improve growth rate and/or lean:fat ratios in the carcass have received considerable adverse publicity and are either banned or have no licence for their use in the EC. This has led to the development of a number of techniques, involving the use of antibodies, aimed at regulating metabolic processes involved in determining growth and body composition.

A different approach to the problem of excess fat deposition involves the use of antibodies directed against the plasma membranes of adipocytes in order to elicit their destruction and thereby limit the storage capacity for fat. This technique has been demonstrated in rats, sheep and pigs in both passive and active immunization techniques. Once again, however, this promising approach is limited by the lack of a commercially suitable vaccine. The identification of individual membrane proteins which are antigenic has been achieved and this provides the prospect of producing recombinant DNA-derived vaccines.

Whether these new approaches will be perceived as acceptable to the general public remains a serious concern and a potential limitation to their development as many would-be sponsors cut back their support for research in these areas.

1. Introduction

Considerable improvements in animal production systems have been achieved through a better understanding of nutritional requirements and through improved disease control. The pursuit of further improvements subsequently led to the development of hormonally-based approaches, principally involving steroid hormones, which have been successfully used world-wide. However, increasing concerns over the use of such compounds from the point of view of safety led to a ban upon their use in the EC. This ban paradoxically served to encourage research into a wide variety of alternative techniques which it was hoped might prove to be more "consumer acceptable".

The object of this review is to discuss a novel approach to manipulate growth and body composition which has developed out of the need to produce long-term effects without the need to use frequent treatment. The technique which will be discussed in detail is an

Y. Kitagawa et al. (eds.), Animal Cell Technology: Basic & Applied Aspects, Vol. 10, 49–54.

approach pioneered by our own research groups involving immunodestruction of adipocytes, the principal lipid storage cells of the body.

Two other points are worthy of mention - firstly that immunomodulation of reproductive process is also an active area of research which will not be discussed in this review (the reader is referred to a review covering this subject (Flint, 1990)) and secondly that all of the areas discussed in this review are potentially amenable to molecular biological approaches, particularly transgenics. The principal distinction to be made is that immunization techniques offer a greater degree of flexibility and speed of response to changing demands but obviously suffer the drawback of the necessity to treat all animals rather than benefit from transmission through the germline.

1.1. ANTIBODIES TO ADIPOCYTE PLASMA MEMBRANES

1.1.1. *Passive immunization*

We have developed a non-hormonal technique which adopts a more direct approach to the problem of excess fat deposition, resulting in a long post-treatment withdrawal period and without the need for slow release formulations such as implants. It involves the production of antibodies to the plasma membrane of adipocytes which are capable of destroying such cells and which leads in many instances to an inability to regenerate lost tissue.

Adipocyte plasma membranes from rats were used to produce antisera in sheep which were highly immunogenic, whilst their cross-reactivity with liver and erythrocytes was low. *In vitro*, such antibodies induced the lysis of collagenase-isolated adipocytes (Flint *et al.* 1986). The antisera were also effective *in vivo* in young rats when administered for 4 days. Within one week of treatment considerable infiltration of fat depots could be seen with histological evidence of adipocyte lysis and cellular infiltration by polymorphs and lymphocytes. This effect disappeared within 2 weeks and resulted in a 75% decrease in adipocyte mass 2 months later (Futter and Flint, 1987). Although not designed to be site-specific the antiserum could act in such a manner; subcutaneous injections producing decreases in subcutaneous fat with lesser effects on internal fat; whilst intraperitoneal injection produced the converse.

Surprisingly appetite and protein deposition increased and feed conversion efficiency improved during the period 3-8 weeks after treatment (Panton, Futter, Kestin and Flint, 1990). Unfortunately, there were unwanted side effects during the 24h post treatment, including reduced food intake and, with some but not all antisera, transitory effects on locomotor function. These side effects remain unexplained. Considerable activation of the immune system occurs rapidly after treatment and this could be part of the explanation, in which case a slow release formulation might reduce or remove these unwanted effects. Indeed the delivery of antibody by subcutaneous route, with slower uptake of antibody, produces less dramatic side-effects than intraperitoneal treatment.

Studies have been extended to other species, including sheep, pigs, chickens and rabbits. In sheep, passive immunization against adipocyte plasma membranes has produced only small effects, reducing both fat content and liveweight gain (Moloney and Allen, 1989; Nassar and Hu, 1991). Although this may reflect a reduction in food intake, our studies using local subcutaneous injections into the backfat of sheep have demonstrated destruction of adipocytes in vivo supporting an effect of the antibody on adipocyte cellularity. In rabbits, large reductions in adipose tissue mass have been achieved within 1 week of treatment with antibodies to rabbit adipocytes although there was a considerable ability to replete fat stores by 2 months (Dulor *et al.* 1990). Attempts to induce adipocyte lysis in the chicken in vivo have so far proved unsuccessful (Butterwith *et al.* 1989), although this is probably due to the use of mammalian antibodies to chicken adipocytes since they are unable to activate chicken complement, an essential component of the destructive process.

Undoubtedly the most successful studies in agricultural species have been achieved in pigs, where subcutaneous injections produced total fat depletion at injection sites which lasted for over 3 months (Kestin *et al.* 1993). Intraperitoneal injection of antiserum also produced changes in carcass composition with a 30% decrease in backfat thickness, a 25% decrease in fat content of the forelimb joint and a concomitant increase in muscle, resulting in no change in body weight or total joint weight. Subcutaneous administration of antibody did demonstrate a potential limitation of this approach as the vast majority of the antibody appeared to be active close to the site of injection presumably limiting its systemic effectiveness.

Passive immunisation as a single-shot treatment is a highly attractive approach because of the long withdrawal period from treatment to slaughter. Unless given in a delay vehicle, antibody delivery in this fashion tends to produce unacceptable, transitory side effects in some instances. There are obvious advantages in administering antibody to animals at as early an age as possible, since the animal is likely to be handled at this stage anyway and the smaller the body weight the smaller the dose required. In fact, it is possible to administer antibodies orally during the first day of life in species such as the pig, where the neonate absorbs immunoglobulin from its mother's colostrum. Such an approach in neonatal pigs, however, failed to produce any effects on growth or body composition. Although antibody was successfully absorbed, the neonatal pig has very little complement and this probably reduced the effectiveness of the antiserum. In addition, however, when the antiserum was administered locally into subcutaneous adipose tissue, it induced fat depletion at the injection site for about 2 weeks, but was followed by subsequent compensatory growth of adipose tissue. By contrast if treatment is given at 3 weeks of age, when complement levels are higher, there appeared to be no compensatory growth of adipose tissue. It seems therefore that the precise timing for treatment will need to be determined to limit recovery and this is likely to differ in different species.

Alternative approaches could include the use of slow release formulations, although we have also already demonstrated the limited effectiveness of prolonged treatment, because the administration of heterologous antiserum leads to the production of neutralising anti-antibodies (Futter and Flint, 1987). It is feasible, however, that such long term treatment of the neonate may lead to a state of immunological tolerance thereby allowing long-term effectiveness of the antibodies.

1.1.2. *Active immunization*

An alternative approach could involve active immunization using adipocyte plasma membranes as an immunogen to provoke an autoimmune response against adipose tissue. We have used this approach with adipocyte plasma membranes, conjugated to a carrier protein to enhance immunogenicity, in rats, sheep and pigs. Although there were consistent reductions in adipocyte cell numbers after immunisation of rats with rat adipocyte plasma membranes the effects were offset by compensatory increases in adipocyte volume, thereby reducing the overall effect on adipose tissue weight. Two studies in sheep, have been conducted, one failing to produce any significant effects, although the second produced a decrease in liveweight gain, fat and lean tissues. By contrast, in pigs immunized with pig adipocyte plasma membranes there were significant decreases in backfat thickness and in fat dissected from the foreloin joint. There was a small though non-significant effect upon dissected lean tissue and body weight giving some cause for optimism, although these effects are clearly still not as great as achieved by passive immunization. Although these preliminary findings are encouraging it is clear that antibody titres achieved after immunization are low and that effects may be improved via alternative immunization strategies, for example in terms of the immunogen itself or in the use of additional techniques to abrogate the natural resistance to the development of an auto-immune response.

1.1.3. *Mechanism of action of antibodies*

Initial studies demonstrated that antibodies could induce lysis of adipocytes in vitro solely in the presence of complement (ie in the absence of cell-mediated killing) (Flint *et al.* 1986). Studies *in vivo* however demonstrated cellular infiltration of adipose tissue suggesting that cell-mediated responses might be important (Futter and Flint, 1987). We subsequently demonstrated that, in complement-deficient rats, antibodies to rat adipocyte plasma membranes were completely ineffective in vivo (Futter *et al.* 1992). Although this does not exclude cell-mediated cytotoxic effects, it clearly demonstrates that complement activation is a critical first step. In considering the potential use of monoclonal antibodies, consideration should be given to their ability to activate complement, which will vary depending upon their subclass. Several other observations suggested that additional mechanisms might also be in operation since, in early studies, we demonstrated that adipocytes which survived the cytotoxic effects of the antiserum were subsequently less effective at storing triglyceride, suggesting a long-term metabolic defect (Futter and Flint, 1987). This did not seem to be due to any defect in their response to insulin (lipogenesis) or noradrenaline (lipolysis) although in the rabbit, there was a profound decrease in lipoprotein lipase (LPL) activity in adipose tissue after treatment with antibodies to adipocytes, which would reduce the potential to store lipid (Dulor *et al.* 1990). Studies in sheep (Moloney, 1990) demonstrated that treatment with antibodies to adipocytes could produce insulin-resistance which again would be expected to reduce lipid accretion in adipose tissue. Although metabolic effects may play a role immediately after treatment it is most likely that long-term effects are principally dependent on cell killing.

1.1.4. *Antibody specificity*

Although antisera to adipocyte plasma membranes might be expected to cross-react widely with other tissues, the relative specificity for adipose tissue can be surprising although in some cases, for example in the chicken, cross-reaction can be high. In the species we have examined so far the relative specificity towards adipose tissue was pig > rat > sheep > chicken, a pattern which fits the respective effectiveness of these antisera to reduce fat deposition *in vivo*. This suggests that the more specific is the antiserum the more effective it is likely to be.

By using affinity chromatography techniques and Western immunoblotting it has been possible to prepare individual "specific" antigens from adipose tissue, to produce antibodies to them and to demonstrate in vivo destruction of adipocytes in pigs and sheep. Production of antigens for immunisation involving large-scale purification from membranes is not likely to be acceptable so the most attractive strategy would seem to be to produce antigens using recombinant DNA technology.

1.1.5. *Precursor cells*

All of the studies described thus far have concentrated upon the mature adipocyte but if targeting of antibodies to precursor cells could be achieved, additional benefits might accrue. For instance, one of the concerns about producing long-term effects of antisera to mature adipocytes is that eventually the cells which have been destroyed will be replaced by the development and differentiation of precursor cells. Thus by preventing or reducing this possibility, longer-lasting effects may be evident. In addition, it would seem inherently more efficient, energetically, to destroy adipocytes before they store triglyceride, rather than to allow them to store it and then have to break it down and mobilise it again. Precursor identification has been a very active research area and it is evident that there are antigens which appear upon differentiation, disappear upon differentiation or which occur upon both precursor and fully mature adipocytes. This last finding suggests that antisera prepared against mature adipocytes should be capable of destroying precursor cells and this has been demonstrated in the chicken (Butterwith *et al.* 1992) as well as in preliminary observations of our own with rat and sheep. However, if precursor proliferation occurs mainly in the foetus or early neonatal period, these may not be appropriate periods for antibody treatment, if a functional immune system is not fully developed. The current role, if any, of cytotoxic damage to precursor cells after passive treatment with antibodies to mature adipocytes, *in vivo*, in pigs, sheep or rats, is unknown.

2. Conclusions

The practical and political limitations upon the use of hormonal growth promoters such as anabolic steroids, ß-adrenergic compounds and GH have led to the search for alternative strategies which are cost-effective and "consumer acceptable".

The use of antibodies as cytotoxic agents for destruction of adipocytes can obviate the need for active immunization techniques if the effects of passive immunoreaction are effectively irreversible or only slowly reversed. This has been demonstrated for adipose tissue in certain circumstances. The ability to use antibodies passively allows their titre and specificity to be determined so that a more uniform response to treatment can be expected and, because treatment effects can last for months, long withdrawal periods can be achieved - a major advantage over hormonally-based approaches.

A number of significant advances in basic research have occurred in recent times which offer considerable opportunities to develop these techniques to practical viability including recombinant DNA derived antigens, chimaeric proteins and antibody engineering. Whilst these may all be of value in demonstrating feasibility the additional challenge in animal production systems must be to achieve this in a cost-effective way.

3. References

Butterwith, S.C., Kestin, S., Griffin, H.D. and Flint, D.J. (1989). Cytoxic antibodies to chicken adipocytes and their precursors: lack of tissue specificity. *British Poultry Science* **30**, 371—378.

Butterwith, S.C., Kestin, S., Griffin, H.D., Beattie, J. and Flint, D.J. (1992). Identification of chicken (Gallus domesticus) adipocyte plasma membrane and differentiation specific proteins using SDS-PAGE and Western blotting. *Comparative Biochemistry and Physiology* **101B**, 147—151.

Dulor, J.P., Reyne, Y. and Nougues, J. 1990. In vivo effects of a treatment with antibodies to adipocyte plasma membranes in the rabbit. *Reproduccion Nutrition et Developpement* **30**, 49—58.

Flint, D.J. (199). Immunomodulatory approaches for manipulation of animal growth, body composition and fecundity. *Biotechnology and Genetic Engineering Reviews* **8**, 97—131.

Flint, D.J., Coggrave, H., Futter, C.E., Gardner, M.J. and Clarke, T.J. 1986. Stimulatory and cytotoxic effects of an antiserum to adipocyte plasma membranes on adipose tissue metabolism in vitro and in vivo. *International Journal of Obesity* **10**, 69—77.

Futter, C.E. and Flint, D.J. 1987. Long-term reduction of adiposity in rats after passive immunization with antibodies to rat fat cell plasma membranes. In *Recent Advances in Obesity Research V* (ed. E.M. Berry), pp. 181-185. John Libbey, London.

Futter, C.E., Panton, D., Kestin, S. and Flint, D.J. 1992. Mechanism of action of cytotoxic antibodies to adipocytes on adipose tissue, liver and food intake in the rat. *International Journal of Obesity* **16**, 615—622.

Kestin, S., Kennedy, R., Tonner, E., Kiernan, M., Cryer, A., Griffin, H., Butterwith, S., Rhind, S. and Flint, D.J. 1993. Decreased fat content and increased lean in pigs treated with antibodies to adipocyte plasma membranes. *Journal of Animal Science* **71**, 1486—1494.

Moloney, A.P. 1990. Immunizing against adipose plasma membranes to reduce body fat: effects on plasma metabolites and insulin. *Biochemical Society Transactions* **18**, 336—337.

Moloney, A.P. and Allen, P. 1989. Growth and weights of abdominal and carcass fat in sheep immunized against adipose cell membranes. *Proceedings of the Nutrition Society* **48**, 14A.

Nassar, A.H. and Hu, C.Y. 1991. Growth and carcass characteristics of lambs passively immunized with antibodies developed against ovine adipocyte plasma membranes. *Journal of Animal Science* **69**, 578—586.

Panton, D., Futter, C., Kestin, S. and Flint, D. 1990. Increased growth and protein deposition in rats treated with antibodies to adipocytes. *American Journal of Physiology* **258**, E985—E989.

IMMUNOLOGICAL MANIPULATION OF LACTATION

C. J. WILDE, C.V.P. ADDEY, F.M. CAMPBELL and M. PEAKER
Hannah Research Institute, AYR, KA6 5HL, UK

Abstract

The rate of milk secretion in the mammary gland is regulated by a complex interaction of peptide and steroid hormones. These hormones, and growth hormone in particular, have been the target of immunological manipulation in dairy animals aimed at controlling milk production. Recently, it has become clear that milk secretion is also regulated by a local mechanism operating within the mammary gland, which modulates the rate of secretion according to frequency and completeness of milk removal. We have shown that this occurs through feedback inhibition by a milk constituent, and have identified a novel milk protein in goat's milk able to inhibit milk secretion *in vivo* and *in vitro* in a concentration-dependent, reversible manner. The protein has been named FIL (Feedback Inhibitor of Lactation), and since it is synthesised in the mammary epithelial cell along with other milk constituents, feedback inhibition is considered an autocrine mechanism. The autocrine mechanism was targetted in lactating goats, which were immunized against FIL three times in declining lactation. After the third immunization, when antibodies against FIL were consistently detected in milk, the decline in milk secretion was significantly reduced compared with sham-immunized controls. Immunization also reduced significantly the fall in milk yield due to once daily milking. These studies show that feedback inhibition is an important determinant of milk secretion rate in lactating animals, and indicate that this local mechanism can be manipulated immunologically to control the productivity of the dairy animal.

Introduction

Immunological manipulation offers a means of controlling the activity of specific target molecules and their biological functions with the minimum of intervention. In animal production, the targets of immunological manipulation, directly or indirectly, have usually been the metabolic hormones important in growth or milk production. For example, active immunization of lactating sheep against somatostatin release inhibiting factor produced an increase in milk yield (Sun *et al.*, 1990), and the effect of growth hormone administered systemically was enhanced by monoclonal antibody against the hormone when the two were given together (Pell *et al.* 1989). Another approach has been to generate hormone mimics in the form of anti-idiotypic antibodies, albeit with variable results. Anti-idiotypes raised

Y. Kitagawa et al. (eds.), Animal Cell Technology: Basic & Applied Aspects, Vol. 10, 55–61.

against growth hormone antibodies increased body weight gain in hypophysectomised rats in a manner analogous to growth hormone (Gardner *et al.*, 1990), but failed to stimulate milk production in lactating cattle (Schalla *et al.*, 1994).

Applied to endocrine mechanisms, the specificity of immunological manipulation may be compromised because the target molecule is, by its nature, pluripotent both in terms of its target tissues and metabolic activity. This disadvantage may be avoided by generation of antibodies specific for one functional epitope on the hormone molecule. Passive immunization with epitope-specific antibodies should then allow silencing of one of the hormone's several metabolic activities. Alternatively, these antibodies could used in active immunization to create anti-idiotypic antibodies which reproduce the biological activity of that hormone epitope. Immunological manipulation is, however, used to best advantage when the target is a tissue-specific event or a monofunctional molecule. The nature of lactation therefore makes it an eminently suitable target, particularly as, apart from their regulation by systemic hormones and growth factors, mammary development and secretory activity are controlled locally, within each mammary gland (Peaker and Wilde, 1996). In this paper we describe experiments in which lactating goats were actively immunized against a protein identified as an autocrine regulator of milk secretion, with the aim of neutralizing feedback inhibition and increasing milk yield (Wilde *et al.*, 1996).

Results and Discussion

Feedback inhibition of milk secretion

In lactating goats we have shown that the rate of milk secretion is regulated acutely by frequency of milking, and that this control occurs through feedback inhibition by a milk protein. The protein, named FIL (Feedback Inhibitor of Lactation) is synthesised by the mammary epithelial cell and secreted, along with other milk constituents, in milk (Wilde *et al.*, 1995). Once in milk, the protein acts on the same luminal epithelial cell to inhibit its secretory pathway (Rennison *et al.*, 1993). Feedback control of milk secretion is, therefore, an autocrine mechanism.

FIL for immunization of goats was prepared from goat's milk as described previously (Wilde *et al.*, 1995). Briefly, defatted milk was centrifuged to sediment casein micelles and the supernatant containing the whey proteins was ultrafiltered to obtain a fraction containing constituents of nominally 6-30 kDa. This whey fraction was subjected to FPLC anion exchange chromatography and one of eight major fractions eluted was found to inhibit the synthesis of casein and lactose in a mammary tissue bioassay. None of the other fractions had any consistent effect. Analysis of this material by chromatofocusing revealed three components, one of which (FIL) accounted in bioassay for all the inhibitory activity, and represented a 40,000-fold purification of the inhibitory activity compared with unfractionated whey. Goat FIL was recognised by antibodies raised against a similar protein in cow's milk. These antibodies counteracted the protein's ability to inhibit milk constituent synthesis in tissue explant culture, indicating that immunoneutralization of feedback inhibition *in vivo* should influence milk secretion rate.

Immunization strategy

The immunization strategy was based on the prediction that, to influence the rate of milk secretion, antibody against FIL had to be present in milk. This was indicated first by the ability of exogenous FIL to inhibit milk secretion when introduced through the teat canal into milk stored in glands of lactating goats (Wilde *et al.*, 1995). A retrograde action by FIL present in milk was also indicated by pbservation *in vitro* that the protein was secreted preferentially, if not exclusively, in that direction. Vectorial secretion of FIL across the apical cell membrane was shown in goat mammosphere cultures. Cells within mammospheres were polarized with the apical membrane facing a central luminal space and the basolateral surface towards the culture medium. Like mammary cells *in vivo*, mammosphere cells also formed tight intercellular junctions, such that the luminal space was sealed. In these circumstances, FIL was detected in luminal secretion and not in the culture medium. Cell polarization within mammospheres, and the ability to open transiently the mammosphere lumina, also allowed FIL to be added on the basolateral side, on the apical side, or on both sides of the secretory cell. At concentrations which markedly inhibited secretion in unpolarized acini cultures, FIL had little effect when added to mammosphere culture medium. Only when the luminal space of the mammosphere was opened by short-term treatment with EGTA in the presence of FIL was there concentration-dependent inhibition of milk protein secretion (D.R. Blatchford and C.J. Wilde, unpublished work). These experiments in cell culture showed that FIL acts on the apical cell surface, so that antibody present only basolaterally would be ineffective.

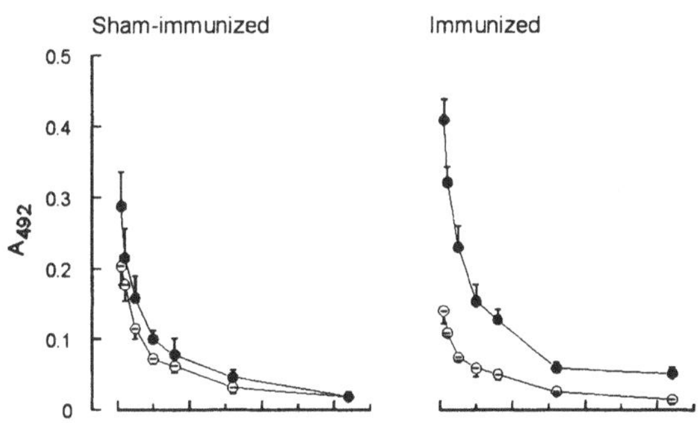

Figure 1. FIL antibody titre in goats immunized or sham-immunized against FIL. Open symbols: pre-immunization. Filled symbols: after third injection.

Anion exchange-purified FIL was conjugated to bovine serum albumin for immunization and administered by intraperitoneal injection (I.P.) to five goats in weeks 10, 18 and 26 of lactation. Six control goats received antigen emulsion without FIL. I.P. immunization was chosed because, unlike most immunization routes, it is known to elicit a local response in the mammary gland (Sheldrake *et al.*, 1985), apparently by stimulating mammary plasma cells (Yurchak *et al.*, 1971; Nickerson *et al.*, 1984). A Pluronic L121-squalene-Tween 80 (12.5:25:1, v/v) adjuvant was tested in preliminary experiments and preferred to Freund's adjuvant's because it caused no inflammation at the injection site.

Blood and milk were sampled at the first morning milking in each week of lactation. Antibody titre in serum and milk was measured by non-competitive enzyme-linked immunosorbent assay. Briefly, microtitre plates were coated with FIL, blocked by standard methods, and then incubated with dilutions of goat serum or a milk immunoglobulin G (IgG) fraction. After extensive washing, binding was detected using donkey anti-goat IgG conjugated to horseradish peroxidase. Antibody titre in serum was assessed by the displacement of pre- and post-immunization sera. By this criterion, the first and second immunizations produced variable and transient increases: a response to immunization was seen in some goats but not in others. After the third immunization, consistent rightward displacement of the dilution curve for post-immunization serum indicated an immune response in all goats (Figure 1). Serum antibody titre remained high for at least four weeks thereafter. Apart from a low and highly-variable response in the milk after the first immunization, antibody was detected at significant levels in the milk only after the third immunization.

Milk yield rsponse to FIL immunization

Goats were milked twice daily until week 33 when one gland of each animal was switched to once daily milking. Single-gland milk yields and time of milking were recorded

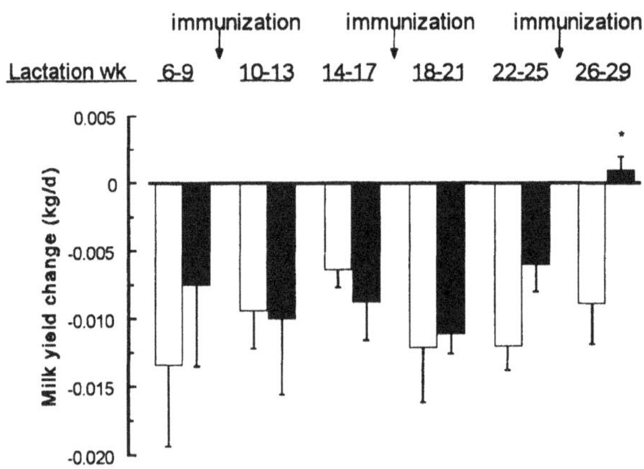

Figure 2. Effect of active immunization against FIL on change in daily milk yield in lactating goats. Mean ± SEM for sham-immunized (open bars) and FIL-immunized goats (filled bars) in successive four-weekly periods. *P< 0.05 compared with sham-immunized animals during the same period. Reproduced from Wilde *et al.*, 1996.

throughout the experiment. Goat milk yield is maximal usually in the second month of lactation and declines thereafter. This decline in yield, which is probably due to a net decrease in secretory cell number (Wilde *et al.*, 1987), was quantified by calculating the mean daily change in yield during successive four-weekly periods, beginning four weeks prior to the first immunization. As expected, the yield of control (sham-immunized) goats declined progressively throughout the experiment. Before the first immunization, and until the third immunization, the decline in milk yield of goats immunized against FIL was similar to that of control animals. After the third immunization, however, when antibody against FIL was present in milk, the decline in milk yield of immunized goats during a four-week period was reduced significantly compared with the preceding four-week period ($P <$ 0.05; Figure 2). During this period, a small increase was recorded in immunized goats (3.0 $\pm$ 0.9%), a value significantly different from the decrease in yield recorded in control animals (11.9 $\pm$ 3.3%; $P <$ 0.05). Therefore, when antibodies against FIL were present in milk, the decrease in the rate of milk secretion with stage of lactation was temporarily reversed.

Switching one gland of control goats to once-daily milking decreased milk yield unitlaterally in that gland, as observed previously (Wilde and Knight, 1990). The change occurred within the first two days, during which time the yield of the once-milked gland was 85% of that in the two days preceding the change in milk frequency. In immunized goats, the decrease in yield with once daily milking was markedly less (92% of previous). This difference between control and immunized goats was significant ($P <$ 0.02) when variation in yield of the contralateral glands was taken into account (by calculation of a relative milk yield quotient (Linzell and Peaker, 1971)). Therefore, the presence of antibody against FIL had protected against the adverse effect of once daily milking, presumably by counteracting the increase in FIL's concentration in milk which, according to the predicted kinetics of autocrine feedback, would be expected with infrequent milking.

Milk composition

Antibodies against FIL influenced milk secretion rate but not milk composition. The concentrations of total lipid, protein and lactose were unchanged in immunized goats after the third injection, and did not differ between glands milked twice and once daily in either control or immunized animals. This is consistent with FIL's postulated role as a mediator of the effect of changing milking frequency on mammary function: when the rate of milk secretion is altered by a change in milking frequency, milk composition is unaffected. Coordinate regulation of milk constituent secretion by FIL may be due to the protein's action on the secretory pathways of the mammary epithelial cell (Rennison *et al.*, 1993). Evidence from primary cell cultures indicates that FIL inhibits constitutive secretion by blocking endoplasmic reticulum to Golgi transport. This is accompanied by dramatic changes in cell ultrastructure, including dispersion of the Golgi apparatus and extensive disruption of the endoplasmic reticulum. There is evidence that the endoplasmic reticulum is involved in lipid secretion as a microlipid progenitor (Keenan *et al.*, 1992), as well as in protein synthesis. FIL's ability to dismantle these intracelluar structures, but in a wholly-reversible manner (Rennison *et al.*, 1993), would thus be expected to inhibit secretion of both the aqueous and lipid components of the cells' secretion.

Conclusion

Isolation of FIL and characterization of autocrine control has provided an opportunity to control milk secretion by means other than hormonal manipulation. It is clear from our studies that active immunization against the FIL protein can mimic the effect on milk yield of an increase in milking frequency. Conversely, immunization can protect milk yield against the adverse effect of once-daily milking. Many questions remain to be answered about the nature of autocrine control in mammary function, and its integration with other intramammary mechanisms, including those activated by mammary distension during prolonged milk accumulation (Peaker, 1980). Immunological manipulation of autocrine control may, nevertheless, offer a novel and effective means of controlling the productivity of dairy animals.

Acknowledgements

This work was funded by the Scottish Office Agriculture, Environment and Fisheries Department. We are grateful to Marian Cartlidge and Shirley Connor for expert technical assistance.

References

Gardner, M.J., Morrison, C.A., Stevenson, L.Q. and Flint, D.J. (1990) Production of anti-idiotype antisera to rat GH capable of binding to GH receptors and increasing body weight gain in hypophysectomised rats. *J. Endocrinol.* **125**, 53-59.

Keenan, T.W., Dylewski, D.P., Ghosal D. and Keon, B.H. (1992) Milk lipid globule precursor release from endoplasmic reticulum reconstituted in a cell-free system. *Eur. J. Biochem.* **57**, 21-29.

Linzell, J.L. and Peaker, M. (1971) The effects of oxytocin and milk removal on milk secretion in the goat. *J. Physiol.* **216**, 717-734.

Nickerson, S.C., Pankey, J.W. and Boddie, N.T. (1984) Distribution, location and ultrastructure of plasma cells in the uninfected, lactating bovine mammary gland. *J. Dairy Res.* **51**, 209-217.

Peaker, M. (1980) The effects of raised intramammary pressure on mammary function in the goat in relation to the cessation of lactation. *J. Physiol.* **301**, 415-428.

Peaker, M. And Wilde, C.J. (1996) Feedback control of milk secretion from milk. *J. Mammary Gland Biology Neoplasia* **1**, 307-315.

Pell, J.M., Johnson, I.D., Pullar R.A., Morrell, D.J., Hart, I.C., Holder, A.T. and Aston, R. (1989) Potentiation of growth hormone activity in sheep using monoclonal antibodies. *J. Endocrinol.* **120**, R15-R18.

Rennison, M.E., Kerr, M.A., Addey, C.V.P., Handel, S.E., Turner, M.D., Wilde, C.J. and Burgoyne, R.D. (1993) Inhibition of constitutive protein secretion from lactating mouse mammary epithelial cells by FIL (feedback inhibitor of lactation), a secreted milk protein. *J. Cell Sci.* **106**, 641-648.

Schalla, C.L., Roberge, S., Wilkie, B.N., McBride, B.W. and Walton, J.S. (1994) Production of anti-idiotypic antibodies resembling bovine somatotropin by active immunization of lactating cows. *J. Endocrinol.* **141**, 203-208.

Sheldrake, R.F., Husband, A.J., Watson, D.L. and Cripps, A.W. (1985) The effect of intraperitoneal and intramammary immunization of sheep on the numbers of antibody-containing cells in the mammary gland, and antibody titres in blood serum and mammary secretions. *Immunology* **56**, 605-614.

Sun, Y.X., Sinclair, S.E., Wynn, P.C. and McDowell, G.H. (1990) Immunization against somatostatin release inhibiting factor increases milk yield in ewes. *Aust. J. Agric. Res.* **41**, 393-400.

Wilde C.J., Addey, C.V.P., Boddy, L.M. and Peaker, M. (1995) Autocrine regulation of milk secretion by a protein in milk. *Biochem. J.* **305**, 51-58.

Wilde, C.J., Addey, C.V.P. and Peaker, M. (1996) Effect of immunization against an autocrine inhibitor of milk secretion in lactating goats. *J. Physiol.* **491**, 465-469.

Wilde, C.J., Henderson, A.J., Knight, C.H., Blatchford, D.R., Faulkner, A. and Vernon, R.G. (1987) Effect of long-term, thrice daily milking on mammary enzyme activity, cell population and milk yield in the goat. *J. Anim. Sci.* **64**, 533-539.

Wilde, C.J. and Knight, C.H. (1990) Milk yield and mammary function in goats during and after once-daily milking. *J. Dairy Res.* **57**, 441-447.

Yurchak, A.M., Butler, J.E. and Tomasi, T.B. (1971) Fluorescent localization of immunoglobulins in the tissue of the cow. *J. Dairy Sci.* **54**, 1324-1325.

LIGHT CHAIN SHIFTING - CHARACTERIZATION OF LIGHT CHAIN REPLACEMENT IN HUMAN PLASMA B CELLS

HIROTAKA HARUTA[1], HIROFUMI TACHIBANA[2] and KOJI YAMADA[2]

[1] Graduate School of Genetic Resources Technology, Kyushu University,
[2] Department of Food Science and Technology, Faculty of Agriculture, Kyushu University, Hakozaki 6-10-1, Higashi-ku, Fukuoka City, Japan

Abstract

Light chain shifting (LCS) is a phenomenon that some human antibody producing plasma B cells lost their original immunoglobulin light chain and begin to express new light chains which are different from original one. Each cells after LCS produced only one type of light chain and cell clone producing more than one light chain protein was not ever found. We have demonstrated that LCS is carried out by a secondary light chain gene rearrangement event and LCS-inducible cells also expressed V(D)J rearrangement-relating factors such as recombination activating genes RAG-1, RAG-2 and terminal deoxynucleotidyl transferase, all of which are usually not expressed in the plasma B cells.

Introduction

B lineage cells recombine immunoglobulin (Ig) gene segments in a highly ordered manner during their differentiation process (1). This mechanism is termed V(D)J recombination and is completed before the immature B cell stage. It was thought that further V(D)J recombination did not to occur after this developmental level. However, recent studies have demonstrated that V(D)J recombination can occur during later stages including immature or even in the mature B cell stages (2, 3). These studies indicate that highly differentiated B lineage cells still have the potential to undergo secondary V(D)J rearrangement under certain conditions.

Light chain shifting (LCS) is a curious phenomenon that plasma B cells, which are thought not able to change their Ig components, replace their Ig light chains (4, 5). NAT-30 (μ, λ) and HB4C5 (μ, λ) cells are human plasma B cell lines which have been already shown to be LCS-inducible (4, 5). NAT-30 is a 6-thioguanine resistant subline derived from Namalwa cells, a human Burkitt lymphoma cell line (6). HB4C5 is a human to human hybridoma which was made using NAT-30 cells as fusion partner. It

Y. Kitagawa et al. (eds.), Animal Cell Technology: Basic & Applied Aspects, Vol. 10, 63–67.

has been shown that Namalwa, NAT-30, and HB4C5 cells produce antibodies of which the heavy and light chains are genetically identical (5). We have previously demonstrated that after concanavalin A (Con A) stimulation, a sub-population of NAT-30 and HB4C5 cells began to produce light chains which are genetically different from original one (4, 5).

The product of recombination activating genes RAG-1 and RAG-2 are essential factors to undergo V(D)J rearrangement and expression of them are generally restricted in the early developmental stages of the lymphopoiesis (7). Terminal deoxynucleotidyl transferase (TdT) is an enzyme which has the ability to add random deoxyribonucleotides to the 3'-OH group of nucleotide fragment in a template independent manner. Expression of TdT is restricted at the pro-B cell stage when Ig heavy chain gene rearrangement occurs (8, 9). TdT introduces short nucleotide sequences termed N region between the D-J or V-DJ segment of the Ig heavy chain genes and are thought to contribute to diversity (9). Ig light chain genes generally lack N region, consistent with the expression pattern of TdT, which is down-regulated in the pre-B cell stage when Ig light chain genes are rearranged (10).

Here, we have characterized LCS phenomenon by assessing the expression patterns of recombination-relating factors and by sequencing the new light chain genes derived from LCS.

Results

LCS-inducible cells were in the plasma stage

Cell surface marker analysis of NAT-30 and HB4C5 cells showed that they have a CD20⁻/CD38⁺ phenotype (Fig. 1), indicating that they are in the plasma B cell stage (11). This result shows that LCS takes place in the plasma stage B cells, which are thought not to change their Ig genes further (1).

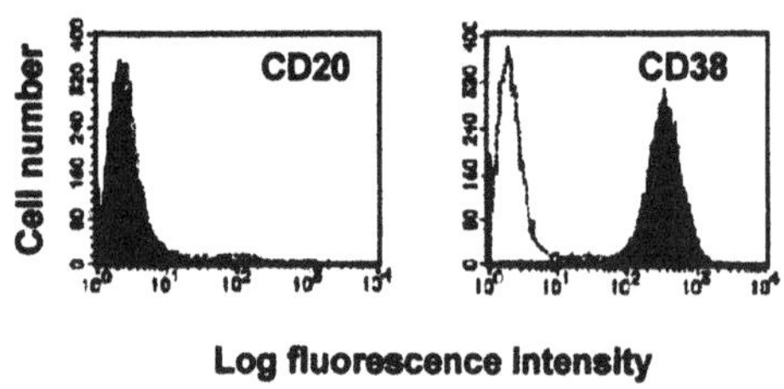

Fig.1 Cell surface analysis of HB4C5 cells. Open area represents negative control. Similar results were obtained for NAT-30 cells (data not shown).

Each light chain shifted cells produced only one light chain protein

New light chain producers were cloned by limiting dilution from Con A-stimulated NAT-30 or HB4C5 cells (4, 5). Immuno blot analysis of supernatants from these new light chain producers revealed that each clone expressed only one type of lambda light chain which was different in size from original light chain (4). New light chain producers were shown to be stable and did not change their light chains further.

LCS was a result of secondary light chain gene rearrangement

Genomic DNA from NAT-30, HB4C5 and new light chain producers CA2 and CA3 was extracted and employed as templates for PCR amplification using the P-CA2V and P-JIR primers followed by Southern hybridization to detect the CA2 VJ coding joint.

Primer and probe positions are shown in the Figure 2. NAT-30 and HB4C5 cells did not have VJ coding joint for CA2 light chain, however, CA2 cells acquired the VJ coding joint (Fig. 2). Similar result has obtained for another new light chain producer CA3 (data not shown), indicating that secondary VJ rearrangement were carried out during LCS. So far as we know, secondary rearrangement in the plasma stage B cells were not reported.

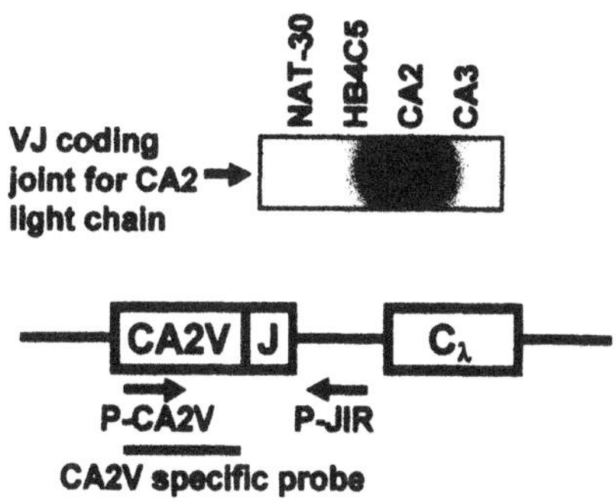

Fig.2 **Detection of VJ coding joints for new light chain produced from CA2 cells, one of new light chain producers. NAT-30 and HB4C5 cells did not have VJ coding joints for CA2 light chain, however, CA2 cells acquired it during LCS.**

LCS-inducible cells were expressing RAG-1 and RAG-2

RAG-1 and RAG-2 are essential factors to carry out V(D)J recombination, however, expression of them are generally suppressed in the plasma B cells (7). We performed RT-PCR followed by Southern hybridization to detect RAG expression and demonstrated that RAG-1 and RAG-2 were expressed in the LCS-inducible cell lines NAT-30 and HB4C5. This result shows that RAG genes were abnormally expressed in the LCS-inducible plasma cells and also supports that LCS is carried out via secondary Ig gene rearrangement (5).

LCS-inducible cells expressed TdT and some new light chains contained N region

Some of the new light chain genes were already been cloned and sequenced (4, 5). Germline V_λ segments, which have a high homology with the new light chain sequences and therefore thought to be the originate segments for them were identified by searching through the DNA sequence database Genbank. Joining and constant (J-C_λ) regions of all new light chains so far cloned were identical except at the beginning point of the J_λ region, indicating that one specific J-C_λ fragment was used predominantly in the secondary rearrangement during LCS. It is known that sterile transcript such as variable region only or J-C_λ region only can be detected before V(D)J recombination (12). Therefore, we have obtained germline sequence of this J-C_λ fragment by cloning the J-C_λ region germline transcript from LCS-inducible cells. Next we aligned new light chain sequences with corresponding germline V_λ and J-C_λ sequences. As shown in the figure 3, some new light chains contained N-region, which is usually absent in the light chain VJ junction. Existence of N-region in some of new light chains indicated the possibility of TdT expression during the LCS process. To ascertain whether TdT is expressed or not, we performed RT-PCR to detect TdT expression in NAT-30, HB4C5 and the T cell leukemia line MOLT-4, which is known to express TdT constitutively (13), as a positive control. Sense primer P-TdTF : 5'-ACT AAA gCT TAT ggA TCC

```
                         Vλ region                          N region          Jλ region

HUMIGL8V       T TAC TGT CAG ACC TGG GGC ACT GGC ATT CA              T TGG GTG TTC GGC GGA GGG ACC AAG C
     clone-1   - --- --- --- --- --- --- --- ---          GC         - --- --- --- --- --- --- --- --- -
     clone-2   - --- --- --- --- --- --- --- ---          GC         - --- --- --- --- --- --- --- --- -
     clone-3   - --- --- --- --- --- --- --- --           AGA            - --- --- --- --- --- --- --- -

V5-4(D87000)   T TAC TGT GAG ACC TGG GAC AGT AAC ACT CA
     clone-4   - --- --- --- --- --- --- --- --- --- --   G            --- --- --- --- --- --- --- --- -
     clone-5   - --- --- --- --- --- --- --- --- --- --   A                --- --- --- --- --- --- --- -
     clone-6   - --- --- --- --- --- --- --- --- ---                 - --- --- --- --- --- --- --- --- -
```

Fig.3 Examples of VJ joining region sequences from some of the new light chains. Upper sequences designate V_λ and J_λ germline fragments. HUMIGL8V and V5-4 are germline V_λ segment registered in Genbank DNA database. The same nucleotide as upper sequence were represented as '-'. Some of the new light chains had N region between V_λ and J_λ regions.

ACC ACg AgC gTC C-3' which is specific to the 5' region of TdT gene and anti-sense primer P-TdTR : 5'-ACT AgC ggC CgC TAg gCA TTT CTT TCC CAC g-3' which is specific to the 3' region of TdT gene was used for amplification. As shown in the figure 4, expression of TdT was detected in the LCS-inducible cell lines NAT-30 and HB4C5. We cloned and sequenced the amplified fragments and concluded that the N-region observed in VJ junctions of new light chains were added by TdT during the LCS process.

Discussion

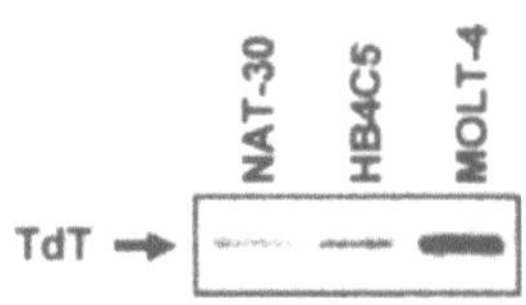

Fig.4 RT-PCR analysis to detect TdT expression. TdT was expressed in the NAT-30 and HB4C5 cells.

We have demonstrated that LCS was carried out by secondary light chain gene rearrangement. Supporting this result, expressions of RAG-1 and RAG-2 were detected in the LCS-inducible cells. TdT was also shown to be expressed and some new light chains contained N region. It has been reported that Epstein-Barr virus (EBV) nuclear antigen 1 (EBNA-1) expression induces RAG-1 and RAG-2 expressions in mature B lymphocytes (14). Namalwa cells, the progenitor line of LCS-inducible cells, is infected with EBV and express EBNA-1 (data not shown). It is possible that RAG expression in Namalwa and LCS-inducible cell lines are induced by EBNA-1 expression. There are increasing reports showing that products of EBV gene modulates expressions of various cell surface antigens, transcription regulators and Ig gene-related molecules of the host cell including RAG-1 and RAG-2 (14, 15). As far as we know, there have been no report describing a relationship between TdT expression and EBV viral product(s), however, there is a possibility that the expression of TdT and other factors needed to carry out LCS as well as RAG-1 and RAG-2 is modulated by EBV gene expression. We speculate that there may be some kind of correlation between EBV infection and acquirement of the ability to induce LCS. Whether or not some EBV-transformed lines other than Namalwa are also able to induce LCS is now under investigation.

Investigating LCS phenomenon may lead to a deeper understanding of the regulation mechanism for immunoglobulin gene rearrangement.

Acknowledgments

This work was supported in part by grants from the Program for Promotion of Basic Research Activities for Innovative Biosciences (PROBRAIN). We thank Perry Seto for reviewing this manuscript.

References

1) Rolink, A., and Melchers, F. (1991) Molecular and cellular origins of B lymphocyte diversity, *Cell* 66, 1081-1094.
2) Tiegs, S. L., Russell, D. M., and Nemazee, D. (1993) Receptor editing in self-reactive bone marrow B cells, *J. Exp. Med.* 177, 1009-20.
3) Papavasiliou, F., Casellas, R., Suh, H., Qin, X. F., Besmer, E., Pelanda, R., Nemazee, D., Rajewsky, K., and Nussenzweig, M. C. (1997) V(D)J recombination in mature B cells: a mechanism for altering antibody responses, *Science* 278, 298-301.
4) Tachibana, H., Kido, I., and Murakami, H. (1994) Heterogeneous expression of human antibody λ chains by concanavalin A-resistant hybridomas lead to changed antigen binding, *J. Biol. Chem.* 269, 29061-29066.
5) Tachibana, H., Ushio, Y., Krungkasem, C., and Shirahata, S. (1996) Inducing the loss of immunoglobulin λ light chain production and the rearrangement of a previously excluded allele in human plasma B cell lines with concanavalin A, *J. Biol. Chem.* 271, 17404-17410.
6) Murakami, H., Hashizume, S., Ohashi, H., Shinohara, K., Yasumoto, K., Nomoto, K., and Omura, H. (1985) Human-human hybridomas secreting antibodies specific to human lung carcinoma, *In Vitro Cell. Dev. Biol.* 21, 593-596.
7) Ma, A., Fisher, P., Dildrop, R., Oltz, E., Rathbun, G., Achacoso, P., Stall, A., and Alt, F. W. (1992) Surface IgM mediated regulation of RAG gene expression in E mu-N-myc B cell lines, *EMBO Journal* 11, 2727-34.
8) Coleman, M. S., Hutton, J. J., De Simone, P., and Bollum, F. J. (1974) Terminal deoxyribonucleotidyl transferase in human leukemia, *Proc. Natn. Acad. Sci. U.S.A.* 71, 4404-4408.
9) Landau, N. R., Schatz, D. G., Rosa, M., and Baltimore, D. (1987) Increased frequency of N-region insertion in a murine pre-B-cell line infected with a terminal deoxynucleotidyl transferase retroviral expression vector, *Mol. Cell. Biol.* 7, 3237-3243.
10) Feeney, A. J. (1990) Lack of N regions in fetal and neonatal mouse immunoglobulin V-D-J junctional sequences, *J. Exp. Med.* 172, 1377-1390.
11) Zandecki, M., Facon, T., Bernardi, F., Izydorczyk, V., Dupond, L., Francois, M., Reade, R., Iaru, T., Bauters, F., and Cosson, A. (1995) CD19 and immunophenotype of bone marrow plasma cells in monoclonal gammopathy of undetermined significance, *J. Clinical Pathology* 48, 548-52
12) Van Ness, B. G., Weigert, M., Coleclough, C., Mather, E. L., Kelley, D. E., and Perry, R. P. (1981) Transcription of the unrearranged mouse C kappa locus: sequence of the initiation region and comparison of activity with a rearranged V kappa-C kappa gene, *Cell* 27, 593-602.
13) Riley, L. K., Morrow, J. K., Danton, M. J., and Coleman, M. S. (1988) Human terminal deoxyribonucleotidyltransferase: molecular cloning and structural analysis of the gene and 5' flanking region, *Proc. Natl. Acad. Sci. U.S.A.* 85, 2489-2493.
14) Srinivas, S. K., and Sixbey, J. W. (1995) Epstein-Barr virus induction of recombinase-activating genes RAG1 and RAG2, *J. Virol.* 69, 8155-8158.
15) Kriauciunas, K. M., Goldstein, B. J., Lipes, M. A., and Kahn, C. R. (1993) Modulation of expression of insulin and IGF-I receptor by Epstein-Barr virus and its gene products LMP and EBNA-2 in lymphocyte cell lines, *J. Cell. Physiol.* 154, 486-495.

CREATION OF HYBRIDOMAS FROM MICE EXPRESSING HUMAN ANTIBODY BY INTRODUCTION OF A HUMAN CHROMOSOME

H. YOSHIDA[1], K. TOMIZUKA[1], H. UEJIMA[2], K. SATOH[1], A. OHGUMA[1], M. OSHIMURA[2] and I. ISHIDA[1]

[1] *Central Laboratories for Key Technology, Kirin Brewery Co., LTD., 1-13-5, Fukuura, Kanazawa-ku, Yokohama, Kanagawa 236, Japan*

[2] *Department of Molecular and Cell Genetics, School of Life Sciences, Tottori University, Nishimachi 86, Yonago, Tottori 683, Japan*

1. Introduction

Monoclonal antibodies are not only used in human diagnostics, but also in human therapy. People cannnot be immunized in vivo with any kind of antigen, and many challenges have been done to obtain human monoclonal antibodies for therapeutic use. Transgenic approach is useful to generate highaffinity monoclonal antibodies. It is known that for each type of Ig chain - kappa light chains, lambda light chains, and heavy chains - there is a separate pool of gene segments from which a single polypeptide chain eventually synthesized. Each pool is on a three different chromosome and contains a large number of genes. But only a limited amount of DNA can be transferred using standard techniques. We have developed a novel procedure to introduce foreign genetic material into mice by using a chromosome itself as a "vector". Human chromosome (hChr.) 14, 22, or hChr.2-derived fragment including Ig heavy, lambda or kappa genes was transferred into mouse ES cells via microcell-mediated chromosome transfer (MMCT). In this study we immunized the chimaeric mice containing hChr.14-fragment by human serum albumin (HSA) and obtained a hybridoma expressing specific antibody to HSA comprised human μ chain.

Furthermore we created the chimaeric mice containing two distinct human chromosomes in order to engineer mice that produce complete human antibody molecules, as an antibody molecule is composed of two identical light chains and two identical heavy chains.

2. Material and Methods

2.1. CONSTRUCTION OF MH(ES) CELLS AND CHIMERA PRODUCTION

Our strategy to introduce human chromosomes into mice is outlined in Fig. 1. MMCT was utilized to introduce a human chromosome mouse ES cells. First, we constructed the libraries of mouse A9 cells containing a hChr. tagged with pST*neo*B suitable for conferring G418-resistance to mouse ES cells. A9/14-C11 and A9/2-W23 were A9 cell lines containing the fragment

Y. Kitagawa et al. (eds.), Animal Cell Technology: Basic & Applied Aspects, Vol. 10, 69–73.

derived from hChr.14 and hChr.2, respectivery. A9/22-G was an A9 cell line containing a intact hChr.22.

Re-tagging of G418^r (G418 resistance)-tagged human chromosomes with puro^r (puromycin resistance) marker was performed as follows. 1 x 10^7 cells of microcell-hybrid A9 cell lines were transferred with pPGKpuro plasmid linearized by SalI digestion. About 200 resultant puromycin-resistant colonies derived from each hybrid A9 cell line were mixed, and microcells prepared from 1x10^8 cells of each clone mixture were fused with wild-type A9 cells. The fused cells were cultured in the presence of both G418 (800mg/l) and puromycin (10 mg/l) for 2 weeks. The resultant double drug-resistant clones were chosen for further analysis. To test whether these two markers are present on the same chromosome, microcells from these four clones were fused with wild-type A9 cells again and the fused cells were first selected with G418 for 2 weeks. G418^r colonies were then cultured in the presence of puromycin for 3 days. By this second round of MMCT, more than 50% of the G418^r colonies obtained from clones (A9/2-W23P originally derived from A9/2-W23 and A9/22-G2, A9/22-G3 from A9/22-G) expected to contain a double tagged human chromosome were resistant to puromycin as well.

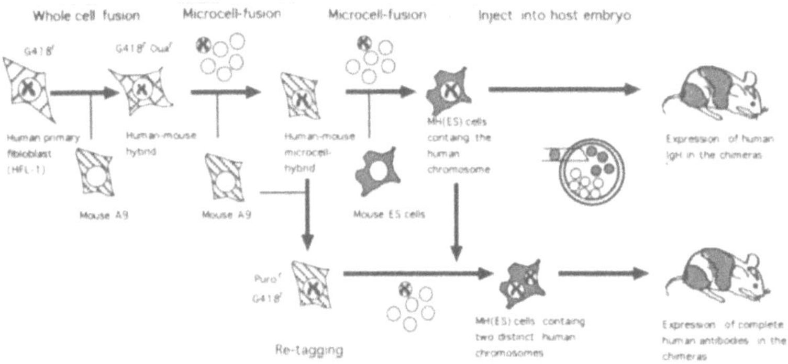

Figure 1. A schematic diagram showing the constract of MH(ES) cells to produce chimeric mice containing human chromosomes. G418^r: G418-resistant, Puro^r: puromycin-resistant Oua^r: ouabain-resistant

Transfer of a second human chromosome into the MH(ES) 14-4 cells, which is derived from TT2 ES cell line and retains a G418^r-tagged hChr.14-fragment, by MMCT was performed as described. 1 x 10^7 cells of recipient MH(ES)14-4 line were used for the fusion with microcells from 1 x 10^8 cells of donor hybrid A9 lines. Puromycin (0.75 μg/ml) selection was done 24 h after microcell fusion. The resultant puro^r colonies were picked between day 7 and day 9 after the selection and cultured in the presence of both G418 (300 μg/ml) and puromycin (0.75 μg/ml). Each clone was analyzed by PCR.

From ten to twenty MH(ES) cells from each cell line were injected into a 8-cell stage embryo derived from Jcl:MCH(ICR) mice (CREA JAPAN, INC.). Injected embryos were then transplanted to the uteri or oviducts of pseudo pregnant recipients and allowed to proceed to term.

Chimerism in a resulting offspring was determined by extent of coat pigmentation. The TT2 line, derived from C57BL/6xCBA-F1 embryo, gives an agouti coat color in an albino MCH(ICR) background.

2.2. HYBRIDOMAS

The chimaeric mouse bearing human chromosome #14 was immunized intraperitoneally with 50 μg of human serum albumin (Sigma) emulsified in RIBI adjubant (RIBI ImmunoChem Research) on days 0, 7, 14. The spleen was removed from the immunized mouse. The spleen cells were fused with a P3X63-Ag8.653 myelomas using PEG. After a fusion, the cells were seeded in 96-well plates in HAT supplemented medium. After HAT selection, G418 was supplemented to select human chromosome containing clones. The number of wells positive for hybridoma growth was determined visually and the human antibody-secreting hybridomas were screened by ELISA.

2.3. ASSAYS

Human immunogloblin were assayed using anti-human antibodies immobilized on ELISAplates and detected with peroxidase-conjugated anti-human immunogloblin antibodies as described before. Hμ/hκ was assayed using mouse monoclonal anti-hk (Kirkegaard & Perry Laboratories) immobilized on ELISAplates (Corning) and detected with peroxydase-conjugated mouse anti-hμ (The Binding Site Limited). Similarly, hμ/hλ were assayed using mouse monoclonal anti-hλ (Kirkegaard & Perry Laboratories) immobilized on ELISAplates and detected with peroxydase-conjugated mouse anti-hμ (The Binding Site Limited). Human IgM (Organon Tekinka) and human IgM/λ (Caltag) were used for the standard. Anti-HSA human immunogloblin μ were assayed using HA immobilized on ELISAplates and detected with peroxidase-conjugated anti-human immunogloblin antibodies (The Binding Site Limited) using human IgM (Organon Tekinka) for the negative control. The samples, standard, and antibody conjugates were diluted with mouse serum (Sigma) supplemented PBS. TMBZ (Sumitomo bakelite) was used for substrate, and absorbance at 450 nm was measured using a spectrophotometer (Bio-tek instrument).

3. Results

3.1. CREATION OF HYBRIDOMAS FROM THE CHIMERA CONTAINING A HUMAN CHROMOSOME 14 FRAGMENT

The spleen was removed from chimera containing hCHr.14 fragment and were fused with a P3X63-Ag8.653 myelomas. The number of growth positive wells in HAT medium was 670. The number of growth positive wells in G418 medium, it was supposed to contain human chromosome, was140. The frequency of G418 resistant clones was similar to that estimated from coat color of mouse. Six human antibody positive clones was obtained. The anti-HSA human Ig μ positive wells were cloned by limiting dilution culture. The clone was named H4B7 (Table 1).

The amino acid sequences were deduced from variable region of human antibody cDNA derived from clone H4B7. It was revealed that the H4B7 hybridomas contained a combination of genes for VH4 family and JH2. These results show that the chimaeric mouse retaining human #14 partial fragment produced a functional human antibody heavy chain protein (data not shown).

TABLE 1. Human antibbody positive wells after fusion of B cells with murine myeloma cells

	positive wells
Growth (HAT selection)	670
Growth (G418 selection)	140
Human Igμ	6
Anti-HSA human Igμ	1

3.2. VIABLE CHIMAERIC MICE PRODUCED FROM MH(ES) CELLS RETAINING TWO INDIVIDUAL HUMAN CHROMOSOME FRAGMENTS

The double drug-resstant (G418^r and puror) clones, A9/2-W23P derived from A9/2-W23 and A9/22-G2 from A9/22-G, were shown to retain double tagged chromosomes by a second round of MMCT and two-color FISH analysis with a human-specific COT-1 DNA and pPGKpuro probes. Each puror-tagged human chromosome was transferred into the MH(ES)14-4 cells. AS a result, puro^r clones were obtained at a frequency (1-8 x 10^{-7} per fused ES cell) comparable to those of previous MMCT experiments. MH(ES)14.2-5,6 and MH(ES)14.22-5 were obtained by using microcells prepared from A9/2-W23P and A9/22-G2, respectivery. For the chimera production,ten to twenty cells of each MH(ES) cell line were injected into 8-cell embryos obtained from albino MCH(ICR) mice. Twenty-six, eight and sixteen mice were born from 160 (MH(ES)14.2-5), 115 (MH(ES)14.2-6), 302 (MH(ES)14.22-5) injected embryos. seven chimeras with 3-15% agouti coats, 2 chimeras with 40-50% agouti coats and 5 chimeras with 3-15% agouti coats were obtained from MH(ES) 14.2-5, 14.2-6, and 14.22-5.

3.3. EXPRESSION OF COMPLETE HUMAN ANTIBODIES IN THE CHIMAERAS

Serum samples obtained from the chmaeras, non-immunized or immunized, were analyzed by ELISA for detection of hμ, hκ, or hλ polypeptides. Two of seven chimaeras from MH(ES)14.2-5, both chimaeras from MH(ES)14.2-6 and all five chimaeras from MH(ES)14.22-5 were tested. As expected, hμ (0.2-5.0 mg/l) was detected in the sera of all tested chimaeras. In addition, hκ (0.6-6.0 mg/l) was detected in three of four chimaeras produced from MH(ES)14.2-5 or MH(ES)14.2-6, and hλ (1.5-22 mg/ml) was detected in all 5 chimaeras from MH(ES)14.22-5. Table 2 shows the results of representative chimaeric individuals that we choose for demonstrating the production of complete human antibodies. As a result, significant levels of complete human antibody molecules (hμ/hκ or hμ/hλ) were detected in all the tested serum samples obtained from the chimaeras.

TABLE 2. Serum titters of human Igs in chimaeric mice

MH(ES)cell lines	Chimeras	Chimerism(%)	hμ(mg/l)	hκ(mg/l)	hμ/hκ(mg/l)
MH(ES)14.2-5	C14-2-15	10	0.2	1.0	0.075
MH(ES)14.2-6	C14.2-26	40	1.5	1.3	0.018
MH(ES)14-4	C14-15	50	6.4	n.d.	<0.002
MH(ES)2-1a	C2-8	70	n.d.[b]	25	<0.002
				hλ(mg/l)	hμ/hλ(mg/l)
MH(ES)14.22-9	C14.22-9	25	0.3	1.5	0.050
	C14.22-102-3	10	0.8	4.4	0.094
	C14.22-10		5.0	22	0.33
	C14.22-12	30	0.8	3.6	0.051
	C14.22-12		3.7	7.0	0.050
MH(ES)14-4	C14-15	50	6.4	n.d.	<0.002
MH(ES)22-1a	C22-2	50	n.d.	18	<0.002
	C22-15+C22-2[a]		n.d.	n.d.	<0.002

[a]An equal volume of serum samples prepared from the two chimaeras(C14-15 and C2-8, or C14-15 and C22-2) was mixed together, and used as negative controls.

[b]n,d., not determined.

4. Discussion

We demonstrated that conventional hybridoma technology can be used to obtain a human sequence antibody from chimaeric mouse bearing a human chromosome. Furthermore complete human antibody molecules (hμ/hλ or hμ/hκ) were detected in the sera of the chimeras produced from MH(ES) cells containing both the hIgH and hIgL losi.

On the other hand, the frequency of human antibody producing hybridoma clones as well as the levels of complete human antibodies in the chimaeras were low. Improving the levels of complete human antibodies in the chimaeric mice is required to obtain human monoclonal antibodies specific to a variety of antigens. We are now constructing such an ES cell line by multiple rounds of gene-targeting.

5. Acknowledgements

We thank K.Hanaoka for technical advice and valuable discussions; S. Watanabe, T. Yagi and P.W. Laird for giving pPGK*puro*.

TRANSGENICALLY PRODUCED BIOPHARMACEUTICALS

M.W. YOUNG
Genzyme Transgenics
5 Mountain Road
Framingham, MA 01701-9322, USA

1. Transgenically Produced Biopharmaceuticals

The goal of transgenic biopharmaceutical production is the efficient production and recovery of commercially important proteins from the milk of transgenic animals. A transgenic animal carries specifically altered DNA, usually from a different species ("heterologous DNA"), that is stably integrated into its genome. In the case of biopharmaceutical production, this foreign DNA contains the genetic information that specifically directs secretion of a target recombinant protein into milk, along with the host animal's own milk proteins, during lactation after the birth of offspring.

Goats are well suited for biopharmaceutical production because of their reasonably short gestation and maturation times, high milk yield and protein content, and general acceptance as a source of dietary milk. Goat milk has been extensively characterized biochemically, which facilitates development of downstream processes to recover the transgenically produced recombinant protein. Genzyme Transgenics also is developing other transgenic species for selected applications, such as cows for very large quantities (tonnes) or rabbits for small quantities (grams) of highly valuable proteins.

Genzyme Transgenics has produced a broad range of peptides and proteins for a variety of applications, including:

(a) *Parenteral drugs*, such as antithrombin-III (rhATIII) for restoring ATIII levels in patients with acquired ATIII deficiency, alpha-1 proteinase inhibitor (rhA_1PI) for treating inherited A_1PI deficiency and cystic fibrosis, and recombinant monoclonal antibody against Lewis antigen (BR96) for targeting anti-cancer drugs to breast cancer cells *in situ*.

Y. Kitagawa et al. (eds.), Animal Cell Technology: Basic & Applied Aspects, Vol. 10, 75–79.

(b) *Orally active drugs,* such as glutamic acid decarboxylase (GAD) for oral tolerization in treating insulin-dependent diabetes mellitus.

(c) *Second generation versions* of approved recombinant drugs such as insulin, human growth hormone, blood clotting factors and interferon. These generally are products with known markets and proven efficacy, for which transgenic production offers advantages such as safety, cost, or large-volume availability.

(d) *Nutraceuticals,* which are foods with a claimed health benefit, such as nutritional supplements and infant formula additives.

2. The Transgenic Production System

1. BIOLOGICALLY ACTIVE PROTEINS AND ANIMAL HEALTH

Both animal welfare considerations and regulatory requirements specify that transgenic animals used for biopharmaceutical production be normal and healthy, especially when they are expressing biologically active human proteins in their milk. The mammary gland is an exocrine organ that generally sequesters its heterologous transgene products away from the circulation and then removes them from the animal. Genzyme Transgenics has successfully expressed high levels of biologically active monoclonal antibodies and rhA_1PI in goats, as well as the lactation-regulating hormone prolactin in mice (3), with no untoward effects on animal health or milk production.

To reduce the risk of potential side effects, such as physiological or reproductive problems resulting from low-level leakage of recombinant protein from the milk into the circulation (2). Genzyme Transgenics has developed new milk-specific promoters that exert tighter tissue and temporal regulation of transgene expression. These promoters permit high-level expression of biologically active proteins without affecting the health or lactation of the production females. Another strategy is to design special constructs to secrete an inactive or precursor form of the target protein, thereby eliminating the risk from particularly potent bioactive recombinant proteins.

2. NON-SECRETED PROTEINS

Genzyme Transgenics has developed proprietary strategies for producing proteins that normally are not secreted by the cells that make them. Intracellular or membrane-bound proteins are integral to the cell.. Ordinarily, in cell culture systems, they must be harvested by cell lysis and protein extraction. Such proteins are not likely to be secreted into milk either, even from targeted transgene constructs. However, specific regulatory DNA sequences may be incorporated into the construct to redirect intracellular protein trafficking. This approach was used successfully to produce transgenic mice secreting high levels of the intracellular proteins GAD and myelin basic protein (MBP).

3. NON-SECRETED PROTEINS CONTINUED

In another approach, Genzyme Transgenics was the first to take advantage of the targeting of membrane-bound proteins to milk fat globule membranes. The fat globules are secreted into milk as usual and the recombinant proteins are recovered by proprietary methods. This strategy was used successfully with the cystic fibrosis transmembrane regulator (CFTR), a potential therapeutic agent for cystic fibrosis (1).

4. FUSION PROTEINS

For some applications, the desired end-product is a fusion protein that combines the active region of a therapeutic protein with sequences that target a particular cell or organ in the body. Such combinations may be produced by splicing the relevant gene sequences from two different proteins together in a single transgene construct and expressing this in milk. For example, under a collaborative agreement, Genzyme Transgenics has produced transgenic mice expressing gram-per-liter amounts of a fusion protein combining portions of an anti-tumor antibody with a biologically active enzyme. This structure is designed to target the enzyme to cancer cells *in situ*, where it will activate a "pro-drug" form of an anti-cancer drug. The goal is to administer effective anti-cancer drugs to patients in a less toxic form to reduce the devastating side effects.

3. The Economics Of Transgenic Production

Biopharmaceutical production in transgenic animals currently is driven by the need for recombinant proteins in quantities that cannot be provided economically from other sources. The worldwide markets for blood-derived proteins such as recombinant human ATIII approach 100 kg per year, and the market for human serum albumin could exceed 500 metric tonnes if it were available in such quantities.

Production costs will depend in part on herd size.

Capital costs for raw material production will be significantly less for transgenic production than for cell culture (Table 1). A state-of-the-art farm and dairy facility may be established for several million dollars, which is an order of magnitude less than the cost of a cell bioreactor with comparable capacity. This reduces the capital commitment for production facilities, especially early in the product development life cycle before clinical efficacy is established. In addition, raw material production capacity may be expanded as needed by incrementally increasing the size of the production herd. This is considerably faster and less expensive than building and validating a new cell culture facility.

Protein recovery costs also have the potential to be significantly less for transgenic production systems. Transgenics offers three relevant technical advantages. First, milk as a raw starting material may be more consistent from batch to batch compared to cell lysate or cell culture supernatant.

Second, the recombinant protein concentration in the milk is much higher than for cell culture supernatant or lysate -- often up to tens of grams per liter. This minimizes protein losses and reduces raw material input volume, which in turn decreases the need for

Table 1. Comparison of monoclonal antibody productivity in cell and transgenic goat milk for 100 kgs purified protein

	CHO cells	Goat
Synthesis	Batch culture	Mammary gland
Expression level	1 g/L, 10 day cycle	8 g/L per day, 300 days
Process Recovery	60%	60%
Annual output for 100kg of purified protein	170,000	21,000 liters
Days per year	200 (= 20 runs)	300 @ 2 liters/goat/day
Reactor capacity	8,500 Liter*	35 goats (= 21,000)
Cost/gram	\$300 - \$3,000**	\$105***

* Requires seed culturing.

** Some groups have reported the capability to consistently obtain large scale yields of 1 g/L. For proteins other than monoclonal antibodies, there is little reported evidence in support of this level of productivity. Industry figures vary widely being dependent on scale of operation, expression levels, capital cost and batch failure rates.

*** Cost calculated by Genzyme Transgenics with certain assumptions: see section on economics.

water, buffer salts and chromatography media, and reduces the size of processing vessels. Finally, large-scale production lends itself to the use of efficient industrial process engineering and technologies that are already established in the dairy industry and in other large-scale process industries.

Cost effective transgenic production of recombinant proteins at a very large scale -- exceeding 500 kg per year -- creates a new opportunity to develop highly cost-effective protein recovery processes. Genzyme Transgenics has formed several collaborations to investigate a variety of combinations of existing purification approaches, large-volume techniques from the dairy industry, modern chromatography methods that address complex feedstreams, and new highly selective ligand chemistries for protein recovery. Ultimately, the Company expects to reduce overall production costs to levels never before achievable. Although unit production costs are likely to be correspondingly higher for proteins produced in smaller quantities, these costs are at least competitive with other production methods at small scales.

4. References

1. DiTullio, P., S. H. Cheng, J. Marshall, R. J. Gregory, K. M. Ebert, H. M. Meade, and A. E. Smith. Production of cystic fibrosis transmembrane conductance regulator in the milk of transgenic mice. Biotech 10: 74-77, 1992.

2. Echelard, J. Recombinant protein production in transgenic animals. Curr Opinion Biotech 7: 536-540, 1996.

3. Wilburn, B., L. Woodworth, A. Gronbeck, J. Lewis-Williams, M. Harrington, D. Pollock,S. M. Richards, H. Meade, and Y. Echelard. High-level expression of recombinant human prolactin in the milk of transgenic mice. Theriogenology 47: 219, 1997.

OPTIMISATION OF TRANSGENE EXPRESSION IN THE MAMMARY GLAND

C. B. A. WHITELAW and A. J. CLARK
Roslin Institute (Edinburgh) Roslin Midlothian EH25-9PS, UK

1. Introduction

Classical animal breeding regimes rely on the importation of genes by cross-breeding and are therefore limited to those traits already present within a given species. The development of gene transfer technology which enables the generation of transgenic animals - an animal that carries integrated sequences of cloned DNA in its genome - allows these limitations to be overcome. Firstly, the introduced DNA can be derived from species other than the host and, secondly, it can be modified *in vitro* prior to being introduced into the germline.

It has been proposed that transgenesis may allow the generation of farm animals with altered phenotype (e.g. enhanced growth rate), animal models for research and animal bioreactors. It is this last opportunity which has seen the most progress over the last decade, with the expression of a variety of proteins having been targeted to milk of sheep, cattle, pigs, goats and rabbits (Wilmut and Whitelaw, 1994). Indeed, this application of gene transfer in livestock is fast approaching commercial success, with the first products currently in clinical trials.

Research in this topic has been focused on two aspects: the optimisation of transgene expression and the development of methods of DNA introduction. This presentation is primarily concerned with the former but will briefly describe recent advances at the Roslin Institute which concern the development of cell culture based approaches for the generation of transgenic large animals. It should be noted, that in the context of optimising

Y. Kitagawa et al. (eds.), Animal Cell Technology: Basic & Applied Aspects, Vol. 10, 81–89.

transgene expression, many of the problems encountered and which require to be overcome are probably common to all types of gene transfer.

2. Use of Cell Culture Models

The time and expense involved in generating transgenic animals and then evaluating the transgene expression pattern is very restrictive. It follows that if questions about the ability and efficiency of expression could be asked *in vitro* considerable progress could be achieved.

The appropriate transcription of eukaryotic genes requires that the gene sequence is present within chromatin and that it has seen all the relevant developmental cues that it would normally be exposed to. Therefore, different experimental assay systems each have an inherent stringency with regard to identifying transcription control elements. The least stringent assay, transient transfection of cells grown in culture, does not result in the appropriate chromatin structure being formed. The appropriate chromatin structure can be established in stably-transfected cells but the gene does not experience the full repertoire of developmental signals it is normally exposed to. In transgenic animals, however, both criteria are achieved. In addition, there is no apparent selection of integration sites, unlike stable transfection methodology where integration is selected for by expression of a resistance gene. Stably transfection experiments, therefore, may only identify transcriptionally permissive integration events.

Thus, the search for DNA regulatory elements capable of directing appropriate transgene expression *in vitro* is limited by the apparent inability of cell culture systems to reliably model what is required for transcription *in vivo*. This limitation may not include post-transcriptional events (Donofrio *et al.*, 1996), potentially allowing RNA processing and translational enhancing mechanisms to be evaluated *in vitro* prior to incorporation into transgenic studies. We anticipate that the next generation of transgenes will incorporate specific sequence elements, evaluated *in vitro*, that are capable of enhancing post-transcriptional events.

3. Targeting Expression to Milk

With regard to generating animal bioreactors, the most successful application of this technology has been to target expression of biomedically important proteins to milk. This requires the generation of a hybrid transgene. The hybrid transgene comprises DNA control elements from a gene normally expressed in the mammary gland, which direct where and when that gene is expressed, linked to the coding region of the gene of interest.

The first stage in targeting expression to milk entails the characterisation of a mammary specific gene promoter. This element should be evaluated for its ability to and the efficiency by which it can direct expression to the appropriate tissue. In our studies we have utilised the ovine β-lactoglobulin gene promoter, which is expressed efficiently and predominantly in the mammary gland of transgenic mice (Whitelaw *et al.*, 1992). This transcriptional control element is then linked to the coding sequences of interest, generating a hybrid transgene. There are currently several examples of hybrid transgenes which direct high level expression of human pharmaceuticals in milk, e.g. α1-antitrypsin (Archibald *et al.*, 1990; Wright *et al.*, 1991). Indeed, it is now generally accepted that at least most of the milk protein gene promoters are capable of targeting expression to the mammary gland (Wilmut and Whitelaw, 1994). It remains unclear, however, as to the relative efficiency by which each of these promoters accomplishes this.

4. Position-Effects

Most transgenes suffer from position-effects which are due to the conformational demands imposed by the chromatin structure at the site of transgene integration, the complexity of the transgene array and the actual sequences present within the transgene. More and more, it is becoming apparent that the influence position-effects have on expression can vary considerably.

A commonly described example of position-effects concerns transgenes which lack introns (Webster *et al.*, 1997), with several studies indicating that hybrid transgenes based on genomic sequences are expressed much more efficiently that those based on cDNA sequences. In this regard, we have suggested that intronless or prokaryotic sequences may actively serve as foci for gene silencing (Clark *et al.*, 1997). Thus, attempts to efficiently express a cDNA-based transgene provide a stringent test for putative transgene expression strategies.

The expression potential for a given transgene at a given integration site must reside in the chromatin structure the transgene either imposes on itself, regardless of the neighbouring chromatin, or the chromatin structure it finds itself in. As a corollary of this, it is reasonable to propose that strategies which engineer the transgene integration site could enhance the probability of efficient expression.

An obvious route to engineer the site of transgene integration is to insert the transgene at a specific, predetermined, permissive site. This is possible in mice, by utilising the high-frequency of homologous recombination which can be achieved in mouse embryonic stem (ES) cells (Hooper, 1992). Indeed, in the mouse system, this methodology has proven to be a powerful and revealing scientific tool. Unfortunately, even though considerable effort has been applied, no non-murine ES cell lines exists, thus this approach has not been possible for livestock species.

Since the site of integration could not be engineered in transgenic livestock (but see section 5 below), attention was directed to engineering the transgene. A variety of alternative approaches have been pursued.

4.1. DOMINANT REGULATORY ELEMENTS

It is hoped that position-effects could be minimised through the inclusion of dominant regulatory elements within the transgene design and/or through the use of large transgene fragments. It is generally accepted that large genomic transgenes, e.g. yACs (Schedl *et al.*, 1993) are expressed in the expected manner. In most cases, however, the actual sequences and mechanisms which ensure this efficient expression *in vivo* have not be identified. Dominant regulatory elements include locus control regions (LCR), e.g. the LCR of the human β-globin locus (Grosveld *et al.*, 1987), and gene-domain boundary elements (Wolffe, 1994). Unfortunately, although these elements can be effective, they do not always act in a reliable manner (Guy *et al.*, 1996).

4.2. OPTIMISING EXPRESSION OF cDNAs

For many applications, it is desirable to achieve efficient expression of small transgenes, e.g. based on a cDNA. In these cases, efficient expression has been achieved through the inclusion of dominantly acting elements, but this approach has proven to be unreliable. Furthermore, and in relation to our work, no LCR which functions in a mammary-specific manner has been characterised.

Given the lack of a mammary LCR and the desire to express cDNA-based transgenes we have developed a 'transgene rescue' strategy entailing the co-injection of the hybrid transgene with an efficiently expressed β-lactoglobulin genomic transgene (Clark *et al.*, 1992). Co-injection of two transgenes results, in the majority of lines generated, in the co-integration of both transgenes at the same chromosomal site. This can result in a dramatic increase in the expression frequency observed for the hybrid transgene .

The conclusion from these types of studies must be that the overall (trans)gene chromatin architecture is crucial for efficient expression. Therefore, for efficient transgene expression regulatory elements capable of modulating gene architecture must be identified and characterised.

4.3. VARIABLE TRANSGENE EXPRESSION

It is now becoming apparent that position effects can also result in variable levels of transgene expression between (Festenstein *et al.*, 1996) or within lines (Dobie *et al.*, 1996). In some instances this is due to mosaic expression patterns within a tissue, i.e. transgene expression levels vary between adjacent cells within an individual animal. This recently described phenomenon, which in some cases resembles position effect variegation, could have implications for all applications of gene transfer. Importantly, variegated transgene expression is not overcome through the inclusion of dominant control elements nor necessarily evaded through using very large genomic transgenes. This is an issue which is currently attracting considerable interest.

In addition, intra-molecular variation may also occur, i.e. whether all copies within individual transgene arrays are equally active is not known (Whitelaw and Webster, submitted).

5. Nuclear Transfer-the Future for Transgenic Livestock

Until recently all transgenic livestock were produced using the microinjection procedure.

This results in random integration and copy number - the cause of the 'position-effect' - and is limited in that only insertion events can be made. The development of methodology allowing animals to be generated from somatic cells by nuclear transfer (Campbell *et al.*, 1996; Wilmut *et al.*, 1997), which resulted in the generation of 'Dolly', will enable these limitations to be overcome. We are now at the start of an new era in large animal transgenics.

5.1. TARGETED INTEGRATION IN LIVESTOCK

The important step in nuclear transfer procedure is the use of cells grown in culture as the source of donor genetic information. This, in theory (and currently under considerable investigation), allows the selection of specific integration events in an analogous manner to gene targeting in murine ES cells. Once achieved, this technology will dramatically increase the potential use of transgenic livestock for both agricultural and biotechnological applications.

In the context of enhancing protein production in milk, several approaches can be envisaged. For example, it should be possible to target the integration of transgenes into permissive loci within the genome, thereby ensuring reproducible expression patterns from a variety of transgene designs. Alternatively, proteins which impede the downstream isolation process or interfere with the biological activity of the desired transgene product could be eliminated using gene knock-out strategies.

5.2. CONCLUSION

Since the first demonstration that transgenic livestock could produce human proteins in their milk (Clark *et al.*, 1989), considerable progress has been achieved. A variety of milk protein gene based vectors have been developed, which combined with expression systems, e.g. transgene rescue, has resulted in the expression of a large variety of proteins in the milk of transgenic animals (Wilmut and Whitelaw, 1994). Nevertheless, transgene expression has yet to be optimised: many transgene designs still result in a high frequency of non-expressing (silent) animals (Clark *et al.*, 1997).

It is probable that the development of the procedures which resulted in the generation of 'Dolly' will overcome many of the problems currently encountered (using microinjection technology). Not only will nuclear transfer enable targeted transgene integration, once appropriate animals have been identified nuclear transfer will allow the rapid expansion of production flocks or herds.

We can confidently await the more efficient production of proteins in the milk of transgenic livestock through combining gene targeting and nuclear transfer. This will lead in turn, to the generation of a variety of transgenic animals addressing a range of issues, e.g. animal disease models, donor animals for xenotransplantation therapy.

6. Acknowledgements

We thank the present and past members of the Roslin Institute for their help in various aspects of the projects discussed. The authors are supported by the Biotechnology and Biological Sciences Research Council (UK); some of the work described was partially funded by PPL Therapeutics.

7. References

Archibald, A.L., McClenaghan, M., Hornsey, V., Simons, J.P., and Clark, A.J. (1990) High-level expression of biologically active human αl-antitrypsin I milk of transgenic mice. *Proc. Natl. Acad. Sci. USA* **87**, 7303-7310.

Campbell, K.H.S., McWhir, J., Ritchie, W.A., and Wilmut, I. (1996) Sheep cloned by nuclear transfer from a cultured cell line. *Nature* **380**, 64-66.

Clark, A.J., Bessos, H., Bishop, J.O., Brown, P., Harris, S., Lathe, R., McClenaghan, M., Prowse, C., Simons, J.P., Whitelaw C.B.A., and Wilmut, I. (1989) Expression of human anti-hemophilic factor IX in the milk of transgenic sheep. *Bio/Tech.* **7**, 487-492.

Clark, A.J., Cowper, A., Wallace, R., Wright, G., and Simons, J.P. (1992) Rescuing transgene expression by co-integration. *Bio/Tech.* **10**, 1450-1454.

Dobie, K.W., Lee, M., Fantes, J., Graham, E., Clark, A.J., Springbett, A., Lathe, R., and McClenaghan, M. (1996) Variegated transgene expression in mouse mammary gland is determined by transgene integration locus. *Proc. Natl. Acad. Sci. USA* **93**, 6659-6664.

Donofrio, G., Bignetti, E., Clark, A.J., and Whitelaw, C.B.A. (1996) Comparable processing of β-lactoglobulin pre-mRNA in cell culture and transgenic mouse models. *Mol. Gen. Genet.* **252**, 465-469.

Festenstein, R., Tolaini, M., Corbella, P., Mamalaki, C., Parington, J., Fox, M., Miliou, A., Jones, M., and Kioussis, D. (1996) Locus control region function and heterochromatin-induced position effect variegation. *Science* **271**, 1223-1225.

Grosveld, F., van Assendelft, G.B., Greeves, D.R., and Kollias, G. (1987) Position independent, high level expression of the human beta-globin gene. *Cell* **51**, 975-985.

Guy, L.G., Kothary, R., DeRepentigny, Y., Delvoye, N., Ellis, J., and Wall, L. (1996) The β-globin locus control region enhances transcription but does not confer position-independent expression onto the lacZ gene in transgenic mice. *EMBO J.* **15**, 3713-3721 .

Hooper, M.L. (1992) 'Embryonic stem cells introducing planned changes into the animal germline.' Harwood Acadmic, Chur, Switzerland.

Schedl, A., Montoliu, L., Kelsey, G., and Schutz, G. (1993) A yeast artificial chromosome covering the tyrosinase gene confers copy number-dependent expression in transgenic mice. *Nature* **362**, 258-261 .

Webster, J., Donofrio, G., Wallace, R., Clark, A.J., and Whitelaw, C.B.A. (1997) Intronic sequences modulate the sensitivity of β-lactoglobulin transgenes to position-effects. *Gene* **193**, 239-243.

Whitelaw, C.B.A. Harris, S., McClenaghan, M., Simons, J.P., and Clark, A.J. (1992) Position-independent expression of the ovine β-lactoglobulin gene in transgenic mice. *Biochem. J.* **286**, 31-39.

Whitelaw, C.B.A., and Webster, J. (submitted) β-lactoglobulin transgene arrays may be heterogeneous with regard to chromatin structure.

Wilmut, I., Schnieke, A.E., McWhir, J., Kind, A.L., and Campbell, K.H.S. (1997) Viable offspring derived from fetal and adult mammalian cells. *Nature* **385**, 810-813.

Wilmut, I., and Whitelaw, C.B.A. (1994) Strategies for the production of pharmaceutical proteins in milk. *Reprod. Fert. Dev.* **6**, 625-630.

Wolffe, A.P. (1994) Insulating chromatin. *Curr. Biol.* **4**, 85-87.

Wright, G., Carver, A., Cottom, D., Reeves, D., Scott, A., Simons, J.P., Wilmut, I., Garner, I., and Colman, A. (1991) High level expression of active human alpha- l-antitrypsin in the milk of transgenic sheep. *Bio/Tech.* **9**, 830-834.

Figure 1. *Stringency of expression systems to evaluate transcriptional control elements.*

expression assay	likelihood of expression	formation of chromatin	passage through germline	selection for expression
in vitro (cell culture)				
- transient transfection	Most constucts work	-	-	-
- stable transfection	Some constrcuts work	+	-	+
in vivo				
- transgenic mice	Few constructs work	+	+	-

The appropriate transcription of eukaryotic genes requires that the gene sequence is present within chromatin and that it has seen all the relevant developmental cues that it would normally be

Figure 2. *Conceptual hybrid transgene design.*

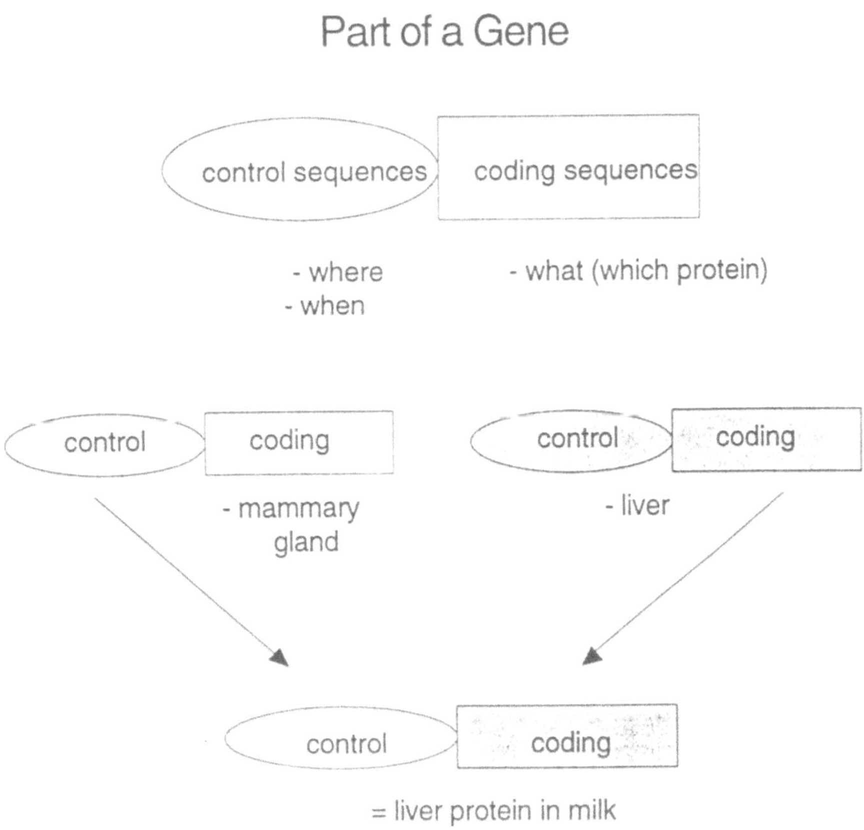

INTRODUCTORY REMARK:
A DECADE OF TRANSGENIC FISH

K. OZATO

Bioscience Center, Nagoya University

Furo-cho, Chikusa-ku, Nagoya 464-1601, Japan

Great progress has been made in the study of transgenic fish, which began in the mid. 1980s, with about 100 publications during this decade. However, it has been proved that some problems inherent in transgenic fish are not found in mammalian models. Although these problems remains to be resolved, the technology of transgenic fish has been standardized in recent studies on model fish and economically important fish. In this symposium, the present status of research of transgenic fish is briefly reviewed and three studies worthy of note, that have contributed to standardization of the technology, are presented.

1. FEATURES OF TRANSGENIC FISH

1) Introduction of DNA is performed by microinjection into nuclei of fertilized eggs in mammalian species. However, application of this method to fish is difficult because fertilized eggs are covered with tough chorion in many fish species and the nuclei cannot be recognized optically. Therefore, various methods have been developed for the introduction of DNA, such as microinjection into oocyte nucleus, microinjection into the cytoplasm of fertilized eggs in which the chorion has been softened by chemical means, electroporation which is efficient in zebrafish, and sperm vectors carrying DNA introduced by electroporation have been used, depending on the properties of the eggs of various species.

2) Extensive extrachromosomal replication of injected genes occurs in the mid-blastula

Y. Kitagawa et al. (eds.), Animal Cell Technology: Basic & Applied Aspects, Vol. 10, 91–92.

stage, inducing strong transient expression. Most of the replicated genes are lost during embryonic development, without integration into chromosomes. This reduces the rate of germ line transmission of foreign genes to a small percentage of the number of eggs injected with DNA.

3) Introduction of DNA is mosaic among cells in one individual. This results in a population of F1 offspring without transgenes. Mosaicism is the most difficult problem encountered for efficient production of transgenic fish lines.

2. STANDARDIZATION OF TECHNOLOGY

1) Any method for DNA introduction into eggs, adapted to the properties of the eggs of various species, can be used.

2) Use of fish promoters results in good expression of transgenes. In the early stages of research, heterologous promoters derived from viruses, mammalian species or humans were used. The rapid progress in the field of fish molecular biology during this decade has made the use of many promoters from various species of fish, possible.

3) GFP is a more sensitive reporter gene than CAT and lacZ.

4) Only a small percentage of eggs injected with DNA contributes to the production of transgenic fish lines.

5) Several transgenic lines expressing physiologically functional genes have been produced using the tyrosinase gene, growth hormone genes, melanin- concentrating hormone gene, in model fish and economically important fish. Several fish lines expressing the GFP gene fused to various promoters have been also produced.

3. ES CELLS AND NUCLEAR TRANSPLANTATION

ES cell lines have been cultured in medaka and zebrafish. However, formation of germ line chimeras has not been successful, using these cell lines. Nuclear transplantation has been attempted in some fish species from the 1970s, by transplantation of embryonic nuclei into eggs with or without nuclei. No fish carrying diploid chromosomes of the original donor nucleus has been obtained. Transplantation of nuclei of culture cells, achieved recently in mammalian species, remains to be studied further.

TRANSGENIC MEDAKA FISH STRAINS WITH LIGHTENED BODY COLOR AND BIOLUMINESCENCE

M. KINOSHITA

Division of Applied Biosciences, Graduate School of Agriculture,

Kyoto University

Sakyo Kyoto 606-01, Japan

1. INTRODUCTION

Aquatic animals have been subjected to transgenic techniques since the early 1980's (Hew, C. L. and Fletcher, G. L. 1992). Although there is no doubt that the production of transgenic fish strains is essential to the study of basic and applied biosciences, only a few studies concerning the fixation of the functional transgene on the host fish have been reported. In this paper, I report the establishment of two medaka transgenic strains : one has lightened body color and the other is bioluminescent.

2. MATERIALS AND METHODS

2. 1. CONSTRUCTS OF TRANSGENES

To establish a strain of medaka with light body color, I constructed the expression vector

Y. Kitagawa et al. (eds.), Animal Cell Technology: Basic & Applied Aspects, Vol. 10, 93–96.

pCMV-MCH, which contains the salmon melanin-concentrating hormone (MCH) gene (Takayama, Y., et al., 1989) driven by cytomegalovirus immediately early promoter. MCH is a peptide hormone (17 amino acids) that is secreted into blood system by pituitary. Its function is to concentrate melanin granules and lighten body color. For the establishment of a bioluminescent medaka strain, I constructed the expression vector pEF-GFP, which contains the green fluorescent protein (GFP) gene driven by medaka polypeptide elongation factor-1 α (EF) promoter. EF is essential for peptide chain elongation and exists in all cell so it is thought to be one of the most powerful promoters. GFP emits green fluorescence when exited by blue light.

2.2. MICROINJECTION OF TRANSGENE INTO MEDAKA EGGS

The wild-type medaka strain (HNI-I) was maintained at 26 ℃ for 14 hours light per day. The expression vector described above (0.0125 mg/ml in PBS) were injected into the cytoplasm of fertilized eggs of HNI-I before the first cleavage (Kinoshita, M. et al., 1996). The injected eggs were incubated at 26 ℃.

3. RESULTS

3.1. ESTABLISHMENT OF A LIGHTENED BODY COLOR STRAIN OF MEDAKA

Among 100 eggs injected, two individuals showed lightened body color (Table 1). One individual carrying the transgene, CMV-MCH, in its germ cells (Table 1) and was mated with wild-type medaka to obtain next generation (F1). Sixteen percent of F1 individuals expressed CMV-MCH and showed lightened body color. By mating transgenic individuals, a lightened body color strain was established. In this strain, the MCH gene was expressed in all tissues tested (lever, muscle, ovary, eye). The secretion of MCH into the blood system in this strain was confirmed by the fact that the serum of this strain could concentrate melanin granules on the scales of wild-type medaka. At the present time, CMV-MCH has been transmitted to the 7th generation, which has lightened the body color.

TABLE 1. Result of microinjection of CMV-MCH into medaka egg

	Eggs injected	Adult	Lightened color	Germ-line transformed
No.	100	50	2	1

3.2. ESTABLISHMENT OF BIOLUMINESCENT MEDAKA STRAIN

Among 150 eggs injected, three individuals (tentatively named No. 1, No. 2, and No. 3) were found to carry the transgene, EF-GFP, in their germ cells and were mated with wild-type medaka to obtain the next (F1) generation (TABLE 2). The green fluorescence derived from the introduced GFP gene was detected in 83%, 37 %, and 61 %, respectively, of F1 progeny of No. 1, No. 2, and No. 3. By mating EF-GFP transformed fish with wild-type medaka, three independent bioluminescent medaka strain were established. The GFP gene was expressed in all cells of an embryo, and the expression pattern was the same in the three strains.

TABLE 2. Result of microinjection of EF-GFP into medaka egg

	Eggs injected	Adult	Germ-line transformed
No.	150	48	3

4. DISCUSSION

In this study, two transgenic medaka strains were established: one had lightened body color produced by introduction of the foreign MCH gene, the other was a bioluminescence strain produced by GFP gene introduction. This reveals that foreign genes inserted into medaka can be transmitted stably to offsprings and express themselves and their function. This also indicates the possibility of producing a model fish for basic sciences, such as the study of gene function, and a commercially available fish strain.

5. REFERENCES

Hew, C. L. and Fletcher, G. L. (1992) *Transgenic Fish*, World Scientific Publishing Co., Singapore.

Takayama, Y., Wada, C., Kawauchi, H. and Ono, M. (1989) Structure of two genes coding for melanin-concentrating hormone of chum salmon, Gene **80**, 65-73.

Kinoshita, M. Toyohara, H. Sakaguchi, M. Inoue, K. Yamashita, S. Satake, M. Wakamatsu, Y. and Ozato, K. (1996) A stable line of transgenic medaka (*oryzias latipes*) carrying the CAT gene, Aquaculture **143**, 267-276.

TRANSGENIC ZEBRAFISH AS A BIOSENSOR

M. YAMASHITA
National Research Institute of Fisheries Science
Fukuura, Kanazawa, Yokohama 236-8648, Japan

1. Introduction

Cells respond to elevated temperature by synthesizing a small number of stress proteins [1]. Such proteins are also synthesized in cells exposed to other stresses, such as heavy metals, toxic compounds, and irradiation. This response has been observed in organisms as diverse as bacteria and humans.

A 70 kDa-stress protein, HSP70, is the most prominent and highly conserved in all organisms. The gene expression of HSP70 is regulated by a heat shock promoter. Its transcription is activated not only by changes in body temperature but also by various stresses. For this reason, HSP70 is considered to be a useful biomarker of stress condition of animals. The molecular basis for this induction has been identified to its regulatory regions, effectively direct the production of heterologous proteins under stress condition.

The present study deals with a transgenic zebrafish expressing the bacterial β-galactosidase gene (*lacZ*) under the control of the HSP70 gene promoter responsive to heat shock and toxic chemicals. The stress inducible expression of the transgene was easily detected *in situ* by staining with chromogenic substrate, X-gal.

2. Generation of Transgenic Zebrafish

The stress-inducible HSP70 gene has been cloned from rainbow trout recently [Yamashita, M., *et al.*, unpublished]. Its promoter region was found to contain three heat shock elements, HSE1, HSE2, and HSE3. Heat shock element corresponds to inverted multimers of the sequence nTTCn or nGAAn which binds with heat shock factor. Each heat shock element of the rainbow trout HSP70 gene also consists of four or three 5-bp units. The CAT assay with expression CAT vectors containing deletion mutant of rainbow trout HSP70 promoter showed only HSE1 has high heat inducible

Y. Kitagawa et al. (eds.), Animal Cell Technology: Basic & Applied Aspects, Vol. 10, 97–100.

transcriptional activity in cultured trout cells.

Since the HSE1 contains three miss-match nucleotides by a comparison of the HSE sequences between the trout and the consensus promoters, we generated the minimum HSP promoter containing the consensus HSE sequence by substitution of the three miss-match nucleotides to the consensus sequence, and fused to the bacterial *lacZ* (Fig. 1).

Transgenic zebrafish were produced by microinjecting one-cell stage embryo with the circular form of plasmid DNA of the *lacZ* construct. Fish were screened by PCR performed on tail fin DNA. The integrated transgene were detected in 32 F0 individuals among 120 DNA-injected fish, and germ line transmission was found in fifteen F0 individuals. Among them a F1 transgenic founder, P7 strain, showed high β-galactosidase activity only under stress-condition. The inheritance of the transgenic offspring was consistent with Mendelian inheritance pattern. The cross between a F1 female and a wild type generated about 50 % positive transgenic F2 progeny.

heat shock element (HSE)

```
        10         20         30         40         50         60
TGGTTGTTAT TCAAGAATAT TCAGGAAAAC ATCTCTGTTG GTTGAGATTG ACATTACTTG

        70 TATA box 80         90        100        110        120
CACCTAGGTT TTCATAGAAT AATATCAATT TCTAAAATAT TTCATTATTT TGTAGGCAAC

       130        140
TAACTCCAAA CCTACAATCT AGA
```

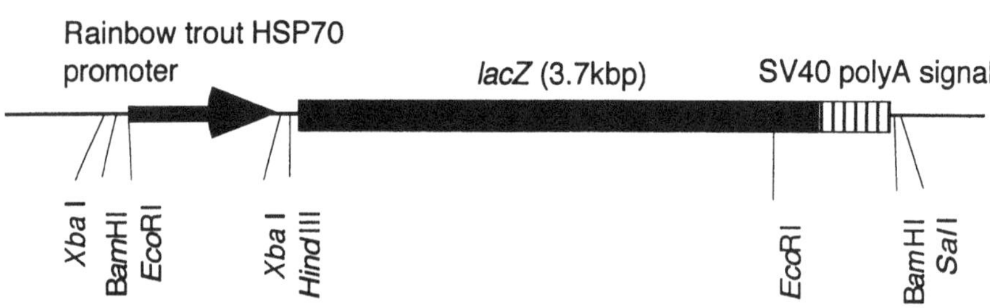

Figure 1. Gene construct of the *HSP-lacZ* gene.

3. Response of the Transgenic Fish to Stresses

Figure 2 shows heat-inducible expression of the *lacZ* transgene of P7 strain in 3 day-old larval fish before and after heat shock stress by X-gal staining. After heat shock treatment by the temperature shift from 28.5 °C to 37 °C for 6 hours, high levels of the *lacZ* expression were observed in every tissue of the body. In comparison, a kind of chemical stressor, sodium arsenite, affect the *lacZ* expression in tissue specific manner. Eyes and intestinal organ showed high β-galactosidase activity. The β-galactosidase expression after the exposure of larval fish to arsenite for 8 hours was found at 5−10 mM, while lethality (LD50) that scored after a 24 hours arsenite-treatment was at 560 mM. This finding indicates that the transgene expression is more sensitive marker than lethality in about 100-fold. This finding shows that heat shock stress affects all of the tissues in the body, while the chemical stressor affects only specific target organs.

By Northern blot analysis of the transgenic zebrafish strain, heat-inducible gene expression was found in both the HSP70 gene and the *lacZ*-transgene. This finding shows that the heat-shock response develops at the gastrula stage 6 to 10 hours after fertilization.

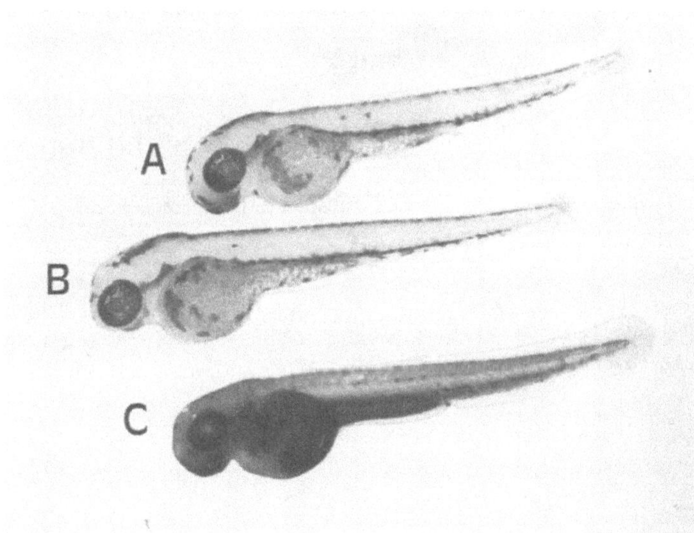

Figure 2. Expression of the β-galactosidase gene under the control of HSP70 promoter. A: normal temperature. B: sodium arsenite. C: heat shock.

Our study shows that the transgenic fish strain provides a bioassay for the stress response at the organismal level (Fig. 3). Aquatic environments are very stressful for living animals. Various physical environments, such as heat shock, irradiation and osmotic stress, chemical stressors such as heavy metals and contaminant, acid and superoxide, and starvation and infection, are serious for the survival of fish in the natural environment and aquaculture. Such various stresses affect the embryogenesis and physiology of fish. The transgenic fish carrying a reporter gene driven by a stress-inducible promoter should be a powerful tool for assay of biological effects of various stress conditions in aquatic environments.

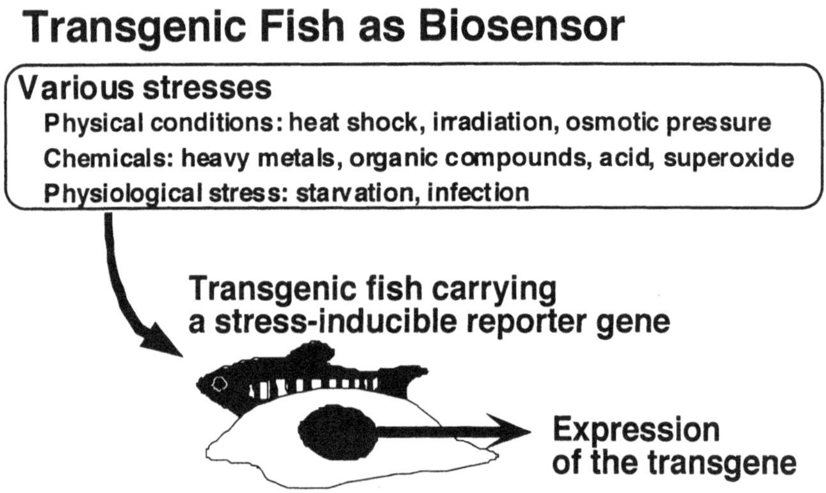

Figure 3. Development of transgenic fish as biosensors.

4. References

1. Morimoto, R. I., Tissieres, A. and Georgopoulos, C.: *Stress Proteins in Biology and Medicine*, Cold Spring Harbor Laboratory Press, Cold Spring Harbor, 1990.

TRANSGENIC SALMON FOR AQUACULTURE

GARTH L. FLETCHER[1] AND CHOY L. HEW[2]
[1]*Ocean Sciences Centre, Memorial University of Newfoundland and A/F Protein Canada Inc., St. John's, Newfoundland, Canada.*
[2]*Research Institute, Hospital for Sick Children and Departments of Biochemistry and Clinical Biochemistry, University of Toronto, Toronto, Ontario, Canada.*

1. Abstract

The potential benefits of transgenic technology to aquaculture are considerable. The isolation and construction of genes responsible for commercially desirable traits and their transfer to aquaculture species could provide a quantum leap in broodstock development over that attainable by traditional selection and breeding methods. Cold and freeze resistance, along with improved growth rates and feed conversion efficiencies are key factors determining the economic viability of many aquaculture ventures. Thus, biotechnologies that could improve these factors must be investigated and evaluated if we are to realize their full potential.

Over the past 12 years, we have generated stable lines of transgenic Atlantic salmon possessing either antifreeze protein (AFP) genes from winter flounder, or a chinook salmon growth hormone (GH) gene construct. The AFP transgene is expressed and AFP secreted into the blood plasma of all generations produced to date. Our GH gene enhanced the growth rates of Atlantic salmon four to six fold on average, and produced smolts within five months of feeding. Feed conversion efficiencies were improved by as much as 20%. Mendelian inheritance of the GH and AFP transgenes and their phenotypic expression has now been demonstrated through the second (F_2), and third (F_3) generations respectively.

2. Introduction

World fisheries are in crisis. Most are exploited to the maximum extent, or overfished, and many are in danger of commercial extinction (Fletcher et al. 1998). As the world population continues to grow exponentially, it is clear that if fish are to maintain their current status as an essential food resource, ways must be found to dramatically improve their production. Aquaculture appears to stand alone as our only major means of meeting demands for fish in the future (New 1997).

A key element to enhanced production of cultured species is the development of genetically superior broodstocks that are tailored to their culture conditions and to the marketplace. Characteristics that are generally desirable include improvements in growth rates, feed conversion efficiencies, disease resistance, cold and freeze resistance, tolerance to low oxygen levels and the ability to utilize low cost, or non-animal protein diets (Hew &

Y. Kitagawa et al. (eds.), Animal Cell Technology: Basic & Applied Aspects, Vol. 10, 101–105.

Fletcher 1997).

Finfish aquaculture is still in its infancy relative to the farming of terrestrial livestock. Despite the acknowledged power of traditional selection and breeding methods, the development of superior broodstock using this process is still relatively slow, and while such broodstock development programs have been underway for salmon since 1971 (Gjedrem 1997), many aquaculture ventures are still reliant on broodstock fish collected from the wild. Therefore, if we are to realize the increased production needed to meet the requirements of the 21st century, a quantum leap in broodstock development is needed.

Transgenic technology provides the only means by which such a quantum leap in production is possible. The identification, isolation and reconstruction of genes responsible for desirable traits and their transfer to broodstock offer powerful methods of genetic/phenotypic improvement that would be difficult, if not impossible to achieve using traditional selection and breeding techniques (Gong & Hew 1995).

This brief communication highlights our progress towards generating genetically engineered Atlantic salmon broodstocks for commercial aquaculture. The issues involved in the production of commercially valuable transgenic fish and their successful integration into industry not only include science but also intellectual property protection, environmental risk assessment, animal welfare, food safety, and consumer acceptance (Fletcher et al. 1998; Entis 1997).

3. Transgenic Salmon

We came into the field of transgenics some fifteen years ago in response to problems faced by the aquaculture industry along the east coast of Canada. Most of these coastal waters are characterized by ice and sub-zero temperatures that are lethal to salmonids. Therefore, sea cage aquaculture of salmon is almost entirely restricted to a relatively small area in the most southerly part of the region (Hew et al. 1995). The challenge for us as scientists was to find a means of producing salmon that would avoid freezing, and thus facilitate the expansion of aquaculture and economic development throughout the entire Atlantic coastal region. The solution was evident from Palmiter et al.'s (1982) clear demonstration of the power of transgenic technologies as a means of genetically improving commercially important animals.

There are two potential ways in which transgenic technologies can be used to solve the problem of overwintering salmon in sea cages in Atlantic Canada, 1) produce freeze-resistant salmon by giving them a set of antifreeze protein genes, and 2) enhance growth rates by growth hormone gene transfer so that overwintering is unnecessary.

3.1 ANTIFREEZE PROTEIN GENES

Our transgenic studies were initiated in 1982 by injecting winter flounder antifreeze protein (AFP) genes into the fertilized eggs of Atlantic salmon. A full length gene encoding the major secretory AFP was used and the AFP transgene was successfully integrated into the salmon chromosomes, expressed, and found to exhibit Mendelian inheritance (Table 1) (Shears et al. 1991). Results to date indicate that the expressed levels of AFP in the blood of these fish are quite low (approximately 50 μg/ml) and are insufficient to confer any significant increase in freeze resistance to the salmon (> 5 mg/ml needed). However, there is little doubt that protective levels could be attained by increasing AFP gene dosage by crossbreeding transgenic salmon with different chromosomal integration sites, use of a strong promoter/enhancer, and designing a gene encoding a more powerful AFP (Hew et al. 1995).

3.2 GROWTH HORMONE GENES

Growth hormone genes are normally expressed in the pituitary gland under the control of the central nervous system (CNS). In order to remove the CNS control, it is necessary to modify the tissue specific elements of the gene so that expression can take place elsewhere. Since the AFP genes are expressed predominantly in the liver, we designed our gene construct using the ocean pout AFP promoter (opAFP) linked to the chinook salmon GH cDNA or a chimeric cDNA genomic GH gene (Hew et al. 1995).

3.2.1. *Growth and Inheritance of the GH-transgene*

The GH transgene genomic integration frequency was similar to that observed for the AFP genes (2-3 %) (Table I). All of the GH-transgenic founder fish were germ cell mosaics, and approximately half of them failed to pass on the GH transgene to their offspring. Approximately 40% of the founder transgenic fish exhibited growth rates that were, on average, several times that of standard (control) salmon over a 30 month period (Fig. 1).

Crosses between F_1 transgenics and standard (wild type) salmon demonstrated Mendelian inheritance of both the GH transgene and the rapid growth phenotype. Three GH transgene inheritance frequencies have been observed to date: 50%, 73%, and 86%, indicating that the transgene is present on one, two, and three chromosomes respectively (Table 1). However, inheritance of the rapid growing pheotype was approximately 50% in the offspring from all three cases. Since 50% inheritance indicates that only one chromosome carries a fully functional GH transgene, it would appear that the transgenic parents of two of the families carry inactive transgenes in addition to the functional transgene.

An illustration of the enhanced growth observed for F_1 transgenics cultured at 16°C is presented in Fig. 2. Similar growth enhancement has been observed for F_2 transgenics (Saunders et al. 1997). In addition to enhanced growth rates, transgenic Atlantic salmon are also considerably more efficient (>20%) at converting dry feed to body weight than are non-transgenic controls (Fletcher et al. 1997).

The results of these studies clearly indicate that stable lines of GH transgenic salmon can be generated that exhibit greatly improved growth rates and feed conversion efficiencies. Such improvements will result in considerable savings to the grower, both in terms of time from egg to market size, and reduced feed costs.

Acknowledgements

The authors gratefully acknowledge NSERC, MRC, IRAP-NRC, ACOA, DFO, and A/F Protein Inc. for providing funding for this research.

Table 1. Inheritance and expression of antifreeze protein and growth hormone transgenes in Atlantic salmon.

Gene	Year	Generation	Cross	Transgenic (% offspring)	Expression (% offspring)
	1985	P_1	Founder	3 %	1%
AFP	1989	F_1	P_1 x Wild (1)	17 %	17%
	1990	F_2	F_1 x Wild (4)	51-54 %	All transgenics
	1992	F_3	F_2 x Wild (1)	52 %	All transgenics
	1992	F_3	F_2 x F_1 (4)	69-77 %	All transgenics
	1989	P_1	Founder	2-3 %	1 %
GH	1992	F_1	P_1 x Wild (1)	19 %	8 %
	1994	F_2	F_1 x Wild (1)	73 %	~ 50 %
	1995	F_2	F_1 x Wild (1)	50 %	~ 50 %
	1995	F_2	F_1 x Wild (1)	86 %	~ 50 %

Eggs were injected with the flounder AFP gene in 1985, and the GH gene in 1989. Transgenics were detected by PCR on red blood cells. AFP gene expression was determined by immunoblotting procedures on blood plasma. Rapid growth phenotype was taken as an indicator of GH gene expression. In all cases but one, crosses were made between a transgenic and a wild type (non-transgenic) individual, the exception being the 1992 cross between two hemizygous AFP transgenic parents (F_2 x F_1). Numerals in parentheses indicate number of crosses using different transgenic parents. Families with 50-54%, 69-77%, and 86% inheritance indicate that the transgene was present on one, two, and three parental chromosomes respectively. The 50% expression in the GH group indicates that only one chromosome possesses a functional transgene.

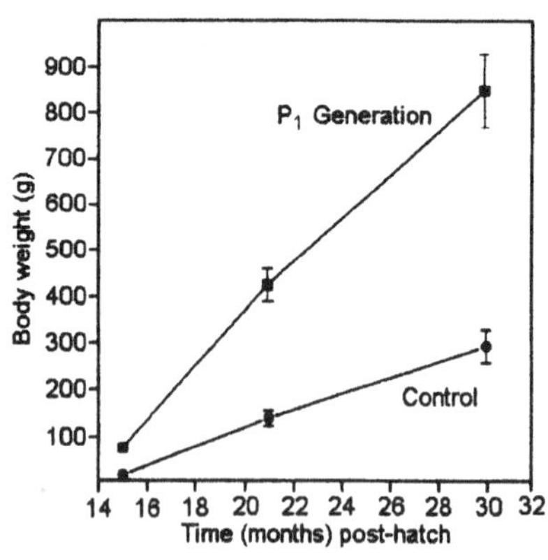

Figure 1. Growth of 6 founder generation (P_1) GH transgenic, and 15 control Atlantic salmon at seasonally ambient temperatures (from 0.5°C to 16°C, mean = 9.5°C). Weights are means ± 1 SE.

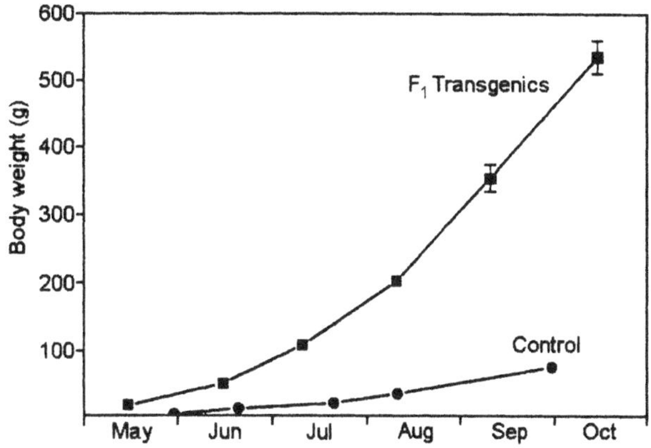

Figure 2. Growth of F_1 generation GH transgenic, and control Atlantic salmon at 16°C (Feb hatch, March first feed). Weights are means ± 1 SE, n = 5-10.

References

Entis, E. (1997) AquAdvantage Salmon: Issues in the introduction of transgenic foods. *Proceedings, International workshop on transgenic animals and food production*, May 22-24, Stockholm, Sweden, (in press).

Fletcher, G.L., Goddard, S.V., Davies, P.L., Gong, Z., Ewart, K.V. and Hew, C.L. (1997) New insights into antifreeze proteins in fish: physiological significance and molecular regulation. In: H.O. Portner and R. Playle (eds.), *Cold Ocean Physiology*, Cambridge University Press (in press).

Fletcher, G.L., Alderson, R., Chin-Dixon, E.A., Shears, M.A., Goddard, S.V., and Hew, C.L. (1998) Transgenic fish for sustainable aquaculture. *Proceedings of the 2nd International Symposium on Sustainable Aquaculture*, A.A. Balkema, Rotterdam (in press).

Gjedrem, T. (1997) Selective breeding to improve aquaculture production. *World Aquaculture* 28 (1), 33-45.

Gong, Z. and Hew, C.L. (1995) Transgenic Fish in Aquaculture and Development Biology. *Current Trends in Development Biology* 30: 177-214.

Hew, C. L., and Fletcher, G. L. (1997) Transgenic fish for aquaculture. *Chemistry and Industry* April 21, 311-314.

Hew, C.L., Fletcher, G. L. and Davies P. L. (1995) Transgenic salmon: tailoring the genome for food production. *Journal of Fish Biology* 47 (Supplement A), 1-19.

New, M. B. (1997) Aquaculture and the capture fisheries - balancing the scales. *World Aquaculture* 28 (2): 11-30.

Palmiter, R.D., Brinster, R.L., Hammer, R.E., Trumbauer, M.E., Rosenfeld, M.G., Brinberg, N.C., and Evans, R.M. (1982) Dramatic growth of mice that develop from eggs microinjected with metallothionein-growth hormone fusion genes. *Nature* 30, 611-615.

Saunders, R.L., Fletcher, G.L. and Hew, C.L. (1997) Smolt development in growth hormone transgenic Atlantic salmon. *Aquaculture* (in press).

Shears, M.A., Fletcher, G. L., Hew, C.L., Gauthier, S. and Davies, P. L. (1991) Transfer, expression, and stable inheritance of antifreeze protein genes in Atlantic salmon (Salmo salar). *Molecular Marine Biology and Biotechnology* 1, 58-63.

FOOD COMPONENTS WITH POTENTIALITIES TO SUPPRESS PROLIFERATION AND INVASION OF CANCER CELLS

KAZUMI YAGASAKI AND YUTAKA MIURA
Department of Applied Biological Science, Tokyo Noko University
Fuchu, Tokyo 183-8509, Japan

ABSTRACT. Endless proliferation and metastasis are two biological properties of cancer cells. The invasion of cancer cells is an important and characteristic step of metastasis. The inhibition of these biological events will prolong the life span of hosts with cancers. Thus, the effects of food extracts and components on the proliferation and invasion of hepatoma (AH109A) cells were investigated using *in vitro* proliferation and invasion assay systems. Some vegetable foods have proved to contain effective and bioavailable components suppressing the proliferation and/or invasion of the hepatoma cells *in vitro*.

1. Introduction

Cancer cells have two biological properties: endless proliferation and metastasis. The best way to make a conquest of cancer is to prevent its occurrence. But once cancer is present, suppression of these two properties of cancer cells seems to be the next best policy.

Cancer cells are believed to arise out of normal cells through three phases called initiation, promotion and progression. During proliferation, some cells get the ability to secede from primary tumor tissues. They are then transported in vessels to a target tissue or organ, adhere to endothelial cells, invade and proliferate again. The invasion of cancer cells is an important and characteristic step of metastasis [1].

We have attempted to identify food components that have the potential to suppress the growth and metastasis of cancer cells by using assay systems for both proliferation and invasion of these cells. As model cancer cells, we employed a rat ascites hepatoma line of AH109A. The cells grow rapidly both *in vitro* [2] and *in vivo* [3], and form solid tumors as well as hyperlipidemia 2 weeks after subcutaneous implantation in the backs of Donryu rats [4]. We describe here that some vegetable foods have proved to contain effective and bioavailable components which suppress the proliferation and/or invasion of the hepatoma cells *in vitro*.

Y. Kitagawa et al. (eds.), Animal Cell Technology: Basic & Applied Aspects, Vol. 10, 107–111.

2. Mateials and Methods

Cell culture and invasion assay. Using AH109A cells, we constructed assay systems for both proliferation and invasion of cancer cells [2]. The hepatoma cells were provided by SRL, Tokyo, Japan. Proliferation of hepatoma cells was estimated by either [^{3}H]-thymidine incorporation method, MTT method, or WST-1 method. The effect on the invasion of AH109A cells was measured by the co-culture system of Akedo *et al.* [5] with slight modifications as described previously [2]. Briefly, primary cultured mesothelial cells (M cells) from mesentery of rats were seeded at a density of 1.5 x 10^5 cells/60 mm ϕ culture dish with 2 mm grids and cultured in 3 ml of DM-160 containing 10% calf serum (10% CS/DM-160) for 7-10 days to attain a confluent state. AH109A cells (2.4 x 10^5) were then seeded on the monolayers of M cells and cultured without or with test substances for 48 hr. Invading AH109A cells and colonies underneath M cell layers were counted with a phase-contrast microscope. Usually at least 10 areas were counted and the invasive activity of AH109A was indicated by the number of invading cells and colonies/cm^2.

Preparation of food extract and food extract-loaded rat serum. Foods were washed and homogenized in water, centrifuged and the supernatant fluids were obtained. The components of beverages such as tea and coffee were extracted with boiled water. After being sterilized by membrane filtration, each food extract was added to experimental media. To obtain food extract-loaded serum, rats fasted overnight were given oral intubation of food extract (1 ml/100 g body weight), blood was withdrawn 0, 0.5, 1, 2, 3 and 5 hr after the intubation, serum was prepared, and the effects of the food extract-loaded rat serum on the proliferation and invasion of AH109A were tested *in vitro* after sterilization [2,6].

Statistical analysis. Data were expressed as means ± standard errors. Statistical analysis was done using a one way analysis of variance, followed by Tukey's Q test.

3. Results and Discussion

Using two *in vitro* assay systems, we investigated the effects of various food extracts and food components on the proliferation and invasion of hepatoma cells. Table 1 shows the summary of results obtained where foods are classified into three groups. The first group is foods containing components with potential to suppress both the proliferation and invasion, such as three kinds of teas, coffee, cabbage, clam and so on. The second group is foods containing components with potential to suppress proliferation such as leek, garlic, broccoli and yacon. The third is foods that contain components with potential to suppress invasion: spinach, carrot, ginger, perilla, and the simplest amino acid glycine.

TABLE 1. Foods that contain components suppressing proliferation and/or invasion of AH109A cells

Proliferation	Invasion	Foods
⬇	⬇	Teas (green, oolong, black), Coffee, Cabbage, Bitter cucumber, Clam, Fresh-water mussel
⬇	→	Leak, Garlic, Broccoli, Yacon
→	⬇	Spinach, Carrot, Parsley, Ginger, Perilla, (Glycine)

⬇ ; suppressed, →; not suppressed

Several examples are as follows. The water-soluble fraction of cabbage suppressed weakly but significantly the proliferation, and it also inhibited the invasion significantly and dose-dependently. In a separate experiment, cabbage extract was found to induce cytolysis of AH109A cells, when the hepatoma cells were co-cultured with whole splenocytes. Cabbage extract was found to activate macrophages with the aid of T cells. Thus, cabbage extract seems to suppress the hepatoma proliferation by affecting AH109A cells both directly and indirectly [7]. The water-soluble fraction of leek dose-dependently inhibited the hepatoma proliferation without affecting the invasion of AH109A at the same concentrations. The spinach extract dose-dependently suppressed the hepatoma invasion, while it did not affect the proliferation at the same concentrations,.

It was important to learn whether or not the bioactive components in foods were effective when orally given, because foods should be *orally* ingested. Donryu rats were fasted overnight, and at 10 o'clock the next morning were given oral intubation of samples; blood was obtained 0, 0.5, 1, 2, 3 and 5 hrs after administration. Serum prepared from blood was used instead of calf serum in the proliferation and invasion assays. Hot water extracts from powdered green, oolong and black teas suppressed both the proliferation and invasion of hepatoma cells when added to medium. We therefore examined the effects of the three kinds of powdered tea extract-loaded rat sera: all three inhibited the proliferation and invasion of hepatoma cells. The inhibitory actions were the strongest 2 hrs after oral intubation and weakened as time proceeded [8]. Likewise, instant coffee powder (ICP)-loaded rat serum also inhibited the proliferation and invasion. Unlike the tea-loaded serum, ICP-loaded serum still maintained its inhibitory action 5 hrs after oral intubation. Leek extract-loaded rat serum also inhibited the hepatoma proliferation, although the inhibitory action was weak. We also examined the effect of spinach extract-loaded rat serum on the proliferation and invasion of AH109A cells and, as expected, it inhibited the invasion; this serum also inhibited hepatoma proliferation. Spinach extract did not inhibit the proliferation, so these results suggest the possibility that spinach affects the hepatoma proliferation by way of host functions such as cytokine production. In any event, foods screened by the two *in vitro* assay systems seem more or less effective when they are orally ingested.

What are the chemical entities of effective food components? Soybean isoflavone genistein, for example, has been reported to be excreted in urine of Japanese males living in Kyoto and consuming a traditional Japanese diet 38-times more than in urine of Caucasian males consuming a Western type diet [9]. Japanese consume a large quantity of soybeans and soybean-related foods. Moreover, genistein is known to have chemopreventive action on carcinogenesis [10]. This component suppressed both the proliferation and invasion of AH109A cells. Genistein is known as an antioxidant [11] and an inhibitor of tyrosine kinase [12]. Teas, especially green tea, are popular beverages in Japan. Theophlline, (-)-epicatechin (EC), (-)-epicatechin-3-gallate (ECG), (-)-epigallocatechin (EGC) and (-)-epigallocatechin-3-gallate (EGCG) are common constituents in green, oolong and black teas, while theaflavin and theaflavin gallate (Tfs) are specific components of black tea. Theophylline and EC had little or no effect on either the proliferation or invasion of AH109A cells, while other catechins, ECG, EGC, EGCG and Tfs suppressed the proliferation and invasion to a greater or lesser degree. These effective tea components are also known to have antioxidative actions. Caffeic acid in coffee inhibited the hepatoma invasion, but did not affect the proliferation when used at the same concentrations. Curcumin is a major component in termeric and inhibited the invasion of hepatoma cells, while it did not affect the proliferation. Caffeic acid and curcumin are also known to have antioxidative activity.

Glycine is the simplest and nonessential amino acid, but suppressed the hepatoma invasion without notable effect on the proliferation. This inhibitory effect of glycine on the invasion was cancelled in the co-presence of glycine transporter antagonist sarcocine. This suggests an involvement of glycine transporter in the invasion of AH109A cells [13].

The modes of actions of the above-mentioned components are unclear at present. However, the following factors appear to be involved in the suppression of the proliferation and invasion of AH109A. Cell cycle arrest, induction of apoptosis or necrosis, and other factors may act to inhibit proliferation, and to the inhibit invasion, suppression of adhesion and suppression of motility of cancer cells, inhibition of expression and activities of matrix metalloproteinases and other enzymes are suggested. Further extensive studies are needed to clarify these aspects.

4. References

1. Liotta, L., Rao, C.N., and Barsky, S.H. (1983) Tumor invasion and the extracellular matrix, *Lab. Invest.*, **49**, 636-649.

2. Miura, Y., Shiomi, H., Sakai, F., and Yagasaki, K. (1997) Assay systems for screening food components

that have anti-proliferative and anti-invasive activity to rat ascites hepatoma cells: *In vitro* and *ex vivo* effects of green tea extract, *Cytotechnology,* **23**, 127-132.

3. Yagasaki, K., Tanabe, T., and Funabiki, R. (1994) Reduction of tumor growth and hypercholesterolemia by arginine administration to rats with transplanted hepatoma,*Biosci. Biotech. Biochem.*, **58**, 1139-1140.

4. Kawasaki, M., Yagasaki, K., Miura, Y., and Funabiki, R. (1995) Reduction of hyperlipidemia in hepatoma-bearing rats by dietary fish oil. *Lipids*, **30**, 431-436.

5. Akedo, H., Shinkai, K., Mukai, M., Mori, Y., Tateishi, R., Tanaka, K., Yamamoto, R., and Morishita, T. (1986) Interaction of rat ascites hepatoma cells with cultured mesothelial cell layers: a model for tumor invasion, *Cancer Res.,* **46**, 2418-2422.

6. Miura, Y., Furuse, T., and Yagasaki,K. (1997) Inhibitory effect of serum from rats administered with coffee on the proliferation and invasion of rat ascites hepatoma cells, *Cytotechnology,* **25**, 219-223.

7. Komatsu, W., Miura, Y., and Yagasaki, K. *In vitro* stimulation of tumor necrosis factor production from rat splenocytes by cabbage extract, *Animal Cell Technology: Basic & Applied Aspects,* **10**, in press.

8. Zhang, G.-Y., Miura, Y., and Yagasaki, K. Effects of green, oolong and black teas and related components on the proliferation and invasion of hepatoma cells, *Animal Cell Technology: Basic & Applied Aspects,* **10**, in press.

9. Fotsis, T., Pepper, M., Adlercreutz, H., Hase, T., Montesano, R., and Schweigerer, L. (1995) Genistein, a dietary ingested isoflavonoid, inhibits cell proliferation and *in vitro* angiogenesis, *J. Nutr.,* **125**, 790S-797S.

10. Lamartinier, C. A., Moore, J., Holland, M., Barnes, S. (1995) Neonatal genistein chemoprevents mammary cancer, *P.S.E.B.M.,* **208**, 120-123.

11. Wei, H., Bowen, R., Cai, Q., Barnes, S., Wang, Y. (1995) Antioxidant and antipromotional effects ofthe soybean isoflavone genistein, *P.S.E.B.M.,* **208**, 124-130.

12. Akiyama, T., Ishida, J., Nakagawa, S., Ogawara, H., Watanabe, S., Itoh, N., Shibuya, M., and Fukami, Y. (1987) Genistein, a specific inhibitor of tyrosine-specific protein kinases, *J. Biol. Chem.,* **262**, 5592-5596.

13. Yagasaki, K., Miura, Y., and Fujii, M. (1996) Actions of amino acids on the proliferation and invasion of hepatoma cells in culture, *Rep. Res. Comm. Essen. Amino Acids (Japan),* No.**146**, 84-87.

DEVELOPMENT OF THROMBOPOIETIN: ITS STRUCTURE AND FUNCTION

K. OHGAMI

Pharmaceutical Research Laboratory

Kirin Brewery Co., Ltd.

3 Miyahara-cho, Takasaki, Gunma 370-1295, JAPAN

Abstract

We recently succeeded to purify thrombopoietin (TPO) from thrombocytopenic rat plasma, and to clone rat and human cDNAs, and human genome. It is now well known that TPO is a ligand of c-Mpl, a member of the hematopoietic receptor superfamily, and that TPO is a megakaryocyte lineage-dominant hematopoietic factor. Highest expression of TPO mRNA is detected in liver. We have produced two types of recombinant TPO. One is expressed in mammalian cells, and the other expressed as truncated protein in bacteria that is further modified by pegylation, termed pegylated recombinant human megakaryocyte growth and development factor (PEG-rHuMGDF). These recombinant TPOs showed both in vitro and in vivo biological activities. Our in vitro studies have demonstrated that TPO enhances the growth of megakaryocyte progenitor cells, and the megakaryocyte maturation. In vivo study of recombinant TPO showed great increase in

Y. Kitagawa et al. (eds.), Animal Cell Technology: Basic & Applied Aspects, Vol. 10, 113–117.

circulating platelet counts, and the number of megakaryocytes and its progenitor cells in the bone marrow of normal animals. Administration of recombinant TPO has effectively decreased the duration of thrombocytopenic state and accelerated platelet recovery, in myelosuppressed animal models. These findings suggest that recombinant TPO has great potential for treatment of thrombocytopenic patients.

1. Introduction

It is well known that hematopoiesis is regulated by various humoral factors. Many cytokines which regulate differentiation of hematopoietic stem cells to peripheral blood cells have been identified. In megakaryocyte lineage, megakaryocyte progenitor cells proliferate and differentiate into megakaryocytes. Megakaryocytes further undergo polyploidization and cytoplasmic maturation characterized by the expression of proteins specific for platelets and the development of the demarcation membrane system, leading to the release of platelets into circulation. Although the presence of a humoral factor termed thrombopoietin (TPO) regulating platelet production was originally proposed more than 30 years ago by Yamamoto and Kelemen, the molecule of TPO has not been identified until 1994, when several groups including ourselves independently reached the cloning of a novel factor regulating thrombopoiesis. By using our original assay system, we have succeeded to purify a factor showing this activity and to clone its cDNA. On the other hand, it has been suggested that c-Mpl played an important role in megakaryocytopoiesis. This factor was also identified as a ligand of c-Mpl, an orphan cytokine receptor, by other groups. Therefore, this novel thrombopoietic factor is variously named the thrombopoietin (TPO), c-Mpl ligand, megakaryocyte growth and development factor (MGDF), or megapoietin. [1]

2. Structure of TPO

Human TPO contains 332 amino acids and shows similarity to EPO especially in its amino-terminal region. Based on this similarity, TPO is divided into two domains, an EPO-like domain and carboxyl-terminal domain. The EPO-like domain contains four cystein residues conserved among various species, which are essential for the tertiary structure formation. Several data of structure-function analyses revealed that only the EPO-like domain showed full TPO activity. Contrarily, the carboxyl-terminal domain rich in sugar chains is not necessary for TPO activity. [1] Although the function of the carboxyl-terminal domain is not well analyzed, it is speculated that this region is important for the secretion from cells or in vivo molecular stability.

3. Gene, expression, and regulation of TPO

Human TPO gene contains six exons and the locus of human TPO gene is mapped to chromosome 3q27. [2] Northern blot analysis revealed that TPO mRNA is dominantly expressed in the liver. To lesser extent, the kidney, brain, intestine, skeletal muscle, and bone marrow express TPO mRNA. Our study using in situ hybridization indicates that hepatocytes express TPO mRNA from fetus to adult. The culture supernatant of primary hepatocytes show TPO activity. These data suggest that circulating TPO is mainly produced by hepatocytes in the liver. [3,4] Unlike the case of EPO that is transcriptionally regulated under anemia, the expression of TPO mRNA in the liver is constitutive under various thrombocytopenic states. It is suggested that circulating TPO level is regulated by the total mass of its receptor, c-Mpl, on platelets and megakaryocytes. Constitutive production of TPO is also supported by the studies of mice deficient in c-Mpl. [1,5]

4. In vitro biological activity of TPO

We have produced two types of recombinant TPO in order to study the biological function. A full-length of recombinant human TPO (rHuTPO) was expressed in mammalian cells, and also a truncated form of human TPO was expressed in bacteria that is further modified by pegylation, termed pegylated recombinant human megakaryocyte growth and development factor (PEG-rHuMGDF). Both of rHuTPO and PEG-rHuMGDF can stimulate the proliferation and differentiation of megakaryocytes in culture, finally leading to proplatelet formation and fragmentation. [6] Furthermore, in concert with other several cytokines, TPO also enhances the proliferation of primitive progenitor cells, indicating that the action of TPO is not restricted to the megakaryocyte lineage. [1,7]

5. In vivo effect of TPO

Administration of either rHuTPO or PEG-rHuMGDF into normal animals resulted in significant increase in platelet count with little or no influence on other blood cells. [1.8] TPO is the cytokine showing highest activity to increase platelets among cytokines ever cloned. When thrombocytopenia was induced in mice and non-human primates by various myelosuppressive treatments such as chemotherapy and irradiation, rHuTPO or PEG-rHuMGDF was effective to reduce the duration and depth of the thrombocytopenia. [1,9,10]

6. Conclusion

Clinical trials of rHuTPO and PEG-rHuMGDF are on-going. Results from clinical studies showed that the cancer patients treated with PEG-rHuMGDF following chemotherapy had less severe falls in platelet count than those in placebo control. [1,12] It is expected that discovery of TPO will provide a benefit for the treatment of various thrombocytopenic conditions.

References

1. Eaton, D.L. and de Sauvage F.J.: Thrombopoietin: The primary regulator of magakaryocytopoiesis and thrombopoiesis, *Exp. Hematol.* 25 (1997),1-7.

2. Sohma, Y., Akahori, H., Seki, N. et al.: Molecular cloning and chromosomal localization of the human thrombopoietin, *FEBS Letters* 353 (1994), 57-61.

3. Shimada, Y., Kato, T., Ogami, K. et al.: Production of thrombopoietin (TPO) by rat hepatocytes and hepatoma cell lines, *Exp. Hematol.* 23 (1995), 1388-1396.

4. Nomura, S., Ogami, K., Kawamura, K. et al.: Cellular localization of thrombopoietin mRNA in the liver by in situ hybridization, *Exp. Hematol.* 25 (1997), 565-572.

5. Kuter, D.J. and Rosenberg, R.D.: Appearance of a megakaryocyte-growth promoting activity, megapoietin, during acute thrombocytopenia in the rabbit, Blood 84 (1994), 1464-1472.

6. Kato, T., Horie, K., Hagiwara, T., et al.: GpIIb/IIIa+ subpopulation of rat megakaryocyte progenitor cells exhibits high responsiveness to human thrombopoietin, *Exp. Hematol.* 24 (1996), 1209-1214.

7. Kobayashi, M., Laver, J.H., Kato, T., et al.: Thrombopoietin supports proliferation of human primitive hematopoietic cells in synergy with steel factor and /or interleukin-3, *Blood* 88 (1996), 429-436.

8. Kabaya, K., Akahori, H., Shibuya, K., et al.: In vivo effects of PEG-rHuMGDF on hematopoiesis in normal mice. *Stem Cells* 14 (1996), 651-660.

9. Shibuya, K., Akahori, H., Takahashi, K., et al.: Multilineage hematopoietic recovery by a single injection of PEG-rHuMGDF in myelosuppressed mice. *Blood* 91 (1998), 37-45.

10. Akahori, H., Shibuya, K., Obuchi, M., et al.: Effect of recombinant human thrombopoietin in nonhuman primates with chemotherapy-induced thrombocytopenia. *Br. J. Haematol.* 94 (1996), 722-728.

12. Fanucchi, M., Glaspy, J., Crawford, J., et al.: Effect of polyethylene glycol-conjugated rHuMGDF on platelet counts after chemotherapy for lung cancer. *N. Engl. J. Med.* 336 (1997), 404-409.

BACULOVIRUS NOT ONLY AS AN INSECT EXPRESSION VECTOR BUT AS A GENE TRANSFER VECTOR

JUN KOBAYASHI
Laboratory of Molecular Bioengineering, Faculty of Engineering, Mie University, Tsu, Mie 514, Japan

1. Introduction

Since the first high-level expression of human ß interferon in cultured insect cells using a genetically engineered baculovirus, the *Autographa californica* nucleopolyhedrovirus (AcNPV), was reported in 1983 [1] and the expression of human interferon-α was expanded to the silkworm using the *Bombyx mori* NPV (BmNPV) in 1984 [2], the baculovirus expression vector (BEV) systems have been widely used in many laboratories throughout the world [3-6]. Many modifications have been added to the systems to improve their productivity as well as make them more user friendly. Recently, other insect cell lines, insect hosts and baculoviruses have been investigated for identifying and developing novel BEV systems with improved recombinant protein production capabilities [4,6]. New applications for the BEV systems have also emerged, such as gene transfer vectors for mammalina cells [7-11] and for constructing transgenic silkworms [12], biological insecticides [13-24] and expression libraries [25].

This paper describes our recent works on constructing two novel BEV systems, *Hyphantria cunea* NPV (HycuNPV)/FRI-SpIm-1229 (SpIm) cell system and *Antheraea pernyi* NPV (AnpeNPV)/giant silkworm system, and also reviews the recent progress in the use of the baculovirus as gene transfer vectors.

2. Novel BEV Systems

Because SpIm cells derived from *Spilosoma imparilis* [26] are susceptible to HycuNPV and have been cultured in a low-cost, simply formulated Mitsuhashi-Maramorosch (MM) medium [27] supplemented with a low concentration (3%) of fetal bovine serum (FBS), HycuNPV/SpIm cell system had been considered as a promising candidate enabling economic scale-up for the large-scale production of recombinant products. We have successfully adapted SpIm cell suspension culture in Sf-900 II serum-free medium (GIBCO) but not in serum-free MM medium, and scaled up it to 50 ml by simple shaker culture without significant reductions in cell growth and virus replication [28]. When the recombinant protein production was compared between HycuNPV/SpIm cell system and BmNPV/BmN4 cell system [2], the accumulation of secreted products,

Y. Kitagawa et al. (eds.), Animal Cell Technology: Basic & Applied Aspects, Vol. 10, 119–123.

prothoracicotropic hormone (PTTH) [29], in the culture supernatant of SpIm cells in the serum-free medium was comparable to that in the BmNPV/BmN4 cell system using TC-100 medium [30] with 10% FBS. It was also found that different N-glycosylation patterns of PTTH molecules expressed in these two systems. Further analysis of sugar chains will reveal properties of SpIm cells in the N-glycosylation pathway, one of the important posttranslational modification abilities of eukaryotic expression system.

AnPe cells derived from *A. pernyi* [31] are susceptible to AnpeNPV and have been cultured in MGM-448 medium [32] supplemented with 10% FBS. As the BmNPV system, the AnpeNPV vector system can be expanded foreign gene expression to *in vivo* using giant wild silkworms, *A. pernyi*. Pupal diapause of *A. pernyi* gives this system an advantage over the BmNPV system, because diapausing pupae can be stored for several months until the use for infection of recombinant viruses. Just recently, we have succeeded to construct a recombinant AnpeNPV expressing the bacterial lacZ gene and are investigating the expression level of lacZ gene in diapausing pupae. We will also evaluate the capability of α-amidation in this system by expressing a precursor of amidated peptide, because it was previously reported that the α-amidation, another important posttranslational modification in eukaryotes, which did not occur in insect cell culture, did occur higher efficiency in insect pupae than larvae [33].

3. Gene Transfer Vectors

3.1. GENE TRANSFER INTO MANMALIAN CELLS

In 1995, Hofmann and coworkers [7] first reported that recombinant AcNPV was efficiently taken up by human hepatocytes through an endosomal pathway and that high-level luciferase gene expression from cytomegarovirus (CMV) immediately early promoter was observed in human and rabbit hepatocytes *in vitro* although mouse hepatocytes and some other epithelial cell types are targeted at a considerably lower rate. In 1996, Boyce and Bucher [9] confirmed this baculovirus-mediated gene transfer into mammalian cells using the recombinant AcNPV carrying the bacterial lacZ gene under control of the rous sarcoma virus (RSV) promoter and SV40 splice and poyadenylation signals. Again high-level expression of lacZ gene was observed in human liver cell line HepG2 and primary culture of rat hepatocytes, while cell lines from other tissues showed less or no expression of lacZ. In 1997, Shoji and coworkers [10,11] have succeeded to expand high-level expression from hepatocytes to non-hepatic cell lines by employing the CAG promoter, which consists of cytomegalovirus immediate early enhancer, chicken ß-actin promoter and rabbit ß-globin polyadenylation signal. Expression levels of reporter genes in various mammalian cells infected with recombinant baculoviruses in these studies were summarized and compared in TABLE 1.

All of these studies suggested that baculovirus vector is a good tool for gene delivery into several mammalian cells, the virus does not replicate or cause cytopathology in mammalian cells and the viral polyhedrin promoter is inactive. Although the exact mechanism for the entry of the baculovirus into mammalian cell is not known, the virus may utilize a common molecule as a receptor on the mammalian cells. These works also caution us against the unrestricted use of baculovirus as vectors for gene expression in insect cells *in vitro* or insect *in vivo*. Further study will be required before using baculovirus for gene therapy.

TABLE 1. Baculovirus-mediated expression of repoter genes in mammalina cell lines

Cell line (origin)	Promoter - Reporter gene		
	CMV - Luciferase	RSV - LacZ	CAG - Luciferase
Liver			
Huh7 (human)	++	ND	++
HepG2 (human)	+	++	++
Fibroblast			
NIH 3T3 (mouse)	±	—	ND
Adrenal			
PC12 (rat)	±	—	+
Kidney			
CV-1 (monkey)	—	—	ND
COS-7 (monkey)	ND	—	++
Lung			
A549 (human)	±	—	ND
Cervix			
HeLa (human)	±	—	++
Lymphocyte			
HL-60 (human)	—	—	ND
Ovary			
CHO (hamster)	±	—	ND
Reference	[7]	[9]	[10]

ND = not done; — = no expression; ± = low expression; + = moderate expression; ++ = high expression.

3.2. GENE TRANSMISSION TO INSECT PROGENIES

In 1995, Mori and coworkers [12] reported that recombinant AcNPV infection to the silkworm larvae has resulted in the transovarian transmission of luciferase gene as well as a viral gene to their F2 progenies. Luciferase activities were detected in the virus-infected larvae and pupae (F0), and in the newly hatched larvae of the next generation (F1). Although the transmission of the luciferase gene was confirmed by the PCR amplification and Southern blot hybridization analysis, no luciferase activity was detected in F2 progenies, probably because of a gene inactivation. Thus improvements of this system may enable us to obtain transgenic silkworms for stable production of recombinant proteins and for basic research on molecular biology of insects.

4. The Future of Baculovirus Engineering

Because of predominant use of AcNPV/Sf 9 (and High Five) cell system only for high level expression of foreign genes, there had been quite a few effort to investigate alternative BEV systems and different applications. However, as described above, recent works have demonstrated that BEV systems still have many potentials not only in recombinant protein production but in other use. Further studies along the lines of these works may bring baculovirus engineering to the forefront of the new industrial biotechnology revolution in near future.

References

1. Smith, G.E., Summers, M.D. and Fraser, M.J. (1983) Production of human β interferon in insect cells infected with a baculovirus expression vector, *Mol. Cell. Biol.*, **3**, 2156-2165.
2. Maeda, S., Kawai, T., Obinata, M., Chika, T., Horiuchi, T., Maekawa, K., Nakasuji, K., Saeki, Y., Sato, Y., Yamada, K. and Furusawa, M. (1984) Characteristics of human interferon-α produced by a gene transferred by a baculovirus vector in the silkworm, *Bombyx mori*, *Proc. Jpn. Acad.*, **60**, Ser. B, 423-426.
3. O'Reilly, D.R., Miller, L.K. and Luckow, V.A. (1992) *Baculovirus Expression Vectors : A Laboratory Manual*, W.H. Freeman and Company, New York, NY.
4. Goosen, F.A., Daugulis, A.J. and Faulkner, P. (1993) *Insect Cell Culture Engineering*, Marcel Dekker, New York, NY.
5. Maramorosch, K. and McIntosh, A.H. (1994) *Insect Cell Biotechnology*, CRC Press, Boca Raton, FL.
6. Richardson, C.D. (1995) *Baculovirus Expression Protocols*, Humana Press, Totowa, NJ.
7. Hofmann, C., Sandig, V., Jennings, G., Rudolph, M., Schlag, P. and Strauss, M. (1995) Efficient gene transfer into human hepatocytes by baculovirus vectors, *Proc. Natl. Acad. Sci. USA*, **92**, 10099-10103.
8. Sandig, V., Hofmann, C., Steiner, S., Jennings, G., Schlag, P. and Strauss, M. (1996) Gene transfer into hepatocytes and human liver tissue by baculovirus vectors, *Human Gene Therapy.*, **7**, 1937-1945.
9. Boyce, F.M. and Bucher, N.L.R. (1996) Baculovirus-mediated gene transfer into mammalian cells, *Proc. Natl. Acad. Sci. USA*, **93**, 2348-2352.
10. Shoji, I., Aizaki, H., Tani, H., Ishii, K., Chiba, T., Saito, I., Miyamura, T. and Matsuura, Y. (1997) Efficient gene transfer into various mammalian cells, including non-hepatic cells, by baculovirus vectors, *J. Gen. Virol.*, **78**, 2657-2664.
11. Matsuura, Y., Shoji, I., Yap, C.-C., Tani, H., Ishii, K. and Miyamura, T. (1997) Gene transfer into mammalian cells by baculovirus vector and its applications, *Virus*, **47**, 247-256 (in Japanese).
12. Mori, H., Yamao, M., Nakazawa, H., Sugahara, Y., Shirai, N., Matsubara, F., Sumida, M. and Imamura, T. (1995) Transovarian transmission of a foreign gene in the silkworm, *Bombyx mori*, by *Autographa californica* nuclear polyhedrosis virus, *Bio/technology*, **13**, 1005-1007.
13. Carbonell, L.F., Hodge, M.R., Tomalski, M.D. and Miller, L.K. (1988) Synthesis of a gene coding for an insect-specific scorpion toxin and attempts to express it using baculovirus vectors, *Gene*, **73**, 409-418.
14. Merryweather, A.T., Weyer, U., Harris, M.P.G., Hirst, M., Booth, T. & Possee, R.D. (1990) Construction of genetically engineered baculovirus insecticides containing the *Bacillus thuringiensis* subsp. *kurstaki* HD-73 delta endotoxin, *J. Gen. Virol.*, **71**, 1535-1544.
15. Tomalski, M.D. and Miller, L.K. (1991) Insect paralysis by baculovirus-mediated expression of a mite neurotoxin gene, *Nature*, **352**, 82-85.
16. Stewart, L.M., Hirst, M., Ferber, M.L., Merryweather, A.T., Cayley, P.J. and Possee, R.D. (1991) Construction of an Improved baculovirus insecticide containing an insect-specific toxin gene, *Nature*, **352**, 82-85.

17. Maeda, S., Volrath, S.L., Hanzlik, T.N., Harper, S.A., Majima, K., Maddox, D.W., Hammock, B.D. and Fowler, E. (1992) Insecticidal effects of an insect-specific neurotoxin expressed by a recombinant baculovirus, *Virology*, **184**, 777-780.
18. Tomalski, M.D., King, T.P. and Miller, L.K. (1993) Expression of hornet gene encoding venom allergen antigen 5 in insects, *Arch. Insect Biochem. Physiol.*, **22**, 303-313.
19. Ribeiro, B.M. and Crook, E. (1993) Expression of full length and truncated forms of crystal protein genes from *Bacillus thringiensis* subspecies kurstaki in a baculovirus and pathogenicity of the recombinant viruses, *J. Invertebr. Path.*, **62**, 121-130.
20. Cory, J.S., Hirst, M.L., Williams, T., Halls, P.S., Goulson, D., Green, B.M., Carty, T.M., Possee, R.D., Cayley, P.J. and Bishop, D.H.L. (1994) Field trial of a genetically improved baculovirus insecticide, *Nature*, **370**, 138-140.
21. Chejanovsky, N., Zilberberg, N., Rivkin, H., Zlotkin, E. and Gurevitz, M. (1995) Functional expression of an alpha anti-insect scorpion neurotoxin in insect cells and lepidopteran larvae, *FEBS Letters*, **376**, 181-184.
22. Eldridge, R., O'Reilly, D.R., Hammock, B.D. and Miller, L.K. (1995) Insecticidal properties of genetically engineered baculoviruses expressing an insect juvenile hormone esterase gene, *Appl. Environment. Microbiol.*, **58**, 1583-1591.
23. Bonning, B.C., Ward, V.K., van Meer, M.M.M., Booth, T.F. and Hammock, B.D. (1997) Disruption of lysosomal targeting is associated with insecticidal potency of juvenile hormone esterase, *Proc. Natl. Acad. Sci. USA*, **94**, 6007-6012.
24. Lee, S.-Y., Qu, X., Chen, W., Poloumienko, A., MacAfee, N., Morin, B., Lucarotti, C. and Krause, M. (1997) Insecticidal activity of a recombinant baculovirus containing an antisense c-*myc* fragment, *J. Gen. Virol.*, **78**, 273-281.
25. Davies, A.H. (1997) Expression libraries in recombinant baculoviruses, *IBC's International Conference on Expression Technologies.*
26. Mitsuhashi, J. and Inoue, H. (1988) Obtainment of a continuous cell line from the larval fat bodies of the mulberry tiger moth, *Spilosoma imparilis* (Lepidoptera: Arctiidae), *Appl. Entomol. Zool.*, **23**, 488-490.
27. Mitsuhashi, J. and Maramorosch, K (1964) Leafhopper tissue culture: Embryonic nymphal and imaginal tissues from aseptic insects, *Contrib. Boyce Thompson Inst.*, **22**, 435-460.
28. Takenaka, Y., Kobayashi, J., Matsuda, Y., Gong, C., Nagaya, M., Miyajima, S. and Yoshimura, T. (1997) Evaluation of SpIm insect cells using a novel baculovirus expression vector system emplying the *Hyphantria cunea* NPV, *JAACT'97 NAGOYA*, P-20.
29. Kawakami, A., Kataoka, H., Oka, T., Mizoguchi, A., Kimura-Kawakami, M., Adachi, T., Iwami, M., Nagasawa, H., Suzuki, A. and Ishizaki, H. (1990) Molecular cloning of the *Bombyx mori* prothoracicotropic hormone, *Science*, **247**, 1333-1335.
30. Gardiner, G.R. and Stockdale, H. (1975) Two tissue-culture media for production of lepidopteran cells and nuclear polyhedrosis viruses, *J. Invertebr. Path.*, **25**, 363-370.
31. Inoue, H. and Hayasaka, S. (1995) A new cell line separated from the contractile muscle cell line of Chinese oak silkworm, *Antheraea pernyi*, *J. Seric. Sci. Jpn.*, **64**, 79-81.
32. Mitsuhashi, J. (1984) Isolation of a continuous cell line from larval fat bodies of an Arciid moth, *Spilarctia seriatopunctata* (Insecta, Lepidoptera, Arctiidae), *Zool. Sci.*, **1**, 415-419.
33. Hellers, M. and Steiner, H. (1992) Diapausing pupae of *Hyalophora cecropia*: An alternative host for baculovirus mediated expression, *Insect Biochem. Molec. Biol.*, **22**, 35-39.

IN VIVO GENE ELECTROPORATION : MANIPULATIONS OF CELLS TO WHOLE BODY OF LIVING ANIMALS

T. MURAMATSU
Department of Biological Resources and Environmental Sciences, School of Agricultural Sciences, Nagoya University, Chikusa-ku, Nagoya 464-01, Japan

Abstract *In vivo* electroporation is currently an unfamiliar nonviral means of gene transfer, accounting for only about 1% of total studies related to *in vivo* gene transfer and gene therapy. In the present study, basic principles and applications of *in vivo* gene electroporation are discussed. Like other nonviral methods, *in vivo* gene electroporation has a variety of advantages over viral vectors : any types of cells and tissues in theory could become a target ; handling is easy and quickly done within a matter of second ; repeated administration of DNA is possible ; no immunogenicity is expected, and ; there is no constraints on amounts and sizes of DNA to be used. Gene transfer efficiency of *in vivo* electroporation was found to be equivalent or even superior to that of *in vivo* lipofection, gene gun and direct DNA injection methods. Although most foreign genes are likely to be present in an episomal form when transferred by *in vivo* electroporation, foreign gene products could be detected for more than 1 month depending on tissues and DNA constructs used. Gene expression generated by *in vivo* electroporation could be controlled to a certain extent in a tissue- or cell-specific manner, and be induced by intention. Undoubtedly, there still exists plenty of room for improvement to apply this nonviral gene transfer method to manipulations of cells and animals, and better appraisal should be brought forth in the near future.

Introduction

The possibility that animal cells could be altered in selected and intended ways by introducing foreign DNA into them has long enchanted scientists. The experimental protocols of gene transfer now become feasible both *in vitro* in cultured cells and *in vivo* in tissues of living animals. Table 1 summarizes currently available gene transfer methods, some of which are applicable to both *in vitro* and *in vivo* situations.

Under the circumstances *in vivo*, gene transfer largely rely on viral vectors hitherto. The viral vector methods based on retrovirus, adenovirus, and others account for more than 60% of the total number of papers concerning *in vivo* gene transfer (Muramatsu et al., 1998). However, there are serious concerns about the use of viral vectors especially when gene therapy for humans is to be attempted. Retroviral vectors, for instance, usually suffer from low titers, oncogenic potential, the requirement of active cell division, transcriptional inactivation caused by methylation, and occasional outbreaks of replication competent retrovirus. Although adenoviral vectors can be recovered in high titers and could be used in both dividing and nondividing cells, host immunogenecity prevents their repeated use (Friedmann, 1997). In addition, there is limitation of DNA size to be inserted in these viral vector constructs, usually up to several kb. For researchers, an

Y. Kitagawa et al. (eds.), Animal Cell Technology: Basic & Applied Aspects, Vol. 10, 125–129.

easy, safe, non-toxic, and possibly efficient delivery of genes to a target tissue has been an attractive tissue. Hence the above limitations of the use of viral vectors lead one to search alternative nonviral means of gene transfer.

In vivo electroporation as a nonviral gene transfer method

Under the *in vivo* condition, electroporation (EP) is still an unfamiliar gene transfer method, accounting for only about 1% of the papers on gene transfer (Muramatsu et al., 1998). However, EP has a variety of advantages over other nonviral as well as viral methods : all tissues and cells in theory could become a target ; handling is easy and quickly completed within a matter of second ; repeated administration of DNA is possible ; no immunogenecity is expected ; no DNA size constraints are imposed, and ; no specialized processes for DNA construction is required.

TABLE 1. Methods of gene transfer in cultured animal cells (*in vitro*) and tissues of living animals (*in vivo*)

In Vitro	*In Vivo*
Biological Means	
Viral vectors	Viral vectors
Receptor mediation	Receptor mediation
Protoplast fusion	
Chemical Means	
Lipofection	Lipofection
Calcium phosphate precipitation	
DEAE dextran mediation	
Polybrene mediation	
Physical Means	
Gene gun	Gene gun
Electroporation	Electroporation
Laserporation	Laserporation
Pricking	Direct injection
Microinjection	

In Fig. 1, a variety of *in vivo* gene EP methods are depicted. Essentially, *in vivo* gene EP is completed first by injecting foreign DNA into a target site of any given tissue, followed by electrical pulse application. In an *in vivo* situation, square pulses should be used as they are far better than exponential decay pulses for higher gene expression. When electrical square pulses are applied to a field, heat will be generated. The quantity of the heat is given by the equation as described in the literature (Hofmann, 1995). Based upon our experience, the gene transfection efficiency, in other words intensity of foreign gene expression, is roughly proportional to the amount of heat generated unless the tissues are irreversibly damaged. For example, EP conditions such as 1,000 V/cm for 100 μsec are equivalent to those of 25 V/cm for 160 msec. The results in the mouse testis given in Fig. 2 appear to support the above hypothesis (Muramatsu et al., 1997a). Decreasing the voltage from 100 to 25 resulted in a remarkable reduction in CAT activity while by increasing loading period from 10 to 50 msec at the same 25 V, CAT activity recovered almost that found at 100 V with 10 msec.

Comparison of EP with other nonviral gene transfer methods

Direct comparison among nonviral gene transfer methods has scantily been documented. Undoubtedly, such comparison is quite difficult as each method has its own characteristics. Nevertheless, attraction to answer the question that researchers often wish to know is overpowering : which nonviral method is the best to choose. At least in the mouse testis (Muramatsu et al., 1996, 1997a), and the chicken embryo (Muramatsu et al., 1997b), *in vivo* EP provides better gene transfection efficiency than *in vivo* lipofection and gene gun, as shown by CAT activity and X-gal staining. In

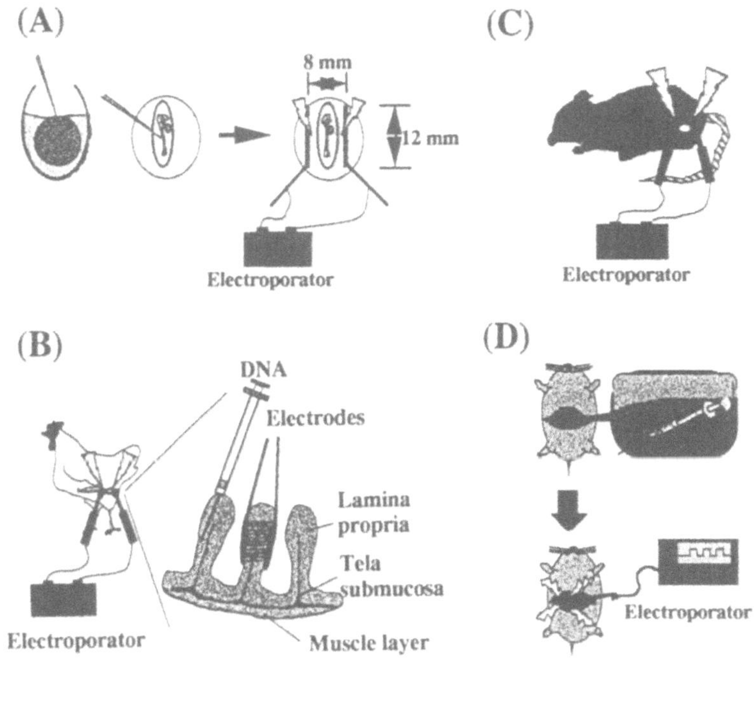

Figure 1. *In vivo* gene electroporation applied to a variety of tissues of living animals : (a) chicken embryos in ovo ; (b) mucosal fold of the oviduct in laying hens ; (c) mouse testis, and ; (d) rat abdominal muscle.

addition, in the chicken testis, *in vivo* EP conferred a stronger firefly luciferase activity than did *in vivo* gene gun (unpublished). When *in vivo* EP was compared with direct DNA injection method in the rat abdominal muscle, gene expression given by *in vivo* EP was approximately 15-50 times stronger than that given by direct DNA injection (unpublished).

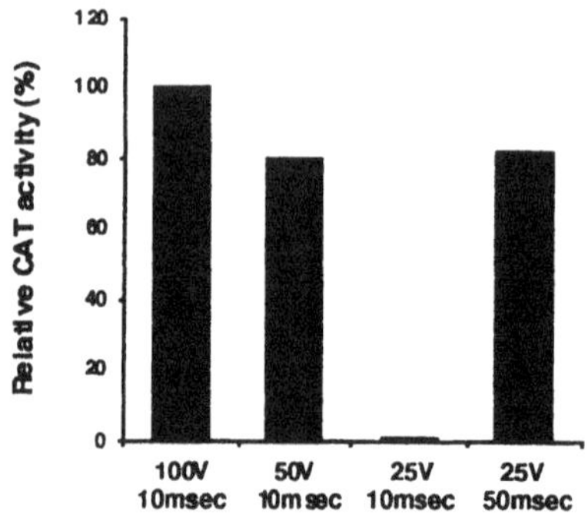

Figure 2. Effects of varying voltage and loading period on CAT expressed as percentage of the mean value for the 100 V/10 msec group. Adapted from Muramatsu et al. (1997a).

In summary, although we have not done extensive comparison of gene transfer efficiency among nonviral methods, our data obtained so far suggest that as far as a localized target area is concerned, *in vivo* EP would be one of the best choices for nonviral gene transfer.

Regulation of gene expression after *in vivo* gene EP

Besides the transfection efficiency, a critical concern is whether or not the expression of transferred genes is properly regulated *in vivo* as was expected. First, duration of gene expression after *in vivo* EP depends on the target tissue, and probably to some extent on the plasmid DNA construct. In the rat liver, reporter gene expression was observed for at least 21 days after transfection (Heller et al., 1996). In skeletal muscles, gene expression of firefly luciferase and mouse erythropoietin was observed at 1 and 3 months after *in vivo* gene EP, respectively (unpublished). Secondly, *in vivo* gene EP may provide not only tissue-specific but also cell-specific gene expression. When spermatid-specific gene promoter was used, substantially stronger CAT gene expression was obtained in the mouse testis than in the leg muscle and liver (Muramatsu et al., 1997a). Thirdly, manipulation of gene expression, either induction or repression, is possible as intended. By using promoters containing steroid response elements, induced gene expression was obtained in the chicken oviduct (unpublished). Moreover, nutritional status, i.e. feeding and starving, could modulate reporter gene expression in the mouse liver when a gene promoter of a key enzyme involved in the gluconeogenesis pathway was used (unpublished).

Conclusion

Despite the poor spread at present, contributing only to about 1% of the total studies, *in vivo* EP will probably be more extensively used in the near future because it is easy, safe, non-toxic, and no limitations in amounts and sizes of DNA applied. There are no constrains on types of cells and tissues targeted, and more importantly, gene transfer efficiency appears to be better in EP than in other nonviral methods. A further appraisal of *in-vivo* gene EP should be brought forth after more detailed analyses are made from a view point of both experimental and therapeutic purposes.

References

Friedmann, T. (1997) Overcoming the obstacles to gene therapy. *Scientific American* **276**, 96-101.

Heller, R., Jaroszeski, M., Atkin, A., Moradpour, D., Gilbert, R., Wands, J. and Nicolau, C. (1996) *In vivo* gene electroinjection and expression in rat liver. *FEBS Letter* **389**, 225-228.

Hofmann, G. A. (1995) Instrumentation, in Nickoloff, J. A. (ed), *Methods in Molecular Biology. Electroporation Protocols for Microorganisms,* Humana Press Inc., Totowa, NJ, vol. 47, pp. 27-45.

Muramatsu, T., Mizutani, Y., Ohmori, Y. and Okumura, J. (1997b) Comparison of three nonviral transfection methods for foreign gene expression in early chicken embryos *in ovo*. *Biochemical and Biophysical Research Communications* **230**, 376-380.

Muramatsu, T., Nakamura, A. and Park, H.-M. (1998) *In vivo* electroporation: a powerful and convenient means of nonviral gene transfer to tissues of living animals. *International Journal of Molecular Medicine* **1**, 55-62.

Muramatsu, T., Shibata, O., Ohmori, Y. and Okumura, J. (1996) *In vivo* electroporation : a convenient method for gene transfer to testicular cells in mice. *Animal Science and Technology (Japan)* **67**, 975-982.

Muramatsu, T., Shibata, O., Ryoki, S., Ohmori, Y. and Okumura, J. (1997a) Foreign gene expression in the mouse testis by localized *in vivo* gene transfer. *Biochemical and Biophysical Research Communications* **233**, 45-49.

IN VITRO STIMULATION OF TUMOR NECROSIS FACTOR PRODUCTION FROM RAT SPLENOCYTES BY CABBAGE EXTRACT

WATARU KOMATSU, YUTAKA MIURA and KAZUMI YAGASAKI
Department of Applied Biological Science, Tokyo Noko University
Fuchu, Tokyo 183, Japan

Abstract. The effect of cabbage extract on *in vitro* production of tumor necrosis factor (TNF) was studied in primarily isolated rat splenocytes and peritoneal macrophages. Cabbage extract was found to stimulate the production of TNF in the cultured splenocytes, but not to do so in the cultured peritoneal macrophages. The production of TNF from peritoneal macrophages was significantly elevated by cabbage extract when these macrophages were co-cultured with isolated splenic T cells. The cytotoxic activities of the splenocytes stimulated with cabbage extract were significantly higher than those with vehicle control when the splenocytes were co-cultured with an ascites hepatoma cell line of AH109A. These results suggest that the stimulatory effect of cabbage extract on TNF production is indirect on macrophages and that T cells are involved in the appearance of the cabbage extract. They also suggest that *in vitro* anti-tumor activity is potentiated by cabbage extract through the stimulation of immune response.

1. Introduction

Tumor necrosis factor (TNF) is the cytokine produced by activated macrophages [1]. TNF has important roles in anti-tumor, immunoregulatory and inflammatory responses [2]. Vegetable extract has been reported to regulate the production of TNF *in vivo* [3,4]. We have found that a cabbage extract also stimulates the production of TNF in resident peritoneal macrophages by its oral administration to rats [5]. The cabbage is thus expected to have an anti-tumor effect with its immunostimulatory action. In the present study, we investigated the effect of cabbage extract on *in vitro* production of TNF in primarily cultured rat splenocytes and peritoneal macrophages, and its anti-tumor effect.

2. Materials and Methods

A cabbage was obtained from a local market and put into a juicer to prepare the cabbage extract. Splenocytes and peritoneal macrophages were isolated from male Donryu rats (8-12 weeks of age), and were cultured in 24-well plates with RPMI 1640 medium supplemented

Y. Kitagawa et al. (eds.), Animal Cell Technology: Basic & Applied Aspects, Vol. 10, 131–134.

with 100 U/ml penicillin, 100 μg/ml streptomycin, 2 mM L-glutamine, 50 μM 2-mercaptoethanol and 10% fetal bovine serum at 1×10^6 cells/well or 1×10^5 cells/well, respectively. To measure the dose-responsiveness of TNF production to cabbage extract, 0, 0.05, 0.1, 0.3, 0.5 and 1% of the extract and 100 ng/ml of lipopolysaccharide (LPS) were added to wells and the cells were incubated for 6 hr. To measure the time-course of TNF production, 1% cabbage extract was added to wells and incubated for 0, 2, 4, 8 and 24 hr. After washing once with RPMI 1640, the cells were incubated for a further 6 hr in the presence of 100 ng/ml of LPS. In both experiments, 100 U/ml interferon (IFN)-γ, which is known to have a priming effect on TNF production [6,7], was used as a positive control. To co-culture peritoneal macrophages with T cells, T cells were prepared from spleen cells and isolated using a nylon wool column for lymphocyte preparation. Isolated T cells were added to the macrophage-culture at 1×10^6 cells/well, and co-cultured for 6 hr in the presence of 0 and 1% cabbage extract and 100 ng/ml LPS. At the end of the incubation, the culture supernatant was harvested and its TNF concentration was determined by bioassay using L929 mouse fibroblasts as described previously [8]. To measure the cytolytic activities, the cultured splenocytes were stimulated by 1% cabbage extract for 2 hr, washed once with RPMI 1640 and co-cultured with the rat ascites hepatoma cell line of AH109A at the effector-target ratios of 10:1 and 20:1 for 24 hr. At the end of the incubation, the culture supernatant was harvested and the cytotoxicity was estimated by measuring activity of lactate dehydrogenase released specifically from dead cells to culture supernatant. Statistical analyses were carried out using Duncan's multiple-range test after one-way analysis of variance or Student's *t*-test.

3. Results and Discussion

The dose-response to cabbage extract of TNF production in splenocytes and peritoneal macrophages was first examined. Cabbage extract stimulated the production of TNF from splenocytes at 0.5 and 1%, and their production was equal to those stimulated by IFN-γ. In contrast, unlike splenocytes, cabbage extract failed to stimulate TNF production by peritoneal macrophages at any dose (TABLE 1). The time-course of TNF production in splenocytes and peritoneal macrophages treated with cabbage extract was next examined.

TABLE 1 Dose-response of TNF production to cabbage extract from rat splenocytes and peritoneal macrophages

	Splenocytes	Macrophages
TNF Production (pg/μg DNA)		
0 (%)	38.1 ± 7.4	87.2 ± 18.7
0.05	52.4 ± 11.5	76.0 ± 14.0
0.1	66.1 ± 17.4	89.7 ± 32.4
0.3	85.5 ± 17.8	96.6 ± 24.8
0.5	161.6 ± 35.7*	123.7 ± 23.0
1.0	148.6 ± 34.3*	86.6 ± 23.4
IFN-γ	157.1 ± 46.5*	212.6 ± 46.0

Each value represents the mean ± SEM for 4 assays. *Significantly different from the control (0%) at $p < 0.05$ by Duncan's multiple-range test.

TABLE 2 Time-course of TNF production from rat splenocytes

	0	2	4	8	24
TNF Production (pg/μg DNA)					
	21.1 ± 5.9	79.9 ± 26.0	60.0 ± 13.2	43.6 ± 10.3	8.1 ± 1.9

Each value represents the mean ± SEM for 4 assays.

The production of TNF by splenocytes peaked at 2 hr, and declined thereafter (TABLE 2). No change was seen in peritoneal macrophages at any period (data not shown). These results suggest that cabbage extract stimulates TNF production *in vitro* and its level is equal to a typical immunostimulator IFN-γ, but its stimulation manner is indirect on macrophage. Cabbage extract might be thought to stimulate macrophage-activating cells such as T cell, because activated T cells were reported to potentiate TNF production in macrophage by direct cell-to-cell contact [9] and by producing the macrophage-activating factors such as IFN-γ [10].

We next examined whether cabbage extract could stimulate TNF production by peritoneal macrophages co-cultured with isolated T cells. In the absence of T cells, TNF production was not changed at any concentration of the extract. In the presence of isolated T cells, however, TNF production was stimulated by cabbage extract, and its production was significantly higher than that in the absence of T cells (TABLE 3). This suggests that cabbage extract stimulates TNF production *via* T cell action.

Because TNF has an important role in anti-tumor effect [2], cabbage extract is expected to have an anti-tumor effect due to its ability to stimulate TNF production; we therefore examined this possibility *in vitro*. Splenocytes treated with cabbage extract almost completely killed AH109A cells (TABLE 4). These results suggest that the extract can kill tumor cells through stimulation of the immune response *in vitro*.

From these results we conclude that cabbage has component(s) which stimulate TNF production, that its effect on macrophage is indirect, involvement of T cell is necessary in the action, and that the cabbage has an anti-tumor effect *in vitro* with actions like a biological response modifier. However, the active component(s) in cabbage remain unidentified. Some flavonoids such as glycoside form [11] or some indole derivatives like 1'-methylascorbigen [12] have been reported to have immunostimulating activity. Therefore, these components in cabbage may be effective to a host defense *in vivo*. Further studies are needed to confirm this.

TABLE 3 TNF production from peritoneal macrophages co-cultured with isolated T cells

	Absence of T cell	Presence of T cell
TNF Production (pg/μg DNA)		
0 (%)	86.5 ± 18.2	97.3 ± 3.2
1.0	75.5 ± 5.5	226.6 ± 64.4*

Each value represents the mean ± SEM for 4 assays. *Significantly different from control (absence of T cell) at $p < 0.05$ by Student's *t*-test.

TABLE 4 Cytolytic activity of rat splenocytes treated with cabbage extract

	Vehicle	Cabbage extract
Cytotoxicity (%)		
10:1	62.9 ± 3.8	91.3 ± 8.7*
20:1	78.7 ± 8.8	100.0 ± 0.0*

Each value represents the mean ± SEM for 4 assays. *Significantly different from control (vehicle) at $p < 0.05$ by Duncan's multiple-range test.

4. References

[1] Carswell, E.A., Old, L.J., Kassel, R.L., Green, L., Fiore, N. and Williamson, B. (1975) An endotoxin-induced serum factor that causes necrosis of tumors, *Proc.Natl.Acad.Sci.USA* **72**, 3666-3670.

[2] Fier, W. (1991) Tumor necrosis factor, *FEBS Lett.* **285**, 199-212.

[3] Yamazaki, M., Ueda, H., Fukuoka, K., Okamoto, M. and Yui, S. (1992) Priming effects of vegetable juice on endogenous tumor necrosis factor, *Biosci.Biotech.Biochem.* **56**, 149.

[4] Yamazaki, M., Ueda, H. and Du, D. (1992) Inhibition by perilla juice of tumor necrosis factor production, *Biosci.Biotech.Biochem.* **56**, 152.

[5] Komatsu, W., Yagasaki, K., Miura, Y. and Funabiki, R. (1997) Stimulation of tumor necrosis factor and interleukin-1 productivity by the oral administration of cabbage juice to rats, *Biosci.Biotech.Biochem.* **61**, 1937-1938.

[6] Gifford, G.E. and Lohman-Matthes, M.L. (1989) Gamma-interferon priming of mouse and human macrophages for induction of tumor necrosis factor production by bacterial lipopolysaccharide, *J.Natl.Cancer Inst.* **78**, 121-124.

[7] Hayes, M.P., Freeman, S.L. and Donnelly, R.P. (1995) IFN-γ priming of monocytes enhances LPS-induced TNF production by augmenting both transcription and mRNA stability, *Cytokine* **7**, 427-435.

[8] Komatsu, W., Yagasaki, K., Miura, Y. and Funabiki, R. (1996) Modification of tumor necrosis factor and interleukin-1 productivity in macrophages from hepatoma-bearing rats by dietary proteins, *Nutr.Res.* **16**, 1699-1707.

[9] Suttles, J., Miller, R.W., Tao, X. and Stout, R.D. (1994) T cells which do not express membrane tumor necrosis factor-α activate macrophage effector function by cell contact-dependent signaling of macrophage tumor necrosis factor-α production, *Eur.J.Immunol.* **24**, 1736-1742.

[10] Stout, R.D. and Bottomly, K. (1989) Antigen-specific activation of effector macrophages by IFN-gamma producing (TH1) T cell clones. Failure of IL-4-producing (TH2) T cell clones to activate effector function in macrophages, *J.Immunol.* **142**, 760-765.

[11] Kunizane, H., Ueda, H. and Yamazaki, M. (1995) Screening of phagocyte activators in plants; enhancement of TNF production by flavonoids, *Yakugaku Zasshi (in Japanese)* **115**, 749-755.

[12] Preobrazhenskaya, M.N., Bukhman, V.M., Korolev, A.M. and Efimov, S.A. (1993) Ascorbigen and other indole-derived compounds from *Brassica* vegetables and their analogs as anticarcinogenic and immunomodulating agents, *Pharmac.Ther.* **60**, 301-313.

EFFECTS OF GREEN, OOLONG AND BLACK TEAS AND RELATED COMPONENTS ON PROLIFERATION AND INVASION OF HEPATOMA CELLS

G. Y. ZHANG, Y. MIURA AND K. YAGASAKI
Department of Applied Biological Science, Tokyo Noko University
Saiwaicho 3-5-8, Fuchu, Tokyo 183-8509, Japan

ABSTRACT. Employing both *in vitro* and *ex vivo* assay systems, we studied the effects of extracts from powdered green, oolong and black teas and related tea components on the proliferation and invasion of an ascites hepatoma cell line of AH109A. The AH109A cells were found to penetrate underneath the monolayer of primary cultured mesothelial cells isolated from Donryu rat mesentery in the presence of 10% calf serum. Each tea extract dose-dependently restrained this invasion and the AH109A proliferation. About 0.04% of each extract had strong inhibitory effects on the invasion and proliferation. When rat serum obtained after oral intubation of each tea extract was added to the culture media instead of calf serum at a concentration of 10%, the invasion of AH109A was significantly inhibited as compared with the control rat serum. The proliferation of AH109A was also inhibited by the tea extract-loaded rat serum obtained after oral intubation of each extract. Furthermore, (-)-epigallocatechin gallate (EGCG), (-)-epicatechin gallate (ECG) and (-)-epigallocatechin (EGC) from green tea as well as the mixture of theaflavin and theaflavin gallates (TFs) from black tea were shown to be the most effective components against the invasion and proliferation of AH109A among all components screened.

These results confirm that not only green tea but black tea, oolong tea and their related components also have an inhibitory effect on the invasion and proliferation of AH109A.

1. Introduction

Tea (*Camellia sinensis*) is one of the most popular beverages consumed worldwide. The history of tea can be traced back more than 5000 years and tea has been used as a drug in traditional Chinese medicine for a long period of time/for much of that time. Many biological activities of teas and tea components have been reported. Its anticancer activity is one of the most important biological activities of tea (1-2). Cancer cells have two biological characteristics, i.e., endless proliferation and metastasis. The invasion of cancer cells is an important characteristic step in metastasis. In the present study, we examined the effects of extracts of green, oolong and black teas and related components on the proliferation and invasion of cancer cells, using hepatoma cells.

Y. Kitagawa et al. (eds.), Animal Cell Technology: Basic & Applied Aspects, Vol. 10, 135–139.

2.Materials and Methods

2.1 *Preparation of extracts from powdered green tea (PGT), powdered oolong tea (POT) and powdered black tea (PBT) and the solution of each tea components*

Two grams of PGT, POT and PBT were respectively extracted by 100 ml of boiling water for 3 minutes. (PGT, POT and PBT were generously provided by Yamato Tea Co., Ltd, Nara, Japan.) After being sterilized by filtration, each tea extract was stored at -20 °C for later experiments. For the *ex vivo* experiment, the obtained solution of each tea extract was evaporated to dryness. The dried materials were reconstituted by 10 ml MQ water (that is, 10 fold concentration). (-)-epigallocatechin gallate (EGCG), (-)-epicatechin gallate (ECG), (-)-epigallocatechin (EGC) (-)-epicatechin (EC) and theophylline (TPL) from green tea as well as the mixture of theaflavin and theaflavin gallates (TFs) from black tea were respectively dissolved in DM-160 medium, while TPL was dissolved in 0.5% DMSO and then mixed with DM-160 medium.

2.2 *Primary culture of mesentery-derived mesothelial cells (M-cells)*

The abdominal cavity of a male Donryu rat (4-6 weeks old, obtained from NRC Haruna, Gunma, Japan) was opened and the mesentery was cut off with scissors. The obtained mesenteries were incubated in 0.25% trypsin (Sigma, St. Louis, MO, USA) in phosphate-buffered saline (PBS) at 37°C for 20 min with gentle stirring. After treatments by digestion, filtration, etc., the M-cells were seeded at a density of $1.2x10^5$ cells/60mmϕ culture dish with 2 mm grids (Costar, Cambridge, MA, USA) in 3 ml of 10% CS/DM-160. The cells were cultured in an atmosphere of 5% CO_2-95% humidified air at 37°C to the confluent state by replacing culture media every other day and were used for the *in vitro* and *ex vivo* invasion assays described below.

2.3 *Culture of AH109A hepatoma cells*

AH109A cells were provided by SRL (Tokyo, Japan) and were maintained in the peritoneal cavities of male Donryu rats. AH109A cells prepared from accmulated ascites could be cultured *in vitro* in 10% CS/DM-160 for up to 2 months. AH109A cultured for at least 1 week after preparing them from rat ascites were used for the assays.

2.4 *In vitro proliferation assay*

Effects of PGT, POT and PBT extracts on AH109A proliferation were examined by the well-known WST-1 method (3). For WST-1 assay, $5x10^3$ AH109A cells were cultured for 44 h in each well of a 96-well-plate containing 10%CS/DM-160 in the absence or presence of PGT, POT and PBT extracts. WST-1 solution was then added and the cells were cultured for another 4 h. Viable cells can make WST-1 produce highly colored formazan dyes and the optical density (OD) was measured through a microplate reader (BIO-RAD, Model 450). The relative proliferation rate of AH109A was calculated by comparing the optical density (OD) between the control group treated with no tea and test groups treated with each tea extract, tea extract-loaded serum or tea component.

2.5 *In vitro invasion assay*

When M-cells became confluent after a week's culture, 2.4x10^5 AH109A cells were seeded on the M-cell monolayer with the 10% CS/DM-160 medium in the absence or presence of different concentrations of each tea extract or tea component and cultured for 48 h. The invasive activity of AH109A was measured by counting the invaded AH109A cells and colonies under a phase contrast microscope. Usually at least 10 areas were counted and the invasive activity of the cells was indicated by the number of invaded cells and colonies/cm^2 (4).

2.6 *Ex vivo proliferation and invasion assay*

Each concentrated PGT, POT and PBT extract was respectively intubated to male Donryu rats (5 weeks old) at a dose of 1.0 ml/100g body weight and blood was collected at 0, 2, and 5 h after oral intubation. The sera were prepared, sterilized by filtration, and added to culture media at a concentration of 10% instead of CS for *in vitro* proliferation and invasion assays of AH109A as described above. The proliferative activities of AH109A in the presence of these sera were measured as the WST-1 method; the invasive activities of AH109A in the presence of these sera were measured in the *in vitro* invasion assay.

2.7 *Statistical analysis*

Data were analyzed by a one-way analysis of variance (ANOVA). When F value was significant ($P<0.05$), differences among the data were evaluated at $P<0.05$ by Tukey's Q test.

3. Results and Discussion

3.1 *Effects of PGT, POT and PBT on proliferation and invasion of AH109A cells*

Table 1. Inhibitory effects of extracts from PGT, POT and PBT on proliferation and invasion of AH109A cells

Teas	Proliferation	Invasion
	(IC_{50}: w/v%)	
PGT	0.037	0.043
POT	0.036	0.040
PBT	0.037	0.029

Table 1 shows the inhibitory effects of PGT, POT and PBT against the proliferation and invasion of AH109A cells in *in vitro* experiment. Each powdered tea extract could significantly inhibit both the proliferation and invasion of AH109A cells. The three kinds of tea extracts had a similar inhibitory effect against the proliferation of AH109A cells with the IC_{50} 0.036% to 0.037%, but PBT showed a stronger inhibitory effect against the invasion of AH109A cells with the IC_{50} 0.029%.

3.2 *Effects of PGT, POT and PBT extract-loaded rat sera on proliferation and invasion of AH109A cells*

Table 2. Inhibitory effects of PGT, POT and PBT extract-loaded rat sera on proliferation and invasion of AH109A cells

Teas	Inhibition rate of proliferation		Inhibition rate of invasion	
	2 h sera	5 h sera	2 h sera	5 h sera
	(% of 0 h)			
PGT	48	22	98	67
POT	35	8	59	43
PBT	59	16	62	34

Table 2 shows the effects of PGT, POT and PBT extract-loaded rat sera against the proliferation and invasion of AH109A cells in the *ex vivo* experiment. Each tea extract-loaded rat serum obtained at 2 and 5 h after oral intubation of each extract was added to the culture media instead of calf serum at a concentration of 10%. Both the proliferation and invasion were significantly suppressed except for the 5 h POT extract-loaded rat serum which did not show a significant effect against AH109A cell proliferation. Each tea extract-loaded rat serum at 2 h after oral intubation had the strongest inhibition and the inhibitory effect lasted 5 h for each one, although only 2 h for POT against proliferation of AH109A cells.

3.3 *Effect of tea components on proliferation and invasion of AH109A cells*

Table 3. Inhibitory effects of tea components on proliferation and invasion of AH109A cells

Components	Proliferation	Invasion
	(IC_{50} : μM)	
EC	>200	—
ECG	99.60	74.25
EGC	75.99	109.29
EGCG	41.28	21.40
TFs	138.00	59.85
TPL	—	—

—: no effect

Table 3 shows the effects of tea components against the proliferation and invasion of AH109A cells in the *in vitro* experiment. EGCG, ECG, EGC and TFs significantly inhibited both proliferation and invasion. EC had weak inhibitory effect against the proliferation but was unable to suppress the invasion of AH109A cells. TPL did not restrain either the proliferation or invasion. The strongest inhibitory effects against both proliferation and invasion of AH109A cells were shown by EGCG.

Based on our results from the present study and those of other studies on the effects of PGT, POT and PBT extracts on normal rat mesentery-derived mesothelial cells (M-cells) and another kind of tumor cell line, the inhibitory effects of PGT, POT and PBT as well as related tea components against the proliferation and invasion of AH109A cells are due to the cell-specific and higher sensitivity of this cell line to the components. The mechanisms of action of these teas and their components against the proliferation and invasion of AH109A cells may be that the components affect a variety of enzyme activities, the signal trans-duction pathways and the relevant gene regulation, although this requires further proof.

References

[1] R.Agarwal and H.Mukhtar: Dietary Phytochemicals in Cancer Prevention and Treatment, Plenum Press, New York, 35-50, 1996.
[2] S. Valcic,et al: Anti-Cancer Drug, 7, 461-468, 1996.
[3] M.Ishiyama, M.Shiga, K.Sasamoto, M.Mizoguchi and PG He: Chem.Pharm. Bull, 41, 1118-1122, 1993.
[4] Y.Miura, H.Shiomi, F.Sakai & K.Yagasaki: Cytotechnology, 23, 127-132, 1997.

MILK SECRETION IN CULTURED MAMMARY EPITHELIAL CELLS

DAVID R BLATCHFORD, COLIN J. WILDE, T. MATSUDA* and N AOKI*
Hannah Research Institute, AYR, KA6 5HL, UK
**Department of Applied Biological Sciences, School of Agricultural Sciences, Nagoya University, Nagoya 464-01, Japan*

Abstract

Mammary epithelial cells secrete milk constituents by several routes. Milk lipid is enveloped by a milk fat globule membrane (MFGM) derived from the apical cell surface, and contains some of its constituent proteins. Soluble milk proteins are secreted by exocytosis. Some studies suggest these pathways are controlled independently, and may therefore provide a way to control milk composition. We propose to study secretion of the aqueous and lipid constituents of milk by comparing the secretion of butyrophilin, the major MFGM protein, with that of casein, the most abundant soluble proteins, in cultures of bovine mammary epithelial cells. Mammary epithelial cells were prepared from tissue of non-lactating, late-pregnant cows by collagenase digestion and density gradient centrifugation, and stored in liquid nitrogen. Cryopreserved cells were cultured initially on tissue culture plastic, then transferred to a reconstituted basement membrane (EHS matrix) and cultured in serum-free medium containing lactogenic hormones. On EHS matrix, cells formed multicellular structures covered in matrix ("mammospheres"), and secreted casein and butyrophilin. Unlike murine mammospheres, bovine mammospheres did not contain one central luminal space, and immunohistochemistry showed milk proteins distributed between the cells. Immunoblotting showed casein and butyrophilin present in both culture medium and an EGTA extract of luminal proteins. However, the results show that bovine mammary epithelial cells differentiate and secrete both soluble and MFGM proteins in primary culture, making this system suitable for study of control of mammary secretory pathways.

Introduction

The rate of milk secretion is controlled by circulating galactopoietic hormones and locally within the gland by frequency and completeness of milk removal. Thus, the productivity of dairy animals can be manipulated by hormone treatment or by husbandry. These manipulations generally influence milk secretion rate but not the composition of milk. Control of milk composition is, however, highly-desirable, since consumer demand is for a high protein, low fat product. The advent of transgenesis may allow milk protein content to be altered by insertion of additional gene copies, or by manipulating the level of

Y. Kitagawa et al. (eds.), Animal Cell Technology: Basic & Applied Aspects, Vol. 10, 141–145.

metabolically-active milk proteins such as α-lactalbumin, a component of the lactose synthetase enzyme (Stacey *et al.*, 1995). Since aqueous and lipid components of milk are secreted by different routes which appear under some circumstances to be independently controlled (Travers *et al.*, 1996), another approach may be to manipulate specifically the secretory pathways for aqueous and non-aqueous milk constituents in the mammary epithelial cell.

Soluble milk constituents, including the majority of milk proteins, are secreted by exocytosis of secretory vesicles (Franke *et al.*, 1976). Secretion of soluble milk proteins is stimulated by raising intracellular Ca^{2+} (Turner et al., 1992) and, conversely, inhibited by a milk-borne feedback inhibitor of secretion (FIL; Rennison *et al.* 1993), indicating that exocytosis is not a constitutive process. Little is known about the regulation of milk lipid secretion, which occurs not by exocytosis, but by extrusion of the apical cell membrane, such that lipid droplets enter milk enclosed by a membrane (Patton and Keenan, 1975). The milk fat globule membrane (MFGM) contains a population of proteins sequestered quantitatively from the apical cell membrane (Mather, 1987). Comparison of MFGM protein secretion with that of soluble milk proteins such as casein or β-lactoglobulin may, therefore, be used to determine the relative activities of the two secretory pathways. To do this, we have developed a culture system in which bovine mammary cells differentiate and secrete milk constituents.

Methods

Mammary cells were prepared from tissue of non-lactating cows in the third trimester of pregnancy by collagenase digestion, and an epithelial cell-rich fraction was obtained by density gradient centrifugation. The cells were snap-frozen until required. After thawing the cells were cultured for 3-4 d on tissue culture plastic in [medium] containing 25% (v/v) serum (?% fetal calf, ?% horse serum)until they were ~90% confluent. The cells were then harvested by trypsin treatment and transferred to a reconstituted basement membrane derived from the Engelbreth-Holm-Swarm tumour (EHS matrix; Kleinman et al., 1986). The cells were cultured for up to 10 d on this matrix in Medium 199 / Ham's F12 medium (50:50, v/v) containing insulin (0.25 μg/ml), hydrocortisone (1μg/ml), triiodothyronine (0.65 ng/ml), epidermal growth factor (10ng/ml) and prolactin (1 μg/ml). Foetal calf serum (10% v/v) was present for the first 24 h to promote cell attachment, and medium (4 ml/well) was changed daily. Cells were fixed *in situ* in 4.0% (w/v) paraformaldehyde in 0.1 M phosphate buffer pH 7.4. The cells and matrix were collected by gentle scraping and frozen as a loose pellet in OCT compound. Air-dried 6 μm cryosections were rehydrated in PBS containing 0.1% (w/v) Triton X-100 and 0.3% (w/v) bovine serum albumin, blocked in PBS containing 25% (v/v) donkey serum, and treated with rabbit polyclonal antiserum against bovine casein (ref) diluted 1:250 in PBS containing 0.1% (w/v) Tween 20. Antibody binding was detected with FITC-conjugated anti-rabbit Ig. Control sections were treated with non-immune serum. Sections were mounted in Vectashield mounting medium containing 4,6-diamidino-2-phenylindole (DAPI) and viewed on a Leica DMRB microscope with epifluorescence optics. Secretory proteins were harvested by collection of culture medium and, after several washes , by treatment of cells *in situ* with EGTA (2.5 mM, 20 min). Culture medium and EGTA extract were dialysed against distilled water for 24 h at 4°C, freeze dried and equivalent amounts were subjected to SDS-polyacrylamide gel electrophoresis (SDS-PAGE) in 12.5% (w/v) gels under non-reducing conditions (Laemmli, 1970). Gels were blotted onto nitrocellulose (Towbin et al.,1979) and probed with antisera against bovine casein (1:500dilution) or butyrophilin (1:500 dilution; Aoki *et al.*, 1994) with alkaline phosphate-conjugated anti-rabbit IgG or anti-mouse IgG as the second antibody. Blots were developed using a colorimetric assay.

Results

Epithelial cells were plated on EHS matrix as small groups of 10-50 cells, and attached within 24 h in the presence of serum. After cell attachment, stress lines were apparent in the substratum as the cells progressively sequestered the surrounding matrix. After 2-3 d of culture, the cell aggregates had become dense structures enshrouded in matrix, such that individual cells were no longer visible. These structures were similar to those formed by mouse mammary cells on EHS matrix, and have been named "mammospheres" (Barcellos-Hoff *et al.*, 1989). Light microscopy suggested that mammosphere integrity was maintained for up to 10 d in culture (results not shown). DAPI nuclear staining showed that the majority of cells were located around the periphery of each mammosphere (results not shown), a phenomenon also observed in murine mammospheres (Barcellos-Hoff *et al.*, 1989; Hurley *et al.*, 1994), and caused by selective apoptosis of centrally-located cells (Blatchford *et al.*, 1998). Unlike murine cultures, however, some cells remained in the centre of the mammospheres, suggesting that this structural reorganisation was incomplete.

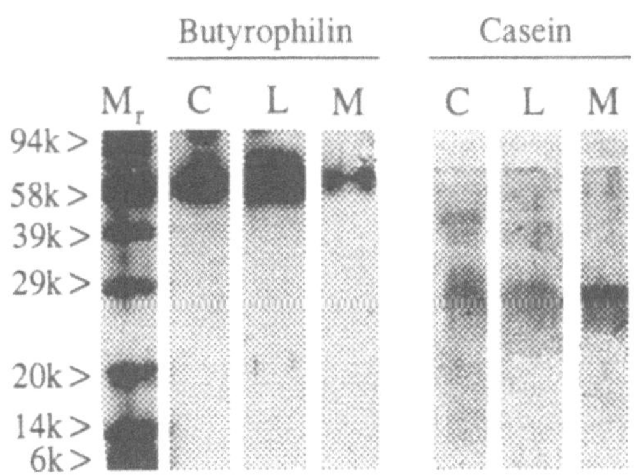

Figure 1. Secretion of butyrophilin and casein by bovine mammary epithelial cells in primary culture. Proteins were detected by immunoblotting in culture medium © and mammosphere lumina (L). M, cow's milk. M_r, molecular weight markers.

Immunohistochemistry of bovine mammospheres on d 7 of culture showed the presence of extracellular casein (results not shown). However, whereas milk proteins secreted by mouse mammospheres typically accumulate within the central luminal space (Blatchford *et al.*, 1995), casein associated with bovine mammospheres was dispersed through the structure and not concentrated in any area. Again, this suggests that bovine mammary epithelial cells, although differentiated and secreting milk proteins, had not formed the hollow luminal structure characteristic of mouse mammary cells on EHS matrix (Barcellos-Hoff *et al.*, 1989; Hurley *et al.*, 1994). Bovine mammary cell differentiation on EHS matrix was confirmed by immunoblotting, which detected casein in culture medium, and also in an EGTA extract of cells *in situ* on matrix (Figure 1). EGTA treatment disrupts tight intercellular junctions and extracts secretory proteins accumulated in the mammosphere lumina (Hurley *et al.*, 1994; Blatchford *et al.*, 1995). Since mammosphere cultures were washed extensively before EGTA treatment, the presence of casein in this extract suggests that some tight junction formation had occurred, with resultant entrapment of secreted protein in luminal spaces. Immunoblotting with monoclonal antibodies against bovine butyrophilin (Aoki et al., 1994) demonstrates that bovine mammary cells also synthesised MFGM proteins. Butyrophilin was, like casein, distributed in approximately equal proportions between culture medium and EGTA extract, indicating that mammosphere formation was incomplete, but that a proportion of cells had established tight intercellular junctions, forming luminal spaces within the structures.

Discussion

The results show that bovine mammary epithelial cells cultured on EHS matrix in the presence of lactogenic hormones undergo functional differentiation and secrete both soluble and MFGM proteins by exocytosis and membrane budding respectively. The system may be improved by optimising mammosphere formation, so that intercellular communication and cell-matrix interactions resemble more closely those occurring *in vivo*. Nevertheless, culture of bovine mammary cells on EHS matrix provides a useful system to study secretion of the aqueous and lipid components of milk and so determine if milk composition may be manipulable by differential control of their secretory pathways.

Acknowledgements

This work was funded by the Scottish Office Agriculture, Environment and Fisheries Department and by The award of a British Council Collaborative Research Grant to CJW is gratefully acknowledged.

References

Aoki, N., Kuroda, H., Urabe, M., Taniguchi, Y., Adachi, T., Nakamura, R. And Matsuda, T. (1994) Production and characterization of monoclonal antibodies directed against bovine milk fat globule membrane (MFGM). *Biochim. Biophys. Acta* **1199**, 87-95.

Barcellos-Hoff, M.H., Aggeler, J., Ram, T.G. and Bissell, M.J. (1989) Functional differentiation and alveolar morphogenesis of primary mammary cultures in reconstituted basement membrane.

Development **105**, 223-235.

Blatchford, D.R., Hendry, K.A.K., Turner, M.D., Burgoyne R.D. and Wilde, C.J. (1995). Vectorial secretion by constitutive and regulated secretory pathways in mammary epithelial cells. *Epithel. Cell Biol.* **4**, 8-16.

Blatchford, D.R., Quarrie L.H., Tonner, E., Flint, D.J. and Wilde C.J. (1998) Influence of micro-environment on mammary epithelial cells survival in primary culture. *J. Cell Sci.* Submitted for publication.

Franke, W.W., Luder, M.R., Kartenbeck, J., Zerban, H. and Keenan, T.W. (1976) Involvement of vesicle coat material in casein secretion and surface regeneration. *J. Cell Biol.* **69**, 173-195.

Hurley, W.L., Blatchford, D.B., Hendry, K.A.K. and Wilde, C.J. (1994) Extracellular matrix and mouse mammary cell function: comparison of substrata in culture. *In Vitro Cell Dev. Biol.* **30A**, 529-538.

Kleinman, H.K., McGarvey, M.L., Hassall, J.R., Star, V.L., Cannon, F.B., Laurie, G.W. and Martin, G.R. (1986). Basement membrane complexes with biological activity. *Biochemistry* **25**, 312-318.

Laemmli, U.K. (1970). Cleavage of structural proteins during the assembly of the head of bacteriophage T4. *Nature* **227**, 680-685.

Mather, I.H. (1987) Proteins of the milk fat globule membrane as markers of mammary epithelial cells and apical plasma membrane. In *The Mammary Gland: Development and Function*. (Neville, M.C. and Daniel, C.W., eds) pp 217-267. Plenum Press, New York.

Patton, S., and Keenan, T.W. (1975) The milk fat globule membrane. *Biochim. Biophys. Acta* **415**, 273-309.

Rennison, M.E., Kerr, M.A., Addey, C.V.P., Handel, S.E., Turner, M.D., Wilde, C.J. and Burgoyne, R.D. (1993) Inhibition of constitutive protein secretion from lactating mouse mammary epithelial cells by FIL (feedback inhibitor of lactation), a secreted milk protein. *J. Cell Sci.* **106**, 641-648.

Stacey, A., Schnieke, A., Kerr, M., Scott, A., McKee, C., Cottingham, I., Binas, B., Wilde, C and Colman, A. (1995) Lactation is disrupted by α-lactalbumin deficiency and can be restored by human α-lactalbumin gene replacement in mice. *Proc. Natl. Acad. Sci. USA* **92**, 2835-2839.

Towbin, H., Staehelm, T., and Gordon, J. (1979). Electrophoretic transfer of proteins from polyacrylamide gels to nitrocellulose sheets:procedure and some applications. *Proc. Natl. Acad. Sci., USA* **76**, 4350-4356.

Travers, M.T., Barber, M.C., Tonner, E., Quarrie, L., Wilde, C.J. and Flint, D.J. (1996). The role of prolactin and growth hormone in the regulation of casein gene expression and mammary cell survival: relationships to milk synthesis and secretion. *Endocrinology* **137**, 1530-1539.

Turner, M.D., Rennison, M.E., Handel, S.E., Wilde C.J. and Burgoyne, R.D. (1992) Proteins are secreted by both constitutive and regulated secretory pathways in lactating mouse mammary epithelial cells. *J. Cell Biol.* **117**, 269-278.

HYPERTHERMIC INDUCTION OF GENE EXPRESSION USING STRESS INDUCIBLE PROMOTER

I. A. Bouhon, A. Ito, M. Shinkai, H. Honda and T. Kobayashi.
Department of Biotechnology, Graduate School of Engineering, Nagoya University, Chikusa-ku, Nagoya 464-8603, Japan.

1.Introduction

The relevance of gene therapy for cancer treatment is directly related to two parameters. First the gene delivery method must satisfy both conditions of targeting and efficiency. Secondly the regulation of the expression system should enable the optimum tuning according to the specific conditions encountered in the tumor tissue. Indeed, tumor tissues have been shown to exhibit different characteristics or sensitivity than normal tissues.

This observation has led several researchers to investigate so called tissue specific vector housing a tumor specific promoter for driving the expression of therapeutic gene (Foster BJ *et al*, 1997; Bowman T*et al*, 1996; Jamora C *et al*, 1996). Nowadays, tumor specificities associated to targetable vectors are expected to yield the ultimate stringent tools in the gene therapy. Besides, cancer treatment by itself also raises the alteration of tumor cellular conditions. It engenders biological responses within the tackled cells apart from impeding the tumor cell growing process. Several examples of such responses have been intensively studied for treatment with cytokine (Goossens V, 1996), cisplatin and taxol (Gately DP *et al*, 1996) and X-ray (Boothman DA *et al*, 1994). The alterations of cellular environment can be regrouped under the appellation of stress.

Then we can elaborate a strategy appealing to treatment-specific promoter, that is, promoter induced by biochemical alteration in the tumor tissue consecutive to the treatment. Our laboratory has been investigating the hyperthermia treatment of glioma by incorporating nanoparticles of magnetite and irradiating the cells with high frequency magnetic field. Considering that several reports mention acidity increase (Kiang JG *et al*, 1990) and generation of radical oxygen intermediates (Grabb PA *et al*, 1995) in conditions of thermotherapy, we have managed to turn these phenomena into induction of gene expression.

In the present paper, we used the gadd153 promoter named from the fact that its expression is stimulated by <u>g</u>rowth <u>a</u>rrest or <u>D</u>NA <u>d</u>amage. DNA damaging agents as MMS, UV and heavy metals are known to have the stronger effects on gadd transcription level (Luethy JD *et al*, 1992). Our preliminary experiments obtained by induction of a luciferase reporter gene driven by gadd promoter examine both the relative induction and

Y. Kitagawa et al. (eds.), Animal Cell Technology: Basic & Applied Aspects, Vol. 10, 147–152.

time course of the luciferase expression under heat stress.

2.Materials and methods

2.1 Plasmid construction

The gadd 153 promoter fragment (799 pb) was isolated by restriction at *Cla*I and *Hind* III sites from Jym CAT0 (kindly provided by Prof. Nikki Holbrook) and inserted into the MCS of pGL3-Basic Vector (Promega, Madison, WI, USA). The promoter controls the transcription of a luciferase reporter gene and has any other promoter or enhancer sequences.

2.2 Cell cultures

The U87MG Glioma cell line and the NIH/3T3 were grown in complete Minimal Essential Medium and Dubbelco Modified Essential Medium respectively, supplemented with 10% fetal calf serum, 10 mM non essential amino acids, and 100 $\mu g\ ml^{-1}$ streptomycin sulfate+100 U ml^{-1}potassium penicillin G as antibiotics. The cells were plated one day before transfection at 2×10^5 cells per 35 mm diameter dishes (Iwaki, Tokyo, Japan) and incubated at 37°C in a humidified atmosphere of 5% CO_2 and 95% air.

2.3 Transfection method

The cells were transfected by liposomes. DLPC (Dilauroyl phosphatidylcholine) and DOPE (Dioleoyl phosphatidylethanolamine) were purchased from Sigma Chemical Co.(St. Louis, MO, USA) and TMAG (N-(α-trimethylammonio-acetyl)-didodecyl-D-glutamate chloride) was provided by Sogo Pharmaceutical Co. (Tokyo, Japan). These lipids were used at the molar ratio of 1 : 2 : 2 as TMAG : DLPC : DOPE as described by Koshizaka *et al* (1989). After formation of a lipidic membrane in a conical tube, 20mg of plasmid was added in PBS (pH 7.4), then the lipids were dispersed by vortex. The suspension was further diluted to the final concentration of 50 mg ml^{-1} in DNA.

2.4 Heat stress treatment

Water bath was used to simulate hyperthermia and heat treatment was performed as follows. The dishes were wrapped individually with hermetic film, placed in a metallic container and were immersed in water bath at 43°C. The temperature of the medium increased quickly and reached equilibrium within 5 min.

2.5 Evaluation of cell viability and luciferase expression

The toxicity of the transfection method and the heat stress were evaluated by daily cell counting. After transfection, at time intervals, cells were harvested until the fourth day after transfection and luciferase activity was determined by luciferase assay kit (Promega).

3. Results and discussion

At first, a glioma as well as a fibroblast cell line were used to test the induction of gadd promoter under condition of thermotherapy (Fig. 1A and B). We could observe that while the expression in fibroblast was slightly induced by heat stress at 43°C for 1 hour, surprisingly an opposite effect was seen for glioma cell line where a moderate repression was noticed.

The cell growth of treated cells was not significantly different from the control for both cell lines (data not shown). The cell division process being not impeded and therefore, the results were consistent with the literature in terms of induction cascade of gadd promoter. Nevertheless, the concept of growth arrest induction of gadd and other related promoter of the CAAT transcription factor family is still under investigation (Jackman J *et al*, 1994) and it is likely that the control of expression is under the transcriptional as well as the post transcriptional process.

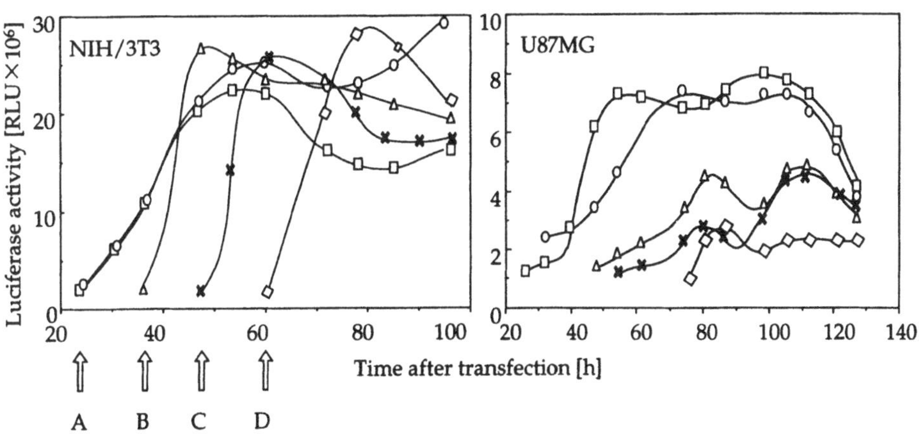

Fig 1. Comparison between glioma (U87MG) and fibroblast (NIH/3T3) cell line for gadd 153 induction under heat stress. □ :Basal level; ○ :Water bath stress applied as indicated by [A] arrow; △:Heat stress applied as indicated by [B] arrow; ✖ :Heat stress applied as indicated by [C] arrow; ◇ :Heat stress applied as indicated by [D] arrow

Then, the difference between normal and cancer cell line can be assumed to be due to differences in these processes. Another plausible explanation is the scavenging of heat generated radicals, responsible for the induction, by enzymes from dismutase or peroxidase family, which would be more active in neoplastic cells (Taniguchi, 1992). Consequently, the influence of the cell line should be further investigated among cancer cells.

Compared with the basal level, the luciferase assay showed a diminution of activity immediately after the heat treatment. This may be explained by instability of the luciferase enzyme under temperature increase.

Next, we have further studied the results obtained with fibroblast by changing the duration of stress attempting to emphasize the induction level (Fig. 2A and 2B).

We could clearly achieve an inductive response correlated to the duration of stress. This response is especially marked when heat stress was applied at later stage after transfection as seen on the Fig. 2B. At 24 hours after stress, the augmentation of activity reached 5 or 7 fold for 2 or 3 hours of heat shock treatment respectively. These results attested the effect of temperature and the duration of heat stress on the expression rate of gadd 153, nevertheless, we cannot determine the intracellular events which triggered directly or indirectly this effect. Also, as mentioned in the above experiment, a correlation between cell viability and expression can be drawn. The mechanism of cell death is essentially due to apoptosis instead of necrosis as cells death was delayed relatively to the stress period. Consequently, DNA repair process or on the contrary, apoptosis itself is the possible cause of induction.

It is interesting to recall that transient transfection is known for a global diminution of inducibility. The construction under episomal form is likely to escape partially from the cellular control and the induction cascade events. However as our final purpose implies the injection of plasmid-entrapped liposomes inside the tumor site, the relevance of stable transformant could not be envisaged in the present therapeutic context. Therefore, our preliminary results instead of stressing the level of induction, are representing significant and realistic phenomenon to be further exploited.

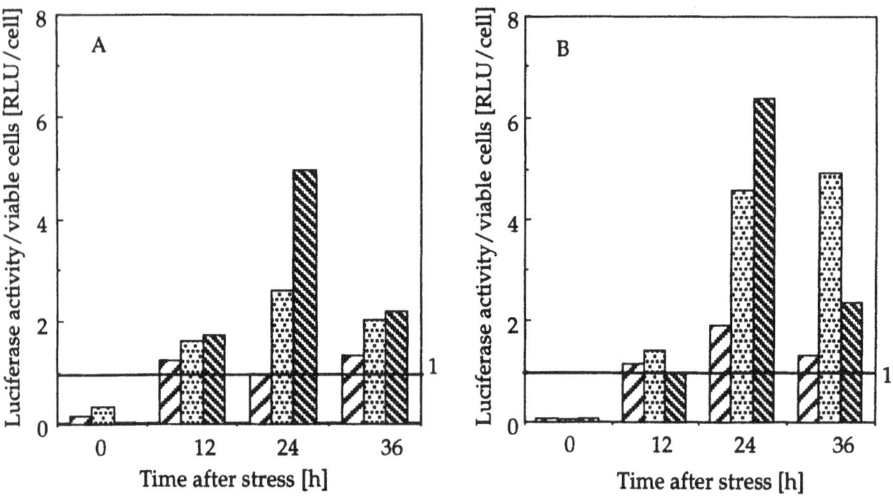

Fig 2. Comparison between duration of heat stress for gadd 153 induction at two different moments; A: Heat stress applied at 36 hours after transfection, B: Heat stress applied at 60 hours after transfection. 1h 2h 3h

Other promoter candidates should equally be investigated, preferably acting independently from the radical oxygen intermediates scavenging process. This way, the possible discrepancy of biochemical pathways exhibited by cancer cell lines could be eluded.

4. Conclusions and prospects

In a different approach from the temperature inducing directly heat sensitive promoter, we have aimed at showing the possibility to use the general biochemical response to heat shock for the induction of stress sensitive promoter.

Our approach proved to be successful in that the expression of a reporter gene driven by gadd 153 increased significantly under heat shock treatment. The level of induction was likely to be proportional to the duration of the stress. This duration was in turn proportional to the cell damages and consequently, the growth arrest and related events are involved in this process, consistently with reported results studying the biochemical pathway of this family of promoters.

Prospectively, we intend to test the feasibility of using a stress inducible promoter to drive the expression of cytokine and improve treatment of brain tumor in combination with local hyperthermia. Previously, we have reported an efficient combination therapy using mild hyperthermia and IFN-β gene therapy in which the expression was driven by the mouse mammary tumor virus promoter (Bouhon IA *et al*, 1997). Also, the expression of cytokine like TNF-α or IL2, exhibiting direct synergy with hyperthermia is likely to support the strategy. Therefore, our approach still offers multiple possibilities of investigation of properly responsive elements which would show desirable characteristics of inducibility in the thermotherapy and immunotherapy framework.

Acknowledgements

We are grateful to Prof. Nikki Holbrook (National Institute on Aging, Baltimore, USA) for providing the JymCAT0 plasmid including the Gadd 153 promoter. We thank very much Profs J. Yoshida, M.Mizuno, and T. Wakabayashi for their helpful advice. This study was partially funded by a Grant-in-aid for Scientific Research (No. 09305055) from the Ministry of Education, Science, Sports and Culture of Japan.

6. References

Boothman, DA. Lee, IW. and Sahijdak, WM. (1994) Isolation of an X-ray-responsive element in the promoter region of tissue-type plasminogen activator: potential uses of X-ray responsive elements for gene therapy. Radiat res. 138: S 68-71.

Bouhon, IA. Shinkai, M. Honda, H. and Kobayashi, T. (1997) Enhancement of cytokine expression in transiently transfected cells by magnetoliposome mediated hyperthermia,. Cytotechnology 25(1), 231-234.

Bowman, T. Symonds, H. Gu, L. Yin, C. Oren, M. and Van Dyke, T. (1996) Tissue specific inactivation of p53 tumor suppression in the mouse. Genes & development 10(7): 826-35.

Foster, BJ. and Kern, JA. (1997) HER2-targeted gene transfer. Human Gene Therapy 8(6): 719-27.

Gately, DP. Sharma, A. Christen, RD. and Howell, SB. (1996) Cis platin and taxol activate different signal pathways regulating cellular injury-induced expression of GADD153. British J. Cancer 73 (1): 18-23.

Goossens, V. Grooten J and Fiers, W. (1996) The oxidative metabolism of glutamine. J.Biol.Chem. Vol.271, No 1, 192-196.

Grabb, PA. and Gilbert, MR. (1995) Neoplastic and pharmacological influence on the permeability of in vitro blood-brain barrier. J. Neurosurg. 82(6): 1053-1058.

Jackman, J. Alamo, I. Fornace, Jr. and Fornace, AJ. (1994) Genotoxic Stress confers preferential and coordinate messenger RNA stability on the five gadd genes. Cancer Res. 54, 5656-5662.

Jamora, C. Dennert, G. and Lee, AS. (1996) Inhibition of tumor progression by suppression of stress protein GRP78/BiP induction in fibrosarcoma B/C10ME. Proc. Nat. Aca. Sci. USA. 93(15): 7690-4.

Kiang, JG. Mc Kinney, LC. and Gallin, EK. (1990) Heat induces intracellular acidification in human A-431 cells: role of Na+-H+ exchange and metabolism. Am. J. of Physiol. C28 (1) 727-737.

Koshizaka, T. Hayashi, Y. and Yagi, K. (1989) Novel liposomes for efficient transfection of β-galactosidase gene into COS-1 cells. J. Clin. Biochem. Nutr. 7, 73-77.

Luethy, JD. and Holbrook, NJ. (1992) Activation of the gadd153 promoter by genotoxic agents: A rapid and specific response to DNA Damage. Cancer Res. 52, 5-10.

Taniguchi, N. (1992) Clinical significances of superoxide dismutases: changes in aging, diabetes, ischemia, and cancer. Advances in clinical chemistry 29:1-59.

ENHANCEMENT OF PRODUCTIVITY OF RECOMBINANT α-AMIDATING ENZYME BY LOW TEMPERATURE CULTURE

K. FURUKAWA & K. OHSUYE
Suntory Institute for Medicinal Research & Development,
2716-1 Kurakake, Akaiwa, Chiyoda-machi, Ohra-gun,
Gunma 370-05, Japan

1. Abstract

We have produced a recombinant C-terminal α-amidating enzyme (799*BglII*α-AE) derived from *Xenopus laevis* by culturing a CHO cell line named 3μ-1S. Recently, we demonstrated that culturing 3μ-1S cells at a temperature below 37℃ led to the following phenomena: inhibited cell growth with high viability, enhanced cellular productivity (maximally at 32℃), and suppressed medium consumption and release of impurities from the cells. Therefore, it is suggested that the 799*BglII*α-AE production will be increased by culturing a sufficient number of the cells at a low temperature (especially at 32℃). Application of the low temperature culture to batch and perfusion cultures was effective in comparison with the culture at 37℃: the productivity per medium and the productivity per time were increased severalfold with enhanced cellular productivity at a low culture temperature. The low temperature culture also increased the relative content of 799*BglII*α-AE in the supernatant and reduced the glucose consumption.

2. Introduction

Currently, many bioactive proteins are produced by culturing recombinant cells on an industrial scale. Several groups have reported the advantages of culturing recombinant cells at low temperature (below 37℃) [1,2,3]; however, knowledge of the effects of culture temperature on recombinant cells is insufficient. We have studied the effects of culture temperature on a recombinant CHO cell line producing a C-terminal α-amidating enzyme (799*BglII*α-AE) and demonstrated that culturing the cells below 37℃ enhanced the cellular productivity while suppressing the cell growth with

Y. Kitagawa et al. (eds.), Animal Cell Technology: Basic & Applied Aspects, Vol. 10, 153–157.

high viability, medium consumption, and the release of impurities from the cells [4]. These phenomena strongly suggest that the production and productivity of 799*BglII*α-AE will be enhanced by culturing at a low temperature (especially 32℃) after obtaining a sufficient number of the cells by culturing at an adequate temperature to promote the cell growth (36 - 37℃). In this report, we describe that applying this method to batch and perfusion cultures is useful for efficient production of 799*BglII*α-AE.

3. Methods

A recombinant CHO cell line, 3μ-1S, secrets a recombinant C-terminal α-amidating enzyme (799*BglII*α-AE) derived from *Xenopus laevis* and can grow in suspension with a serum-free medium [5]. This cell line was maintained as described previously [4,5]. The enzyme catalyzes the conversion of glycine extended prohormone substrates to bioactive C-terminal α-amidated peptide hormones such as human calcitonin [6,7,8].

For culture experiments, we used 1-liter cell culture system which were established as described in our previous report [4]. This system enabled the culture temperature, pH, and DO to be controlled. For perfusion culture, the circulation and perfusion lines were attached to the system. To separate the cells, we used a centrifuge, Centritech Lab System (Sorvall).

3μ-1S cells were inoculated to the 1-liter culture systems at 2 x 10^5 cells/ml and cultured with the serum-free medium without MTX. In the case of batch culture, the culture temperature was shifted from 37℃ to 32℃ at 48 hours. The pH was controlled to 7.2, and DO was kept above 60% air saturation by continuously supplying 100 ml/min of air. In the case of perfusion culture, the cells were precultured at 37℃ until the cell density reached 6 x 10^6 cells/ml, and the culture was then divided into two portions. The culture temperature of one of the vessels was shifted to 34℃ when the cell density again reached 6 x 10^6 cells/ml. The pH and DO were controlled to 7.2 and 60% air saturation, respectively. In each vessel, the perfusion rate was adjusted to 4.5 volume/day after the point of the temperature shifting. In these experiments, the total and viable cell densities were determined daily by the trypan blue dye exclusion method in a hemacytometer. The enzymatic activity of 799*BglII*α-AE was determined as reported previously [7]. The glucose concentration in the supernatant was measured with a Glucose analyzer ST-1 (Oriental Electric, Japan).

4. Results

Figure 1 shows the effects of the temperature shift on batch culture. In the temperature shift-culture, the cell growth was nearly equal to that in the 37℃-culture until 24 hours after the temperature shift, and then it stopped (Fig. 1A). Although the cell density then tended to decrease, the viability was maintained at 80% or above. The glucose consumption was reduced by the temperature shift (Fig. 1B). As shown in Figure 1C, although the production of 799*BglI*Iα-AE in the 37℃-culture leveled off at 1500 units/ml in 96 hours of the culture, that in the temperature shift-culture reached 4900 units/ml on day 9, and it still maintained an increasing trend. In parallel, the cellular productivity in the temperature shift-culture was elevated to about 2-fold after the temperature shift (data not shown).

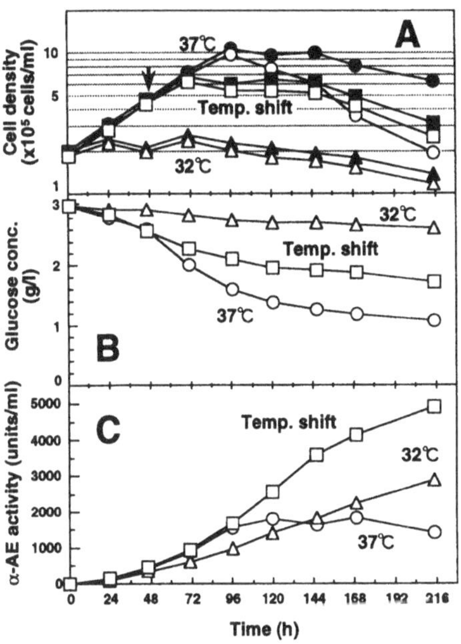

Fig. 1. Effects of temperature shift on batch culture. (A) Growth curves of 3μ-1S cells. Closed and open symbols respectively indicate the total and viable cell densities. The arrow indicates the point of the temperature shift. (B) Time course of glucose concentration. (C) Accumulation of 799*BglI*Iα-AE in the medium.

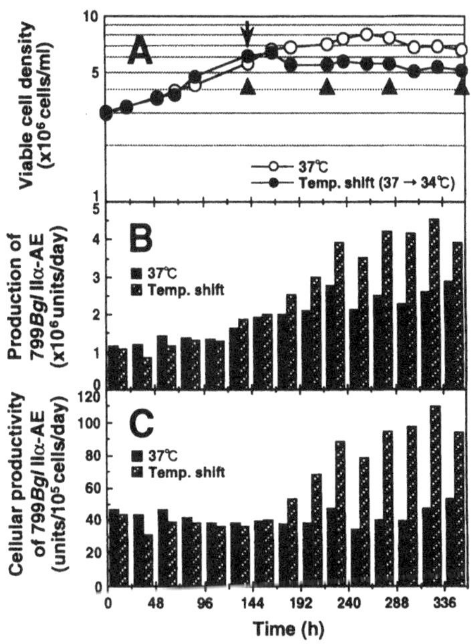

Fig. 2. Effects of temperature shift on perfusion culture. (A) Transition of viable cell density. The arrow indicates the point of the temperature shift. The black triangles are the sampling points of the supernatant for SDS-PAGE. (B) Daily production of 799*BglI*Iα-AE. (C) Cellular productivity of 799*BglI*Iα-AE.

The results of the perfusion culture are shown in Figure 2. Before the point of shifting temperature, the viable cell density, the production per vessel per day and the cellular productivity in each vessel varied in the same way. After that point, the viable cell density in the temperature shift-culture became steady at a slightly lower level than that in the 37℃-culture. On the contrary, the production and the cellular productivity of 799*BglI*Iα-AE were enhanced 1.6-fold and 2.0-fold, respectively (Fig. 2B, C). In addition, the glucose consumption during the culture at 34℃ decreased in comparison with that at 37℃ (data not shown). However, it should be mentioned that the cells suffered significant physical damage in this experiment because the circulation rate became too high. As a result, the growth rate declined during the culture at 37℃. If the temperature was shifted to 32℃ (optimum for the cellular productivity), there was a possibility of a decrease in the viable cell density. Thus, the temperature was shifted to 34℃, not to 32℃.

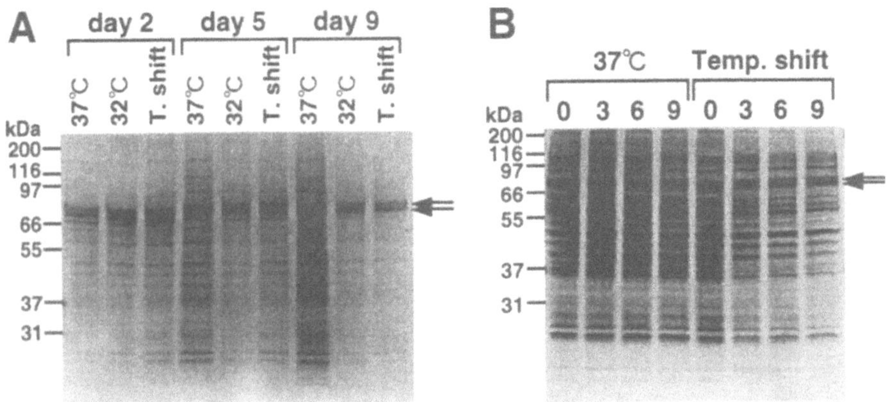

Fig. 3. SDS-PAGE analysis of 799*BglI*Iα-AE secreted into the culture medium. Each supernatant containing 45 units of enzymatic activity from the batch culture (A) and the perfusion culture (B) was subjected to this analysis. The proteins in the gel were visualized by silver staining. The arrows indicate the size of the 799*BglI*Iα-AE protein (75 and 81 kDa). In (B), the numbers above the lanes indicate the days after the point of the temperature shift.

To access the purity, *i.e.*, the relative content of 799*BglI*Iα-AE secreted in the medium, we performed SDS-PAGE analysis. In the case of the batch culture (Fig. 3A), the relative content of 799*BglI*Iα-AE from the temperature shift-culture remarkably increased. On day 9, 799*BglI*Iα-AE became the main protein in the supernatant. The result from the perfusion culture is shown in Figure 3B. In this figure, there are many cellular proteins due to the cell damage mentioned above. However, in the 34℃-culture, the intensities of these impurity bands were apparently reduced.

5. Discussion and Conclusions

We demonstrated here that low temperature culture of 3μ-1S cells could be successfully applied to batch and perfusion cultures, although optimization of the culture should be needed in the future. The total production, as well as the cellular productivity, of 799*Bgl*IIα-AE was significantly increased by shifting to the low temperature. In parallel, the low temperature culture of 3μ-1S cells reduced the glucose consumption and the release of impurities from the cells. In addition, it has been demonstrated that the consumption of most of the amino acids in 3μ-1S cells is also reduced by lowering the culture temperature [4]. These observations indicate that the low temperature culture greatly contributes to an efficient process for the production of 799*Bgl*IIα-AE. The method presented here would be useful for production of bioactive proteins using other recombinant cell lines.

6. References

[1] Rasmussen P (1991) Method of producing proteins with FVIII activity and/or FVIII derivatives. Patent WO 91/00347.

[2] Weidemann R, Ludwig A & Krezmer G (1994) Low temperature cultivation – A step towards process optimization. Cytotechnology 15: 111-116.

[3] Chuppa S, Tsai Y-S, Yoon S, Shackleford S, Rozales C, Bhat R, Tsay G, Matanguihan C, Konstantinov K & Naveh D (1997) Fermentor temperature as a tool for control of high-density perfusion culture of mammalian cells. Biotechnol. Bioeng. 55: 328-338.

[4] Furukawa K & Ohsuye K (1998) Effect of culture temperature on a recombinant CHO cell line producing a C-terminal α-amidating enzyme. Cytotechnology 26: 153-164.

[5] Furukawa K, Okuno K, Onai S, Sugimura K, Yoko-o Y, Ishibashi Y, Ohshima T, Tsuruoka N, Magota K, Tanaka S & Ohsuye K (1993) Production of an α-amidating enzyme (α-AE) in recombinant CHO cells. In: Kaminogawa S *et al.* (eds.) Animal Cell Technology: Basic & Applied Aspects, Vol. 5 (pp. 493-499) Kluwer Academic Publishers, The Netherlands.

[6] Eiper BA, Mains RE & Glembotski CC (1983) Identification in pituitary tissue of a peptide α-amidation activity that acts on glycine-extended peptides and requires molecular oxygen, copper, and ascorbic acid. Proc. Natl. Acad. Sci. USA 80: 5144-5148.

[7] Mizuno K, Sakata J, Kojima M, Kangawa K & Matsuo H (1986) Peptide C-terminal α-amidating enzyme purified to homogeneity from *Xenopus laevis* skin. Biochem. Biophys. Res. Comm. 137: 984-991.

[8] Ohsuye K, Kitano K, Wada Y, Fuchimura K, Tanaka S, Mizuno K & Matsuo H (1988) Cloning of cDNA encoding a new peptide C-terminal α-amidating enzyme having a putative membrane-spanning domain from *Xenopus laevis* skin. Biochem. Biophys. Res. Comm. 150: 1275-1281.

DIALYSIS-FIXED-BED CULTURES FOR HIGHLY EFFICIENT PRODUCTION OF CHIMERIC FAB-FRAGMENTS WITH TRANSFEC TOMA CELLS IN PROTEIN-FREE MEDIUM

R. PÖRTNER[1], I. LÜDEMANN[1], K. REHER[2], M. NEUMAIER[2], H. MÄRKL[1]

[1] *Technische Universität Hamburg-Harburg, Bioprozeß- und Bioverfahrenstechnik, Denicke Str. 15, D-21071 Hamburg, Germany*

[2] *Universitätskrankenhaus Eppendorf, Abteilung für klinische Chemie, Martinistr. 52, 20246 Hamburg, Germany*

1. Abstract

The extremely shear sensitive transfectoma cell line FAB#10, producing a chimeric Fab-antibody fragment specific for the human carcinoembryonic antigen (CEA), was cultivated in fixed bed reactors, were the cells were immobilized in macroporous carriers. As the cell line had a very low productivity, the fixed bed reactors were coupled with dialysis membranes, were low molecular weight metabolites are removed through the membrane and high molecular products are enriched. In fixed bed systems the cell line was stable during long term cultures. With the dialysis technique Fab-concentrations up to 8.000 μg l^{-1} could be obtained, appr. 16 times higher compared to fixed bed cultures without dialysis or batch cultures with suspended cells.

If nutrients are supplied through the membrane, a high loss of nutrients has to be considered. In a novel "nutrient-split" feeding strategy concentrated medium is fed directly to the fixed bed unit, whereas a buffer solution is used as dialysis fluid. This feeding strategy reduces the required medium per mg of product significantly.

2. Introduction

In fixed bed reactors with immobilized animal cells high volume specific cell densities can be obtained under low shear conditions, favoring the use of serum-free and protein-free medium (Lüdemann *et al.*, 1996). For process integrated product enrichment dialysis membranes have proven to be very effectiv and reliable. Low molecular metabolites (ammonia, lactate) are removed through the membrane into a dialysis fluid. High molecular components can not permeate through the membrane and accumulate.

Y. Kitagawa et al. (eds.), Animal Cell Technology: Basic & Applied Aspects, Vol. 10, 159–163.

This concept - fixed bed for cell immobilization coupled with dialysis membrane - was investigated with the transfectoma cell line FAB#10, producing a monovalent chimeric human/mouse Fab-antibody fragment specific for the human carcinoembryonic antigen (CEA). In preliminary studies (Lüdemann *et al.*, 1996) this cell line had turned out to be very shear sensitive and only cultures with immobilized cells in a fixed bed had been successful.

3. Materials and Methods

The cell line FAB#10 was cultivated in a standard, protein-free medium described previously (Lüdemann *et al.*, 1996). For the "nutrient-split" feeding strategy the standard medium was split into a nutrient concentrate (10x, purchased by BioConcept, Umkirch, Germany) and a buffer solution containing all inorganic components (Pörtner *et al.* 1997). Glutamine was fed as an extra solution (50x concentrated).

The set up of the axial-flow fixed bed reactor system (purchased by meredos, Bovenden, Germany) has been described previously (Ong *et al.*, 1994). An axial-flow fixed bed was filled with macroporous SIRAN®-carriers (Schott, Mainz, Germany) with a diameter of 3-5 mm to immobilize the cells. The volume of the fixed bed was 176 ml with a length of 90 mm. The fixed bed was connected with a conditioning vessel (working volume 700 - 1.000 ml) for aeration and medium exchange. When the perfusion loop was used for continuous exchange of the medium in the conditioning vessel without dialysis, fresh medium was pumped from the medium bottle into the conditioning vessel and the spent medium into the harvest bottle.

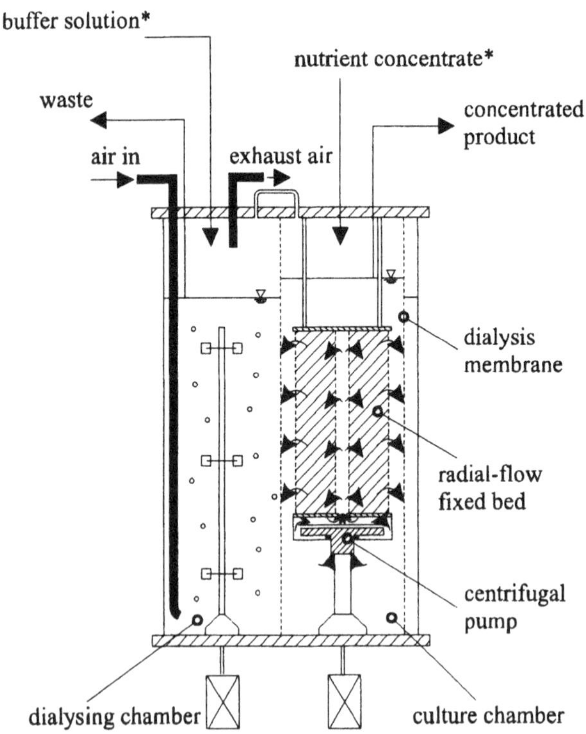

Figure 1. Membrane dialysis reactor with integrated radial-flow fixed bed (* "nutrient-split"-feeding strategy)

For experiments in an axial-flow fixed bed reactor coupled with dialysis module a dialysis loop (dialysis module (BL OG, purchased by Microdyn, Wuppertal, Germany, for specific data compare Ogbonna and Märkl, 1993) and dialysis vessel (working volume 4 l)) were connected with the conditioning vessel. Conditioning vessel as well as dialysis vessel contained standard medium. The reactor system was run in a semi-batch mode, were the Fab fragments should accumulate in the conditioning vessel and nutrients should be supplied through the membrane.

For the "nutrient-split" feeding strategy a membrane dialysis reactor was used (Fig. 1). The reactor consists of two chambers (dialysis chamber 4.9 l, culture chamber 1.2 l) separated by a dialysis membrane (cuprophan, Akzo, Germany) with a cut-off of 10,000. In the culture chamber a radial-flow fixed bed (volume 0.6 l) containing the porous carriers is introduced (for specific data see Bohmann *et al.*, 1995).

4. Results

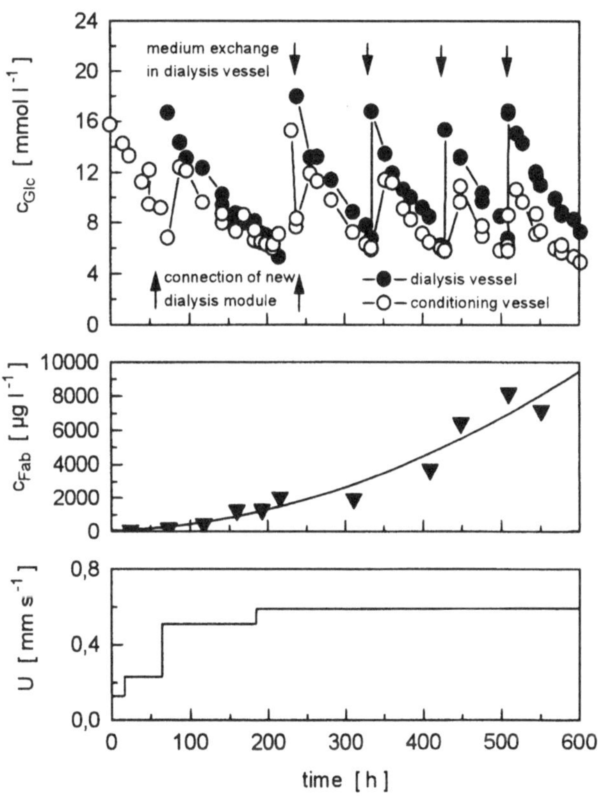

Figure 2 : Experiment in the axial-flow fixed bed with external dialysis loop

Time course of glucose concentration c_{Glc} in conditioning vessel and dialysis vessel, Fab-concentration c_{Fab} and superficial flow velocity U.

In a set of experiments in the axial-flow fixed bed reactor the dilution rate for continuous medium exchange in the conditioning vessel was varied to get different levels of substrate supply. In an experiment with a dilution rate of D = 0.8 d^{-1} the glucose concentration reached a "steady state" value of 3.5 mmol l^{-1} after 340 h, glutamine 0.75 mmol l^{-1}. Lactate had accumulated to 25.5 mmol l^{-1}, ammonia to 2.8 mmol l^{-1}. The Fab-concentration increased to a finally constant value of 510 µg l^{-1}. In a second experiment with a reduced dilution rate of D = 0.2 d^{-1}, the following "steady state" concentrations were reached after 200 h culture time: glucose 3.26 mmol l^{-1}, glutamine 0.28 mmol l^{-1}, lactate 26.7 mmol l^{-1}, ammonia 3.07 mmol l^{-1}, Fab 500 µg l^{-1}. The Fab-concentration was lower compared to the first experiment probably due to the lower substrate supply.

An experiment with the axial-flow fixed bed coupled with dialysis module is shown in Fig. 2. It was started as batch until culture time 75 h. Then the dialysis module and the dialysis vessel were connected. The medium in the conditioning vessel was not changed during culture time 238 h and 600 h to accumulate the Fab fragments. The medium in the dialysis vessel was exchanged batch-wise at culture time 334 h, 428 h and 509 h. At the times of exchange the glucose concentration was in all cases above 4 mmol l^{-1}. The Fab-concentration increased up to appr. 8.000 µg l^{-1}, almost 16 times more than in the experiments without dialysis. But feeding the nutrients through the membrane has a significant drawback: when the substrate concentrations in the conditioning vessel have re-

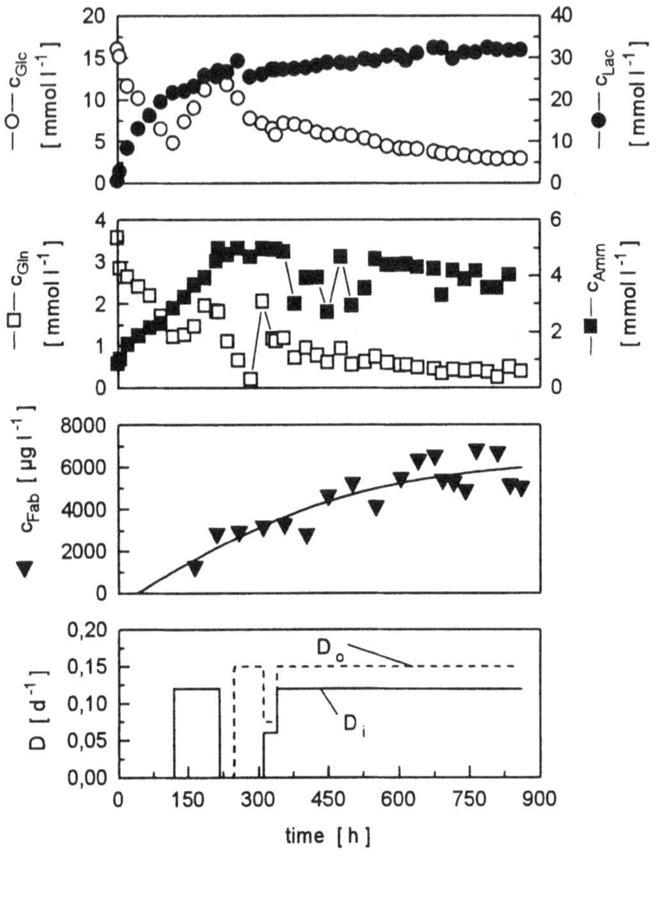

Figure 3 : Experiment in the membrane dialysis bioreactor with integrated radial-flow fixed bed with "nutrient-split" feeding strategy

Time course of the concentrations of glucose c_{Glc}, lactate c_{Lac}, glutaimine c_{Gln}, ammonia cAmm and Fab cFab in the culture chamber as well as dilution rates in culture chamber D_i and dialysis chamber D_o.

ached limiting values, the medium in the dialysis vessel has to be changed, even if this has not been used completely. The efficiency of this operation mode is low.

To solve this problem, the "nutrient-split"-feeding strategy was applied in an experiment in the membrane dialysis reactor with integrated radial-flow fixed bed. The culture was started as batch with complete medium in both chambers of the reactor (Fig. 3). During continuous operation the 10x concentrated nutrient solution was fed directly to the culture chamber, whereas the dialysis chamber was perfused with a buffer solution. Finally "steady state" conditions could be obtained after appr. 600 h of culture time with the following concentrations: glucose 3.18 mmol l^{-1}, glutamine 0.35 mmol l^{-1}, lactate 27 mmol l^{-1}, ammonia 3.6 mmol l^{-1}. From these data it can be concluded, that the nutrient loss through the membrane was significantly reduced. The Fab-concentration increased to 5.800 µg l^{-1}. After 860 h the experiment was stopped.

5. Discussion

Three different reactor set-up's, axial flow fixed bed without and with dialysis as well as radial-flow fixed bed with dialysis, were investigated. A comparision of the volume specific substrate uptake and metabolite production rates as an indicator for the cell activity does not reveal significant differences between the three set-up's (data not shown). A comparision of the different process strategies underline the potential of dialysis cultures run with the "nutrient split" feeding strategy. The antibody concentrations obtained with dialysis cultures were up to 16 times higher compared to those without. The "nutrient-split" feeding strategy leads to a better medium utilization and a significantly reduced medium demand (in this case the sum of concentrated nutrient solution and buffer). The medium demand was almost 50 % lower compared to the other culture methods.

The radial-flow fixed bed in the membrane dialysis reactor has the same geometry as the fixed bed elements of a large-scale version, which has been discussed previously (Bohmann *et al.*, 1995; *Pörtner et al.*, 1997). Therefore design and scale up of industrial scale reactor systems can be done with basic data from small scale reactor systems and data on the properties of the dialysis membrane.

6. References

Bohmann, A., Pörtner, R. and Märkl, H. (1995) Performance of a membrane dialysis bioreactor with radial-flow fixed bed for the cultivation of a hybridoma cell line. Appl. Microbiol. Biotechnol. 43: 772-780

Lüdemann, I., Pörtner, R., Schaefer, C., Schick, K., Šrámková, K., Reher, K., Neumaier, M., Franek, F. and Märkl, H. (1996): Improvement of culture stability of non-anchorage-dependent cells grown in serum-free media through immobilization. Cytotechnol. 19: 111-124

Ong, C.P., Pörtner, R., Märkl, H., Yamazaki, Y., Yasuda, Y. and Matsumura, M. (1994): High density cultivation of hybridoma in charged porous carriers. J. Biotechnol. 34: 259-268

Ogbonna, J.C. and Märkl, H. (1993) Nutrient-split feeding strategy for dialysis cultivation of E. coli. Biotechnol. Bioeng. 41: 1092-1100

Pörtner, R., Lüdemann, I. and Märkl, H. (1997) Dialysis cultures with immobilised hybridoma cells for effective production of monoclonal antibodies. Cytotechnol. 23: 39-45

MONITORING DRUG EFFECTS ON HUMAN PRIMARY TISSUES UTILIZING FIBRE OTICAL SENSORS

H. MAERZ
R. Buchholz, F. Emmrich, F. Fink, L. Pfeifer, U. Marx
University of Leipzig
Insitute for Clinical Immunology & Transfusion Medicine
Department for Medical Biotechnology
Delitzscher St 141
D-04129 Leipzig
Germany

ABSTRACT. The new medical developments create a questionable doubt to the further requirement of animal tests. Instead the call for humanitarian reproductive in vitro models becomes increasingly louder. In cancer research and therapy, the effect of chemostatica in vitro in the so called oncogramm is being tested ; unfortunately, up to now without great success. However, it showed among other things that the dimensional structure of the tissue plays a vital role. We have decided to concentrate our efforts in realising tissue cultivation in hollow fibre bioreactors (HFBR). With two fibre optic sensors - an optical oxygen sensor and an interbolism detecting LASER, we were able to successfully monitor the biological status of tissue culture and the drug or toxic effects of *in vitro* pharmaceutical testing. Several studies on human primary tissues were carried out, whereby the corresponding signals of the oxygen consumption and the NADH content inside the cells showed performance when the tissue culture was exhibited to therapeuticals.

1. Introduction

Pharmaceutical usage *in vitro* has a long proven history. But common drug response assays do not always correlate with *in vivo*-like drug resistance and sensitivity. In contrast, three-dimensional tissue cultivation in hollow fiber bioreactors do, indicating that drug response is a function of tissue architecture[1]. Establishing a test system *in vitro*, feasible for therapeutic drug monitoring, will improve individual patient care, especially in leukæmia therapy. In addition, such a test system will also be suited for toxicological testing of environmental pollution, which today gain more and more signifcance. For detecting drug or toxic effects, suitable sensors have to be found to enlight the black box hollow fibre bioreactor. Therefore, we emphasize on monitoring

Y. Kitagawa et al. (eds.), Animal Cell Technology: Basic & Applied Aspects, Vol. 10, 165–169.

all-purposing interbolism parameters, the oxygen consumption and the NADH content. Both parameters play a major role in the energetic interbolism of cells in general and by detecting both, we can visualize almost all drug or toxic effects.

2. Materials And Methods

Optical Oxygen Sensor MOPS
TCI COMTE Moderne Analysensysteme GmbH, Hanover-Germany
A fluorescent dye is immobilized on the top of the fibre. Blue light (λ=470 nm) is emitted through the fibre exciting the dye to fluores (λ=580 nm). By the Quenching effect, oxygen eradicates the fluorescence dependent to ist concentration. The fluorescence eradication is detected and gives us the current oxygen concentration. Next to its small size, the major advantage of the fibre oxygen sensor is that it does not consume oxygen while measuring. Calibration and sterilization as described by *Maerz et al.*[2]

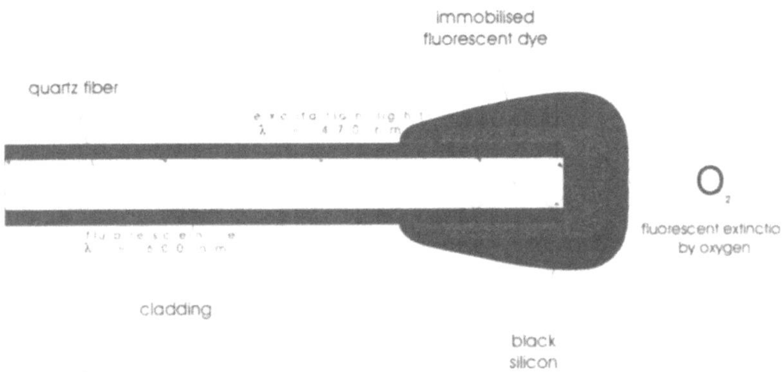

Figure 1: Scheme of the optical oxygen sensor

NADH LASER (Interbolism Detector)
IOM GmbH, Berlin-Germany
The NADH content is measured through autofluorescence of the NADH (λ=580 nm) at ultraviolet light (λ=340 nm). The LASER has two major advantages: First of all, only the intensity and sensitivity of a LASER, enables us to detect the minor NADH content of primary mammalian cells. Secondly, the LASER can detect time-resolved. Measuring time-resolved, enables to detect only the peak of the NADH fluorescence while filtering almost all the other disturbing signals of substances which fluoresce at similiar wavelenghts. With the NADH LASER only the NADH content inside of living cells is detected because dying cells decay their intracellular NADH. Calibration and sterilization as described by *Maerz et al.* [2]

Figure 2: Scheme of the NADH LASER

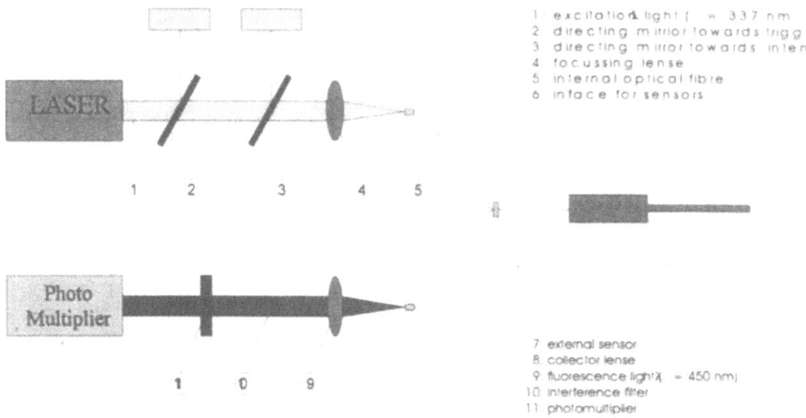

3. Results

*Figure 1.*In this experiment, a therapeutical drug (GM-CSF) was added to human bone marrow. After a short period of time, the NADH signal ascended while the corresponding oxygen signals descended. The drug stimulates hematopoesis, which means cell growth. This figure also exhibits that the NADH signal indicating the drug reaction is detected before the oxygen signal.

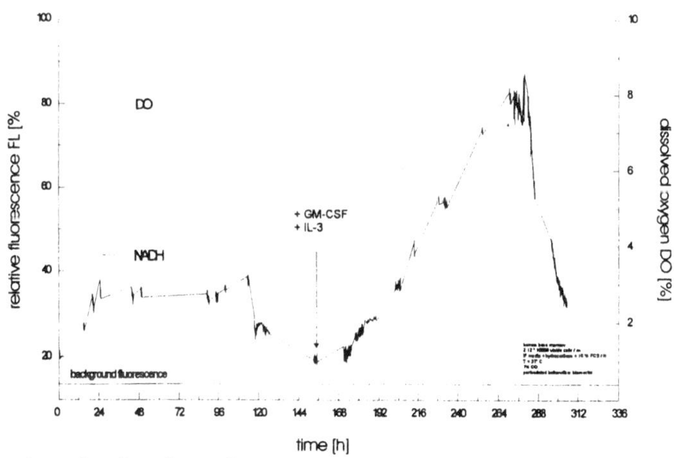

Figure 3: Pharmacological study on human bone marrow in vitro

Figure 4. In this chemotherapy simultion *in vitro Etoposid* which effects mitosis, was added to human bone marrow *in vitro*. After addition of the therapeutical drug, the NADH signal descencded while the corresponding oxygen signal stayed constant.

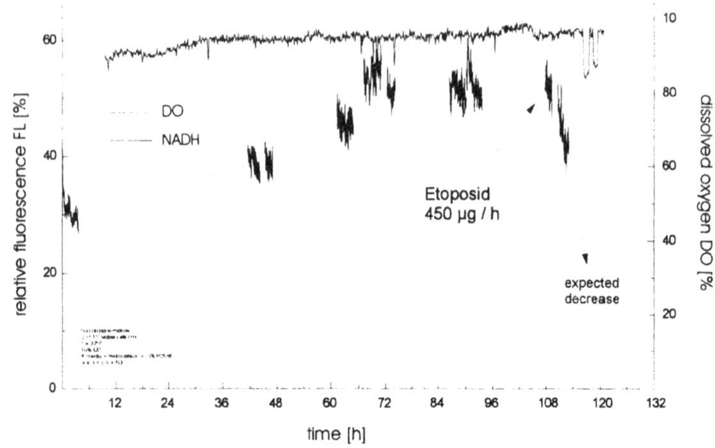

Figure 4: Simulation of a chemotherapy on human bone marrow in vitro

Figure 5. In the last figure the minor toxic effect of a volantile organic carbonaldehyde was tested on human pulmonary cells. The VOC was exposed to the cells via a silicon membrane normally used for aeration. A weak descending signal was detected caused by the toxic effect. After aeration the NADH signal ascended again indicating that the effect was reversible.

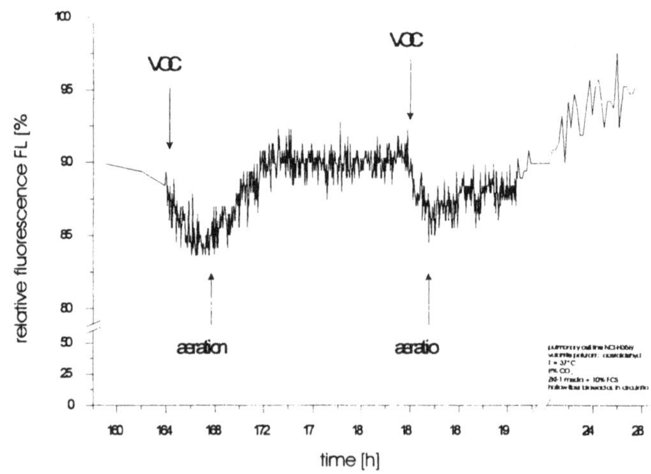

Figure 5: Detection of volantile pollution effect on human pulmonary cells in vitro

Conclusions

Both optical sensors showed perfomance in monitoring pharmaceutical drug effetcs *in vitro*. Especially the NADH LASER is a powerful tool for detecting the actual metabolic state of cells. The corresponding oxygen signal is always detected much later than the more sensitiv NADH signal but is needed for validation. Both sensors together can be used for pharmaceutical studies *in vitro*; but you have to keep in mind that tissue density is a precondition. There is a detection limit for the NADH signal for primary cells at a tissue density of about $5*10^7$ viable cells per mililiter.

5. Acknowledgements

Clive L. Geddes is acknowledged for critical reading.
E. Ulrich and F.Raabe are acknowledged for providing cell material.
U. Sack is acknowledged for last minute technical support.
The Ministry for Science and Art in Saxony (SWMK) is acknowledged for granting this project.

6. References

[1] Hoffman, R.M. (1993) To do Tissue Cultivation in Two or Three Dimensions ? That is the Question., *Stem Cells* **11**,105-11

[2] Maerz, H. et al. (1998) *Microreaction Technology: Development of a Pharmacological Test Sytem in a Miniaturized Hollow Fiber Bioreactor*,W. Ehrfeld (Ed.) / Springer-Verlag Berlin Heidelberg, Berlin, 259-266.

INCREASING ANTIBODY PRODUCTION BY SUPPRESSING CELL PROLIFERATION WITH CONTROLLABLE EXPRESSION OF CYCLIN INHIBITOR P21 AND P27

TETSUO FUJITA , SATOSHI TERADA , HIROSHI UEDA ,
TERUYUKI NAGAMUNE and EIJI SUZUKI*
Department of Chemistry and Biotechnology, Faculty of Engineering, University of Tokyo
Bunkyo-ku ,Tokyo 113, Japan
**Research Institute of Innovative Technology for the Earth, Soraku-gun, Kyoto 619-02, Japan*

ABSTRACT

The authors transfected hybridoma 2E3 cells with cDNA of mouse cyclin inhibitor p21 or p27 in plasmid pMAM-neo, aiming at increasing antibody production by controlling cell proliferation with cyclin inhibitors. The expression of the inhibitor genes was regulated by mouse mammal tumor virus (MMTV) promoter with/without addition of dexamethasone (DEX) in culture medium. Growth rate of the transfected cells was suppressed with DEX, although the transfectant grew slower than the wild type cells even without DEX, presumably because of the leaky transcription from MMTV promoter. The p27 transfectant with DEX produced 3-fold amount of antibody of the wild type or control cells in a batch during 10 days culture, and the p21 transfectant with DEX produced about 1.5-fold during 10 days. The viable stationary phase after the logarithmic growth was prolonged by five days by the expression of p27, although the cell density at the phase was about 60% of that of the wild type and control cells, and during that period the p27 transfectant accelerated antibody production. In the case of p21, the maximum number of viable p21 transfectant cells with DEX was equal to that of the wild type and control cells, and the p21 expression delayed overgrowth of transfectant cells by six days.

KEY WORDS: cyclin inhibitor, hybridoma, cell cycle control, antibody production

1. INTRODUCTION

Hybridoma cells were reported to produce antibodies most effectively in G1 phase in cell cycle (Al-Rubeai et al.), as long as culture medium could afford to support cells in healthy condition. Cells would produce more antibodies, therefore, if they would maintain the longer G1 phase in alive condition, which could be one of the ways for improvement of antibody production. However, when hybridoma cells were arrested in G1 phase by such as several chemicals and cytokines, they often induce apoptosis, that is the cell suicide expressing chromosomal condensation and fragmentation, cell surface blebbling and whole cell shrinkage (Cotter et al.) Thus apoptosis results in the end of expensive cultivation. Moreover, hybridoma cells are sensitive to slight change of culturing conditions and occasionally show down-regulation to the amount of antibody production.

Cyclin inhibitors have several important roles in cell cycle progression, subsequent growth inhibition and differentiation of various cell types. They were

Y. Kitagawa et al. (eds.), Animal Cell Technology: Basic & Applied Aspects, Vol. 10, 171–175.

identified as the protein which could bind to heterodimers composed of cyclins and cyclin-dependent kinases (cdks). The p21 protein, termed Cip-1 (Harper et al.), Waf-1 (El-Deiry et al.) or Sdi-1 (Noda et al.), is the first identified cyclin inhibitor, and overexpression of p21 can arrest several cell lines in G1 phase. The p21 binds to cyclin A-cdc2, cyclin B-cdk2, cyclin D-cdk4/6 and cyclin E-cdk2 complexes, and to the proliferate cell nuclear antigen which is one subunit of DNA polymerase ∂ complexes. The transcription of p21 is regulated by p53, which is one of the most famous oncogene and has many functions suppressing the transformation of normal cells into carcinoma. This regulation between p53 and p21 was recognized as a step in the complex mechanisms during repairing of damaged genomic DNA in nucleus. And p21 has a short half life time, approximately 30 minutes, possibly meaning a temporary stop in cell cycle.

The p27 protein was reported as an inhibitor binding directly to cyclin E-cdk2 complex (Polyak et al.) and cyclin D-cdk4/6 complexes (Toyoshima et al.), and able to arrest several cell lines in G1 phase. The p27 protein has a high homologous region with p21 in N-terminus and is maintained stably in cells during cell cycle. However, the degradation pathway of p27 by conjugation with ubiquitin was reported (Pagano et al.) and p27 protein could have several concerning during the progression of apoptosis (Ezhevsky et al.).

In this study, we attempted to increase antibody production in a batch culturing hybridoma 2E3 cells by arresting cells in G1 phase utilizing the cyclin inhibitors p21 and p27 genetically. And we studied the investigation of the functions of p21 and p27.

2. MATERIALS AND METHODS

Construction of expression plasmid of p21 and p27

The cDNA of p21 derived from pcDSRαΔ-sdi1 plasmid was digested with BamHI and the fragment containing with p21 cDNA was recombinated to pBluescript II KS^+, and inserted between NcoI and XhoI sites in pMAM-neo (Clontech), yielding pMAM-p21 plasmid. The cDNA of p27 derived from pGST-p27 was digested by NotI and inserted at the EcoRV site in pBluescript II KS^+, and thus pMAM-p27 plasmid was prepared.

Culture conditions

Hybridoma 2E3 cells were cultured in Dulbecco's modified Eagle's (DME) medium (Nissui, Tokyo) supplemented with 10% (v/v) fetal calf serum (FCS), 20mM N-(2 hydroxyethyl) piperazine-N'-2-ethanesulfonic acid (HEPES), 0.2% $NaHCO_3$, 2mM glutamine and 0.06mg/ml kanamycin. The cells were cultured at 37°C in humidified air containing 5% CO_2. The viable and dead cells were identified by trypan blue exclusion and the number of cells was measured with haemocytometer.

Transfection procedure

The linear plasmids of pMAM-p21 and pMAM-p27 were prepared and Lipofectin® reagent (Gibco BRL) was used for transfection. The transfection was performed along the recommended protocol with serum free DMEM. Transfected cells were selected in DMEM containing with 400μg/ml of G418 by neomycin resistant gene in pMAM-neo and cloned by limiting dilution.

Incubation under growth suppressed condition

Cell growth was measured for 2E3 transfected with pMAM-p21 and with pMAM-p27. The p27 transfectant was cultivated with or without 0.5μM of DEX for induction of p27 expression, and the p21 transfectant with or without 2μM of DEX, respectively. In the cultivation of transfectant and wild type, 2.0×10^4 cells were initially inoculated in 1ml of complete culture medium with or without DEX.

Measurement of antibody concentration
Culture medium was harvested at several points and the concentration of antibody was measured by ELISA (enzyme linked immunosorbent assay). Produced antibodies were sandwiched between sheep anti-mouse IgG and goat anti-mouse IgG conjugated with horseradish peroxidase. The absorption at 490nm derived from o-phenylenediamine was measured by spectrometer.

3. RESULTS AND DISCUSSION

Preparing 2E3 cells transfected with expression plasmid of cyclin inhibitors
Transfectants were prepared as described in MATERIALS AND METHODS. The concentration of DEX for suppressing proliferation was determined as 0.5μM for the p27 transfectant and 2μM for the p21 transfectant. The p27 transfectant rapidly started to induce cell death over 0.5μM concentration of DEX (data not shown). Contrarily, over 2μM of DEX was little influenced in proliferation of the p21 transfectant. The immediate cell death of the p27 transfectant with DEX suggested that intensive expression of p27 protein induced urgent cell death.

Determining of the amount of antibody
Transfectants were cultured with or without DEX and the number of cells were counted at several points. The wild type 2E3 seemed to have no difference in the growth rate or antibody production, whether they were cultured with or without ~2.5μM of DEX in the medium (data not shown). In common culture condition, the wild type cells produced little antibodies after they passed overgrowth point and started apoptosis. The p21 transfectant without DEX grew slower, and much slower with DEX than the wild type cells (Figure 1-a). The antibody production of the p21 transfectant with DEX was increased to about 1.5-fold amount of that of the wild type cells because of the elongation of culture period (Figure 2). The p21 protein was reported to have the property to suppress apoptosis (Wang et al.), which would help with increasing the antibody production.

The p27 transfectant grew at a rate equal to the wild type in logarithmic growth phase with and without DEX (Figure 1-b). With DEX, the p27 transfectant recorded the maximum viable cell density of about 60% of that for the wild type or control cells, and maintained that density until day 10, while the wild type and control cells rushed into death. The p27 transfectant with DEX accelerated antibody production and produced about 3-fold amount of antibody of the wild type (Figure 2). In the growth-suppressed state, the number of dead cells continuously increased, and on day 10 the total cell number of the p27 transfectant with DEX was almost equal to that of the wild type and control cells (data not shown).

The effect of cyclin inhibitors p21 and p27
The specific antibody production rate was calculated, and plotted against the specific growth rate during the logarithmic growth phase of each cell line (Figure 3). Inducing the expression of cyclin inhibitors p21 and p27 seemed to have no significant effect on improvement of antibody productivity. However, the p27 transfectant cultured with DEX accelerated antibody production during day 4-10 (Figure 2), while the p21 transfectant with DEX increased it at the constant rate. During day 4-10, the dead cell number of the p27 transfectant with DEX increased continuously, while the number of viable cells was stable (Figure 1b). The excess induction of p27 protein by DEX resulted in cell death in hybridoma 2E3, indicating that p27 functioned to induce cell death. Consequently, the p27 transfectant with DEX possessed the stationary phase before overgrowth and accelerated antibody production, resulting in 3-fold production

of antibody compared with the wild type cells.

The expression of p21 was reported to provide a host cell with a resistance to apoptosis (Wang et al.), and in the present study the culture period was prolonged by about five days (Figure 1a). By this prolongation of viable culture period, the expression of p21 contributed to increase antibody production.

Recently, cell cycle inhibitors of p21, p27, and a mutant of p53 were applied for increasing transient protein production by CHO cells with tetracyclin-regulated transcription system (Fussenegger et al.). In their study, the controlled expression of these inhibitory genes could abolish cell cycle progression and did not induce cell death, resulting in 4-fold production of human secreted alkaline phosphatase compared with conventional production system. In the case of hybridoma 2E3, the expression of p27 induced the cell death rather than suppressed the cell proliferation and provided the prolonged stationary phase, and the expression of p21 did not help the cells so much in increasing antibody production.

ACKNOWLEDGEMENT

We wish to thank Dr.Noda for kindly providing us with p21 cDNA and Dr.Toyoshima for p27 cDNA.

REFERENCE

Al-Rubeai, M., Emery, AN., Chalder, S., and Jan, DC. (1992) Specific monoclonal antibody productivity and the cell cycle-comparisons of batch, continuous and perfusion cultures, *Cytotechnology*, **9** (1-3), 85-97

Cotter, TG., and Al-Rubeai, M. (1995) Cell death (apoptosis) in cell culture systems, *Trends in Bio Technol.* **13**, 150-155

El-Deiry, WE., Harper, JAW., O'Connor, PM.,Velculescu, VIE., Canman, CE., Jackman, J., Pietenpol, JA., Burrell, M., Hill, DE., and Wang, Y. (1994) WAF1/CIP1 is induced in p53-mediated G1 arrest and apoptosis, *Cancer Res* **54** (5), 1169-74

Ezhevsky, SA., Toyoshima, H., Hunter, T., and Scott, DW. (1996) Role of cyclin A and p27 in anti-IgM induced G1 growth arrest of murine B-cell lymphomas, *Mol Biol Cell* **7** (4), 553-64

Fussenegger, M., Mazur, X., and Bailey, JE. (1997) A novel cytostatic process enhances the productivity of chinese hamster ovary cells, *Biotechnol and Bioeng* **55** (6), 927-939

Harper, JW., Adami, GR., Wei, N., Keyomarsi, K., and Elledge, SJ.(1993) The p21 Cdk-interacting protein Cip1 is a potent inhibitor of G1 cyclin-dependent kinases, *Cell* **75** (4), 805-16

LaBaer, J., Garrett, MD., Stevenson, LF., Slingerland, JM., Sandhu, C., Chou, HS., Fattaey, A., and Harlow, E. (1997) New functional activities for the p21 family of CDK inhibitors, *Genes Dev* **11** (7), 847-62

Liu, M., Lee, MH., Cohen, M., Bommakanti, M., and Freedman. LP. (1996) Transcriptional activation of the Cdk inhibitor p21 by vitamin D3 leads to the induced differentiation of the myelomonocytic cell line U937, *Genes Dev* **10** (2), 142-53

Luo, Y., Hurwitz, J., and Massague, J. (1995) Cell-cycle inhibition by independent CDK and PCNA binding domains in p21Cip1, *Nature* **375** (6527), 159-61

Noda, A., Ning, Y., Venable, SF., Pereira-Smith, OM., and Smith, JR. (1994) Cloning of senescent cell-derived inhibitors of DNA synthesis using an expression screen, *Exp Cell Res* **211** (1), 90-8

Pagano, M., Tam, SW., Theodoras, AM., Beer-Romero, P., Del Sal, G., Chau, V., Yew, PR., Draetta GF., and Rolfe, M. (1995) Role of the ubiquitin-proteasome pathway in regulating abundance of the cyclin-dependent kinase inhibitor p27, *Science* **269** (5224), 682-5

Polyak, K., Kato, JY., Solomon, MJ., Sherr, CJ., Massague, J., Roberts, JM., and Koff, A. (1994) p27Kip1, a cyclin-Cdk inhibitor, links transforming growth factor-beta and contact inhibition to cell cycle arrest, *Genes Dev* **8** (1), 9-22

Toyoshima, H., and Hunter, T. (1994) p27, a novel inhibitor of G1 cyclin-Cdk protein kinase activity, is related to p21, *Cell* **78** (1), 67-74

Wang, J., and Walsh, K. (1996) Resistance to apoptosis conferred by Cdk inhibitors during myocyte differentiation, *Science* **273** (5273), 359-61

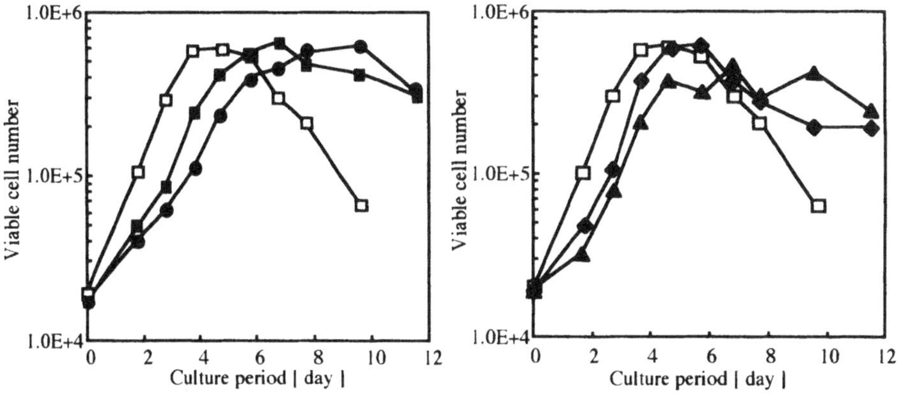

Figure 1. The number of viable cells of the wild type 2E3 and the p21 or p27 transfectant cells. For inducing the expression of p21 or p27 genes, DEX was added in the initial. The 2×10^4 cells were inoculated and the number of viable cells were counted by trypan blue exclusion with haemocytometer. **a**: open square; wild type 2E3: closed square; p21 transfectant without DEX: closed circle; p21 transfectant with 2μM of DEX. **b**:open square; wild type 2E3: closed diamond; p27 transfectant without DEX; closed triangle; p27 transfectant with 0.5μM of DEX. All data represent duplicated experiments.

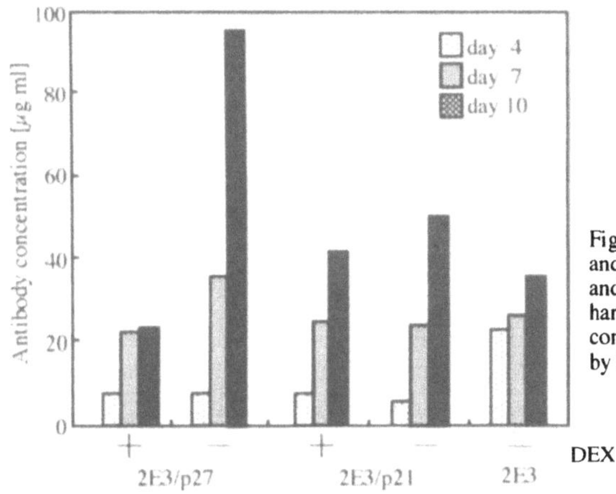

Figure 2. The antibody production by p21 and p27 transfectant with or without DEX and wild type 2E3. Culture medium was harvested on day 4, 7 and 10. The concentration of antibody was determined by ELISA.

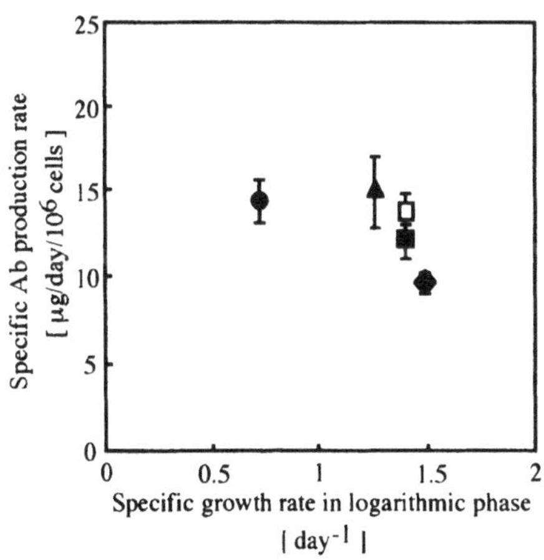

Figure 3. The specific antibody production rate against the specific growth rate was plotted. Each character was as described in Figure 1.

ESTABLISHMENT OF AN APOPTOSIS-RESISTANT AND GROWTH-CONTROLLABLE CELL LINE BY TRANSFECTING WITH INDUCIBLE ANTISENSE c-JUN GENE

YON HUI KIM[1], TAKEHIRO IIDA[1], EDWARD V. PROCHOWNIK[2], EIJI SUZUKI[1]

1. *Department of Chemistry and Biotechnology, Graduate School of Engineering, The University of Tokyo, Bunkyo-ku, Tokyo 113, Japan*
2. *Division of Pediatric Hematology/Oncology, Children's Hospital of Pittsburgh, Pittsburg, PA 15213-2583, USA*

ABSTRACT. By treating the c-*jun* AS cells with dexamethasone (DEX) in DMEM supplemented with 10% serum, the growth of the cells were completely suppressed for duration of 16 days with high cell viability of above 86%. The c-jun AS cells were able to grow well to 10^6 cells/ml without supplementation of any serum components, keeping viability high above 80% for 8 days when cultured in the absence of DEX. The c-jun AS cells stayed at a constant cell density and high viability above 80% for 8 days when they were cultured in the presence of DEX under serum-deprivation. In contrast to the c-*jun* AS cells, the wild type F-MEL cells were unable to grow and they died by apoptosis in 3 days under the serum-deprivation; the internucleosomal cleavage of DNA, a landmark of apoptosis, was clearly detectable. Thus the c-*jun* AS cell line that is resistant to apoptosis induced by serum-deprivation and can reversibly and viably be growth-arrested was established.

1. INTRODUCTION

Cell death may follow two distinct patterns: necrosis and apoptosis. Apoptotic cell death follows a well-defined sequence of events characterized by cell shrinkage, nuclear condensation, and fragmentation of the cell into discrete membrane enclosed bodies. In the early apoptotic death process, an endogenous endonuclease is activated and cleaves DNA within the internucleosomal spacer regions. Hence, a characteristic ladder pattern consisting of multiplies of 180 bp fragments is seen when the DNA from apoptotic cells is electrophoresed on an agarose gel. Reportedly, the cell death in the late logarithmic growth and stationary phases of batch culture is mostly apoptosis (Mercille and Massie, 1994).

Proto-oncogene c-*jun* known as the 'immediate early response' gene is rapidly induced in response to mitogenic stimuli, and it plays a critical regulatory role in the commitment of the cell to proliferate (Ryseck et al., 1988, Smith and Prochownik, 1992). A nuclear

Y. Kitagawa et al. (eds.), Animal Cell Technology: Basic & Applied Aspects, Vol. 10, 177–184.

phosphoprotein c-Jun is a major component of the AP-1 *trans*-acting transcriptional activator complex that promotes transcription of various genes required for progression of the cell cycle (Turner and Tjlan, 1989). The selective inhibition of c-*jun* expression caused exit from the cell cycle, in other words, it blocked cell growth (Smith and Prochownik, 1992; Soprano et al., 1992). The possibility of c-Jun being the inducer for natural cell death, apoptosis, was reported (Colotta et al., 1992; Estus et al., 1994; Rampalli and Zelenka, 1995; Anderson et al., 1995). Interleukin (IL)-6- and IL-2-dependent mouse cell lines underwent programmed cell death after the growth factor deprivation; in this death process, c-*jun* gene was rapidly induced; and the antisense oligonucleotides directed against c-*jun* reduced this c-*jun* induction, while improving survival of the cells after the growth factor deprivation (Colotta et al., 1992). The antisense oligonucleotides for c-*jun* prevented HL-60 cells from apoptosis induced by ceramide (Sawai et al., 1995).

Using this understanding that the c-*jun* gene plays an important role in the progression of the cell cycle and induction of apoptosis or programmed cell death. The objective of the present work is to establish a cell line in which growth can be switched on and off without induction of apoptosis. And we were successful in creating a cell line c-*jun* AS that can be induced by glucocorticoid hormone to block cell proliferation and at the same time suppress apoptosis even under growth factor deprivation. This growth and apoptosis controllable cell line can be utilized to keep high viability in batch, fed-batch, or perfusion culture. The cell line will be also suitable to culture for which use of any serum components is prohibited.

2. MATERIALS AND METHODS

Cell Line and Cell Culture Conditions

Friend murine erythroleukemia (F-MEL) cells were supplied by Riken Cell Bank. F-MEL cells and all the cells derived from F-MEL cells in the present work were cultured in Dulbecco's modified Eagle's medium (DMEM) (Nissui, Tokyo), supplemented with 10% fetal bovine serum (v/v) (JRH, Biosciences), 4 m*M* glutamine, 100 mg/ml kanamycin, 0.2% NaHCO3 (complete medium), unless otherwise specified. Viable cell count was performed by the trypan blue dye exclusion method employing a hemocytometer.

Transfection, Selection, and Gene Amplification

pMS-G c-*jun*AS dhfr vector contains an antisense fragment of c-*jun* gene downstream of a glucocorticoid-inducible murine mammary tumor virus (MMTV) promoter and a dhfr (dihydrofolate reductase) gene. The antisense c-*jun* gene is the inversely inserted sequence that consists of 287 bp of 5`-untranslated region, 48 bp of coding sequence derived from the extreme 3` end of the coding region, and 929 bp of 3`-untranslated region of c-*jun* cDNA. pMS-G c-*jun* dhfr vector contains a complete sense c-*jun* cDNA instead of the antisense c-*jun* gene (Smith and Prochownik, 1992). For transfections, $2x10^6$ F-MEL cells were pelleted at 500 g for 10 min, washed twice in phosphate-buffered saline (PBS), and resuspended in 200 ml of PBS. 20 mg of *Bam*HI-linearized pMS-G c-*jun* AS dhfr DNA or pMS-G c-*jun* dhfr

DNA plus 2 mg of *Nde*I-linearized pSV2neo plasmid DNA in a total volume of 50 ml of sterile water were added to the cell suspension. Transfection was accomplished by electroporation with a homemade apparatus at settings of 650 voltage with 10 pulses at resultant time of 250 msec.

Detection of c-*Jun* Protein by Western Blotting

To assess the effect of c-*jun* antisense transcript induction on c-*jun* protein levels, the c-*jun* AS cells were cultured in the complete medium containing 10^{-6} *M* dexamethasone for 48 h. At size of 39-kDa, c-*jun* protein was detected by rabbit polyclonal antibodies to c-*jun* /AP-1 protein (Medac, Diagnostika) to which peroxidase-conjugated goat anti-rabbit Ig polyclonal antibody (Bio Source International, Inc. Tago Products) bound. F-MEL cells and the c-*jun* AS cells were cultured free from DEX, and treated in the same manner for Western blotting.

DNA Fragmentation Assay by Electrophoresis

$1x10^6$ cells were pelleted and lysed by vortexing vigorously in 0.5 ml of TE buffer (10 m*M* Tris-HCl, pH 7.4 and 1 m*M* EDTA) containing 0.2% Triton X-100, and centrifuged for 10 min at 13,000g at 4°C. DNA was extracted after addition of 0.1 ml of 5 *M* NaCl and 0.7 ml of isopropanol. The supernatant was vortexed, placed at -20°C overnight, and then centrifuged for 10 min at 13,000 g at 4 °C. The DNA pellet was dissolved in TE buffer and treated with RNase prior to electrophoresis on a 2% agarose gel. To visualize the fragmented DNA bands, the gel was stained with 1 mg of ethidium bromide per ml.

3. RESULTS and DISCUSSION

Blocking Cell-Growth by Antisense c-*Jun* Expression

We assessed the effect of c-*jun* antisense transcript induction on c-*jun* protein levels. The viable cell density and viability datas were reported previously. For comparison, we cultured in the same manner the wild type F-MEL cells and those transfected with the complete sense c-*jun* gene. The growth of the c-*jun* AS cells was almost blocked two days after DEX addition. After that the viable cell density was kept almost unchanged for 14 days. A viability greater than 86% was maintained for 16 days. In the absence of DEX, the c-*jun* AS cells grew to the over-growth condition and then started dying. The wild type F-MEL cells also grew to over-growth , and then started dying either in the presence or absence of DEX. Jointly these results indicated that the c-*jun* AS cell line was a desired cell line, that is, a viably growth-arrestable cell line.

Suppressing Apoptosis by Antisense c-*Jun* Expression

At least two groups reported that antisense oligonucleotides directed for *c-jun* gene suppressed apoptosis (Colotta et al., 1992; Sawai et al., 1995). This suggested that the *c-jun* AS cell line is possibly free from apoptosis when cultured in the presence of DEX. Use of oligonucleotides is limited to research purpose, because they are expensive and not stable in culture. Contrarily, the antisense *c-jun* transcripts in the *c-jun* AS cells can be induced with relatively inexpensive DEX when necessary. Therefore, we examined the apoptosis resistance of the *c-jun* AS cells by applying serum-deprivation, one of typical apoptosis inducing conditions. The logarithmically growing *c-jun* AS cells in the complete medium were collected, and washed three time in PBS. Then the cells were placed in the serum-deprived medium, and cultured in the presence or absence of DEX. The wild type F-MEL cells were also subjected to the same procedure. The viable cell density and viability are shown in Fig.1-a and -b, respectively. Under serum-deprivation, the viability of the wild type F-MEL cells decreased to 50% in 3 days, while that of the c-jun AS cells was maintained above 80% for 8 days.

In the absence of DEX, the *c-jun* AS cells, of which *c-jun* expression was partially suppressed by the antisense *c-jun* expression due to the leakiness of MMTV promoter, survived and grew well under serum-deprivation (Fig.1-a and -b). This meant that we could unexpectedly establish a F-MEL cell line that can grow without any serum component. We interpreted this result as *c-jun* expression in the *c-jun* AS cells cultured in the absence of DEX was high enough to drive cell cycling and too low to induce apoptosis.

We examined emergence of the ladder pattern, a land mark of apoptotic cells, on electrophoreses of DNA extracted from the cells of the above experiments. The *c-jun* AS cells cultured with DEX under serum-deprivation were collected on day 4, 6, 8, 10, and 12. DNA samples extracted from the cells were treated as described in Materials and Methods for detecting DNA fragmentation. The wild type F-MEL cells cultured under the serum-deprivation were similarly examined. The clear DNA ladder was detected for the wild type F-MEL cells on day 4, but not for the *c-jun* AS cells treated with DEX (Fig.2). Jointly with the result of Western blotting for detection of *c-jun* protein, this result indicated that suppression of *c-jun* expression inhibits apoptosis.

Reversible Growth Blocking

To apply the *c-jun* As cells to various usages, including protein production and bioassay, it is desirable that the cell cycle arrested *c-jun* AS cells can start growing again after removal of DEX. to examine whether the cell cycle arrest is reversible, we cultured the *c-jun* As cells and the wild type F-MEL cells in the presence or absence of DEX in serum-deprived medium for 4 days and then transferred them into a DEX-free complete medium. The viable cell density is shown in Fig.3a, and the viability is shown in Fig.3b. The *c-jun* AS cells that had been growth arrested by antisense c-jun expression started to grow a day after DEX removal. They grew to a cell density of $7X10^6$ cells/mL, as high as the nontreated cells.

Model for Interlinked Control of Growth and Apoptosis

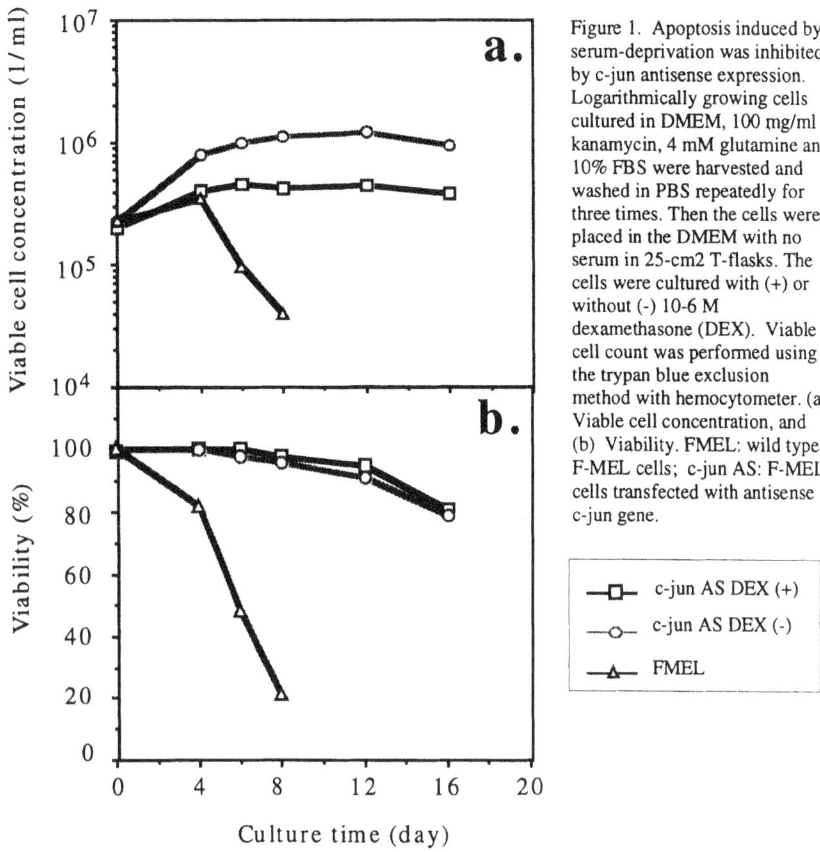

Figure 1. Apoptosis induced by serum-deprivation was inhibited by c-jun antisense expression. Logarithmically growing cells cultured in DMEM, 100 mg/ml kanamycin, 4 mM glutamine and 10% FBS were harvested and washed in PBS repeatedly for three times. Then the cells were placed in the DMEM with no serum in 25-cm2 T-flasks. The cells were cultured with (+) or without (-) 10-6 M dexamethasone (DEX). Viable cell count was performed using the trypan blue exclusion method with hemocytometer. (a) Viable cell concentration, and (b) Viability. FMEL: wild type F-MEL cells; c-jun AS: F-MEL cells transfected with antisense c-jun gene.

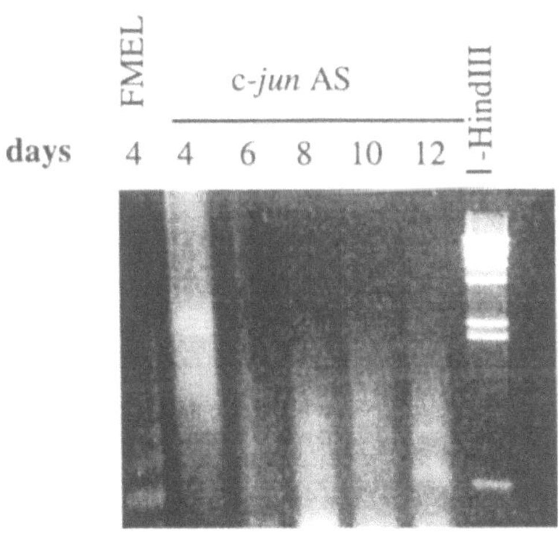

Figure 2. Electrophoresis of DNA of cells cultured under serum-deprivation. The c-*jun* AS cells were cultured with 10-6 *M* dexamethasone (DEX) under serum-deprivation as described for Fig.2. The cells were collected on day 4, 6, 8, 10, and 12 of serum-deprived culture for DNA fragmentation assay, and treated as described in MATERIALS and METHODS. The wild type F-MEL cells cultured under serum-deprivation were collected on day 4 for the assay.

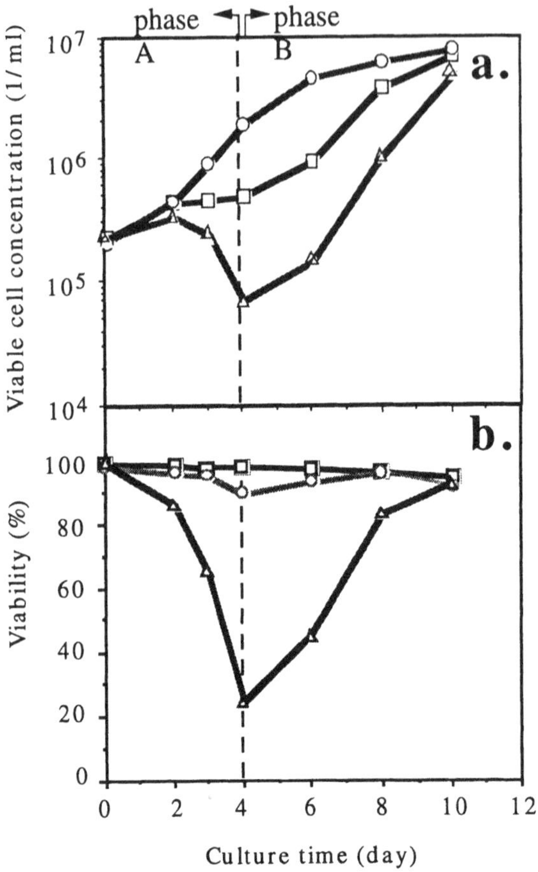

Figure 3. Growth blocking by c-jun antisense expression in serum-deprived culture is reversible. Phase A: The c-jun AS and the wild type F-MEL cells were cultured under the serum deprivation by day 4 in the same manner as described for Fig.3. ( ) c-jun AS cells cultured with 10-6M DEX: ( ) c-jun AS cells cultured without DEX. Phase B: On day 5, the cells were collected, washed with PBS three times, and resuspended in DEX-free medium supplemented with 10% FBS: (a) viable cell concentration; (b) viability

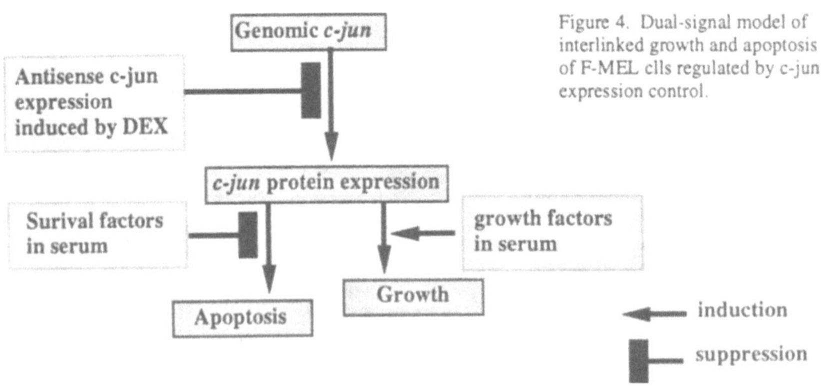

Figure 4. Dual-signal model of interlinked growth and apoptosis of F-MEL clls regulated by c-jun expression control.

We proposed a model Fig.4 of the regulation of growth and apoptosis of F-MEL cells by *c-jun* expression control. There are two hypotheses to be noticed in the model. First, c-jun expression at a level above a threshold induces apoptosis and growth, while growth itself suppresses apoptosis in a growth-rate dependent manner. Slow growth can cancel apoptosis induction due to weak *c-jun* expression. Secondly, serum contains survival factors in addition to growth factors. The group of Evan (Harrington et al., 1994) proposed two models that may explain the induction of apoptosis by c-Myc in serum deprived fibroblasts: one is the conflict model, and other is the dual-signal model of which they were in favor. Our model for *c-jun* is partly similar to their dual-signal model for c-myc and is consistent with our observations listed below, although there are some other models that are also consistent.

1) Suppressing *c-jun* expression of the *c-jun* AS cells by antisense in the presence of serum — no apoptosis, no growth.
2) Expressing *c-jun* of the wild type F-MEL cells in the presence of serum — no apoptosis, fast growth.
3) Expressing *c-jun* of the wild type F-MEL cells in the absence of serum — apoptosis, slow or no growth.
4) Suppressing *c-jun* expression of the *c-jun* AS cells by antisense in the absence of serum — no apoptosis, no growth.
5) Moderately expressing *c-jun* of the *c-jun* AS cells cultured in the absence of serum and DEX — no apoptosis, moderate growth.

A cell line *c-jun* AS that can be reversibly and viably growth-arrested was established by transfecting F-MEL cells with inducible antisense *c-jun* gene. The cell line is resistant to apoptosis induced by serum-deprivation , or by cell cycle arrest. The *c-jun* AS cells can grow even under serum-deprivation when cultured free from dexamethasone, while the wild type F-MEL cells die quick. Jointly with several other reports on the relation between apoptosis and *c-jun* expression, our result strongly suggested that *c-jun* expression is required for onset of at least some types of apoptosis.

References

Anderson, A. J., Pike, C. J. and Cotman, C. W. 1995. Differential induction of immediate early gene proteins in cultured neurons by b-amyloid (Ab): association of c-Jun with Ab - induced apoptosis. J. Neurochem. **65:** 1487-1498.

Colotta, F., Polentarutti, N., Sironi, M. and Mantovani, A. 1992. Expression and involvement of c-fos and c-jun protooncogenes in programmed cell death induced by growth factor deprivation in lymphoid cell lines. J. Biol. Chem. **267:** 18278-18283.

Estus, S., Zaks, W. J., Freeman, R. S., Gruda, M., Bravo, R., and Johnson, E. M. Jr. 1994. Altered gene expression in neurons during programmed cell death: identification of c-jun as necessary for neuronal apoptosis. J. Cell Biol. **127:** 1717-1727.

Mercille, S. and Massie, B. 1994. Induction of apoptosis in nutrient-deprived cultures of hybridoma and myeloma cells. Biotechnol. Bioeng. **4 4**: 1140-1154.

Rampalli, A. M. and Zelenka, P. S. 1995. Insulin regulates expression of c-fos and c-jun and suppresses apoptosis of lens epithelial cells. Cell Growth Differ. **6:** 945-953.

Ryseck, R. P., Hirai, S. I., Yaniv, M. and Bravo, R. 1988. Transcriptional activation of c-jun during the G0/G1 transition in mouse fibroblasts. Nature **334:** 535-537.

Sawai, H., Okazaki, T., Yamamoto, H., Okano, H., Takeda, Y., Tashima, M., Sawada, H., Okuma, M., Ishikura, H., Umehara, H., and Domae, N. 1995. Requirement of AP-1 for ceramide-induced apoptosis in human leukemia HL-60 cells. J. Biol. Chem. **270**: 27326-27331.

Smith, M. J. and Prochownik, E. V. 1992. Inhibition of c-jun causes reversible proliferative arrest and withdrawal from the cell cycle. Blood **79:** 2107-2115.

Soprano, K. J., Cosenza, S. C., Yumet, G., and Soprano, D. R. 1992. Use of antisense oligomers to study the role of c-jun in G1 progression. Ann. N. Y. Acad. Sci. **660:** 231-239.

Turner, R. and Tjlan, R. 1989. Leucine repeats and an adjacent DNA binding domain mediate the formation of functional cFos-cJun heterodimers. Science **243**: 1689-1694

IMPROVEMENT OF CELL SURVIVAL BY INHIBITION OF CASPASES PLAYING KEY ROLES IN APOPTOSIS

S. TERADA[†], T. YATA[†], K. FUKUOKA[†], H. AIZAWA[†],
T. FUJITA[†], T. NAGAMUNE[†] and E. SUZUKI*
[†] *Department of Chemistry and Biotechnology, Graduate School of Engineering, The University of Tokyo ; 7-3-1, Hongo, Bunkyo-ku, Tokyo, 113 Japan*
* *Research Institute of Innovative Technology for the Earth (RITE) ; 9-2, Kizugawadai, Kizu-cho, Soraku-gun, Kyoto, 619-02, Japan*

Improving cell survival at over-growth phase of culture should increase protein production of culture. At the over-growth phase, several adverse conditions such as depletion of nutrients or serum components and accumulation of toxic metabolites trigger apoptosis, a style of cell death. ICE/CED-3 family proteases such as caspase 1 (ICE) and caspase 3 (CPP32) are principal players in apoptosis. The authors aimed at suppressing apoptosis by inhibiting caspases. Two viral proteins, CrmA from cowpoxvirus and p35 from baculovirus are potent inhibitors of the proteases. The CHO cells transfected with p35 gene were named CHO-p35. CHO-p35 and wild type were cultured in serum limited condition (FBS free) for 52 hours. Viability of CHO-p35 was 59 % and that of wild type CHO was 48 % at 52 hours of the culture. However, expression of p35 protein in CHO-p35 was not detected with western blotting method. Method of apoptosis inhibition using ICE/CED-3 tetrapeptide inhibitors was studied. These tetrapeptide inhibitors were designed to involve recognition sequences of the substrates of the proteases. The authors examined effect of a caspase 3 inhibitor, a tetrapeptide Ac-Asp-Glu-Val-Asp-aldehyde (DEVD), on cell survival of hybridoma, myeloma and CHO cells. Caspase 3 inhibitor significantly prolonged by 3 days viable culture period of the bcl-2 transfected cells, while they slightly did that of the untransfected cells. The simultaneous inhibition of multiple pathways of apoptosis could improve cell survival.

1. Introduction

The increasing demand for large quantities of bio-active proteins requires the improvement of productivity of in vitro mammalian cell culture. Monoclonal antibodies (MoAb) are produced by hybridoma culture and recombinant protein are produced by cell, for example, CHO or myeloma, culture. Bio-active protein production of a culture increases when the viable culture period, that is, the growth and viable non-growth culture period extends. However, malignant cells tend to die quickly after reaching the maximum cell density. Therefore, preventing cells from death which starts in the late exponential growth phase and maintaining them viable in batch culture for longer time period should increase bio-active protein production of the culture. Cells die due to depletion of nutrients such as amino acids and glucose, limitation of growth factors such as serum components or some other conditions unfavorable for cells. At least one of these cell-death inducing conditions occurs in the late exponential growth and stationary phases of batch culture. Apoptosis is described as suicidal death and Bcl-2 protein has been found to be functional in suppressing apoptosis [1]. The authors have already reported that overexpression of Bcl-2 in hybridoma prevented cell death and enhanced MoAb

Y. Kitagawa et al. (eds.), Animal Cell Technology: Basic & Applied Aspects, Vol. 10, 185–189.

production [2], [3].

ICE/CED-3 family proteases such as caspase-1 (ICE) and caspase-3 (CPP32, Yama, apopain) play a central effector role in apoptosis [4]. These proteases are produced as inactive precursors and they require cleavage into two subunits approximately 20 and 10 kDa before activation [5]. After activated, they cleave several cellular proteins including poly(ADP-ribose) polymerase (PARP). Cleavage of these proteins appears to be an early event of apoptosis as it occurs before any morphological signs of cell death [6]. Two viral proteins, CrmA from cowpoxvirus and p35 from baculovirus are potent inhibitors of the proteases. In the present study, we focused on a method of inhibiting apoptosis using ICE/CED-3 tetrapeptide inhibitors. These tetrapeptide inhibitors were designed as the appropriate peptide recognition sequences of the substrates of each proteases. A caspase-3 inhibitor [6] contains the P1-P4 amino-acid of the PARP cleavage site (Ac-Asp-Glu-Val-Asp-CHO) and caspase-1 inhibitor [7], Ac-Tyr-Val-Ala-Asp-CHO, contains the amino acid site in proIL-1β that is recognized and cleaved by ICE.

In this study, the authors achieved improvement of cell survival by inhibition of caspases.

2. Materials & Methods

2.1. CELL LINE AND CULTURE CONDITION

A cell line 2E3-O is a mouse hybridoma derived from a mouse myeloma P3X63 AG8U.1 by electric fusion with mouse spleen cells. A cell line X63.653 is a mouse myeloma. The cells were cultured in DME medium (Nissui, Tokyo), supplemented with 10% vol/vol FBS (JRH Bioscience), 20 mM HEPES, 0.2% $NaHCO_3$, 2 mM glutamine, and 0.06 mg ml^{-1} kanamycin. The cells were grown in 6 well plate (Iwaki glass) at 37℃ in humidified air containing CO_2 at 5%.

Chinese hamster ovary (CHO) cells (CRL-9096 ; ATCC) were cultured in α-MEM medium (Nissui, Tokyo), supplemented with 10% vol/vol FBS, 20 mM HEPES, 0.2% $NaHCO_3$, 2 mM glutamine, and 0.06 mg ml^{-1} kanamycin.

Viable and dead cells were determined by counting in a hemocytometer under a phase contrast microscope using trypan blue exclusion.

2.2. TRANSFECTION

The vectors BCMGneo-p35 for expressing p35 and BCMGneo-crmA for expressing CrmA were constructed and prepared in our laboratory by conventional method. The vectors BCMGneo-bcl-2 for expressing human bcl-2 [8] and pZeo-bag-1 for expressing murine bag-1 [9] were also prepared.

CHO cells were transfected with BCMGSneo vector [10], BCMGneo-bcl-2, BCMGneo-p35, and BCMGneo-crmA respectively, and named mock transfectant CHO-mock, bcl-2 transfectant CHO-bcl2, p35 transfectant CHO-p35 and crmA transfectant CHO-crmA, respectively. Myeloma X63-653 was transfected with BCMGSneo vector, with BCMGneo-bcl-2 and pZeo-bag-1, respectively, and named mock transfectant X63-mock, and bcl-2 and bag-1 transfectant X63-bcl2/bag1. Hybridoma 2E3-O was transfected with BCMGSneo vector and BCMGneo-bcl-2, respectively, and named mock transfectant 2E3O-mock, bcl-2 transfectant 2E3O-bcl-2.

The transfection method was electroporation.

2.3. WESTERN BLOT ANALYSIS OF P35-PROTEIN

Single-cell suspensions were lysed in 1% Triton X-100, 0.15 mM NaCl and 10 mM Tris (pH 7.4) with 50 mg/ml PMSF and 2 mg/ml aprotinin at 4 ℃ for 30 min. The cell lysates were boiled in SDS sample buffer for 5 min before being run on a SDS-polyacrylamide gel (13%). The lysate from cells was loaded into each lane. Gels were transferred to nitrocellulose filter overnight. Blots were blocked with 5% skim-milk for 2 hourrs at room temperature. P35 protein was detected by chicken anti-p35 antibody (Promega) to which peroxidase-conjugated anti-chicken Ig antibody bound.

2.4. REAGENTS

As a caspase-3 inhibitor, the authors used Ac-Asp-Glu-Val-Asp-aldehyde (Takara) and as a caspase-1 inhibitor, we used Ac-Tyr-Val-Ala-Asp-aldehyde (Takara). The inhibitors was supplemented to medium at 1 mM concentration.

3. Results & Discussion

3.1. TRANSFECTION OF P35 AND CRMA TO CHO CELL

Expression plasmid encoding p35 and crmA were introduced into CHO cells together with the neomycin-resistance gene. G418-resistant transformants were obtained. Anti-p35 antibody and purified p35 protein were obtained from Promega and the authors measured expression of p35 protein by western blotting method. Though purified p35 was detected in control assay, p35 protein did not detected in the lysate of G418-resistant transformants.

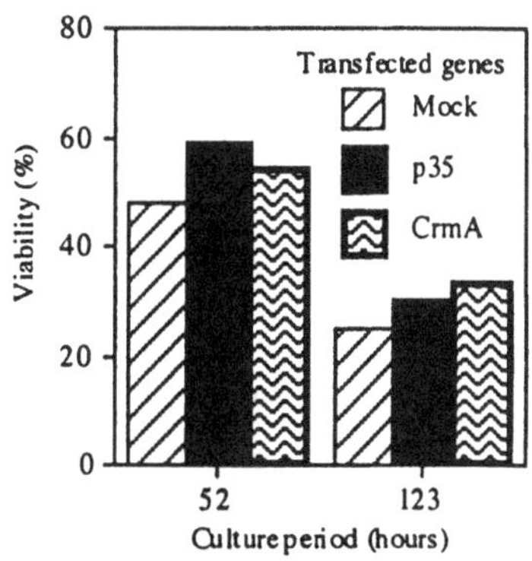

Fig. 1 Viability of CHO cell in serum depletion culture

Though expressions of these viral proteins were not detected, we examined effect of transfection of cells with these viral protein genes on apoptosis in serum depletion culture (Fig. 1). After 52 hour serum depletion culture, the viabilities of CHO-mock, CHO-p35, and CHO-crmA were 47, 59, and 55%, respectively. Viability of CHO-p35 and CHO-crmA were slightly higher than that of CHO-mock. The viral protein, p35 and CrmA might be expressed and they might inhibit apoptosis induced by serum depletion.

3.2. EFFECT OF CASPASE INHIBITOR ON CHO CELL

Effect of transfection of the cells with p35 or crmA was unclear. In order to study inhibition of ICE-Ced-3 proteases, the authors used a tetrapeptide inhibitor (DEVD) which works as specific inhibitor for caspase-3. We examined DEVD effect on CHO cell in serum depletion culture (Fig. 2). After 214 hour serum depletion culture, the viability of CHO cell treated with 1 mM DEVD was 39% and that of untreated cells was 11 %. Treatment with 1 mM DEVD apparently increased viability in serum depletion culture.

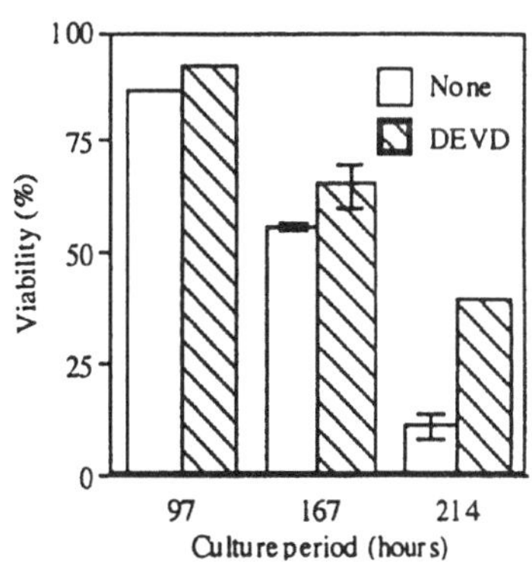

Fig. 2 Effect of Caspase inhibitor (DEVD) on CHO cell

3.3. SIMULTANEOUS INHIBITION OF APOPTOSIS PATHWAYS IN CHO CELL

The tetra-peptide caspase inhibitor was effective against cell death but not enough. We expected that simultaneous inhibition of multiple pathways of apoptosis (Fig. 3) could improve cell survival. Expression plasmid encoding bcl-2 was introduced into CHO cells together with the neomycin-resistance gene. Among G418-resistant transfectants, the high bcl-2 expressing transfectants was selected by ELISA and was named CHO-bcl2. We examined caspase-3 effect on CHO-mock and CHO-bcl2 cell in serum depletion culture (Fig. 4). During 167 hour serum depletion culture, the viabilities of CHO-bcl2 and CHO-bcl2 treated with DEVD were higher than 90 %, that of CHO cells treated with DEVD was 65%, and that of CHO cells untreated was 56 %. After 214 hours, the viability of CHO-bcl2 treated with DEVD was 80% and that of CHO-bcl2 untreated was 33 %. Caspase inhibitor synergestically prolonged culture of bcl-2 transfectant.

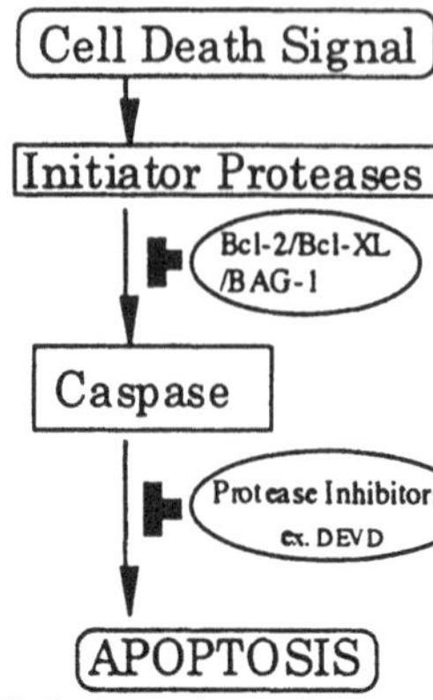

Fig.3 Apoptosis cascade, Modified from D. W. Nicholson, Nature Biotechnology 14, 1996

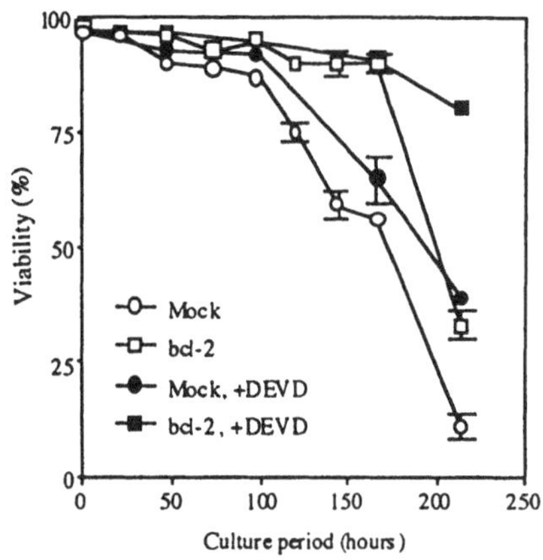

Fig. 4 Effect of Caspase inhibitior & bcl-2 on CHO cells

3.4 SIMULTANEOUS INHIBITION OF APOPTOSIS PATHWAYS IN OTHER CELLS

Synergistical inhibition of apoptosis with bcl-2 transfection and the caspase inhibitor treatment prolonged viable culture period of CHO cells. Hence we investigated simultaneous inhibition of multiple pathways of apoptosis in other cells, hybridoma and myeloma. Bcl-2 and bag-1 genes were transfected to myeloma X63.653. We examined expression of bcl-2 and bag-1 proteins by western blotting method. The transfectant was named X63-bcl2/bag1. We examined effect of bcl-2 and bag-1 expression on cell survival (Fig. 5). After 130 hour culture, viability of X63-mock was 29%, that of X63-mock treated with DEVD was 36 %, that of X63-bcl2 was 43 %, and that of X63-bcl2 treated with DEVD was 51 %. The caspase-3 inhibitor (DEVD) synergestically prolonged culture of bcl-2 and bag-1 co-transfectant myeloma cells.

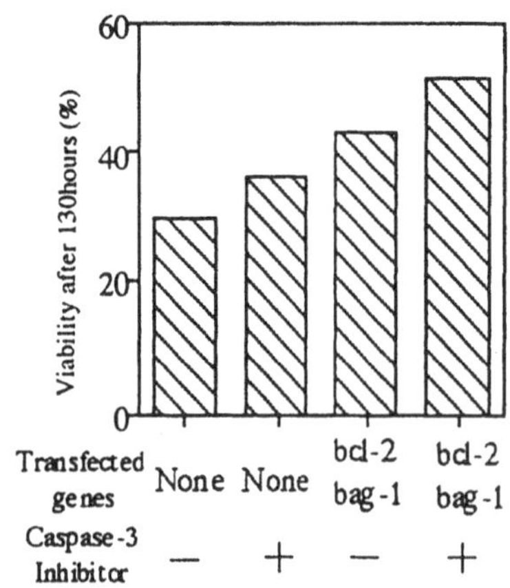

Fig. 5 Effect of Caspase inhibitior & genes transfection on myeloma

Bcl-2 was transfected to hybridoma 2E3-O and was named 2E3O-bcl2. The authors investigated synergistical effect of bcl-2 transfection and treatment with caspase-1 inhibitor (YVAD) and caspase-3 inhibitor (DEVD), respectively (Fig. 6). Neither the caspase-1 inhibitor nor the caspase-3 inhibitor did not prevent decrease cell viability (Fig. 6-a), while the caspase-3 inhibitor amazingly by 4 days and the caspase-1 inhibitor moderately prolonged viable culture period of the bcl-2 transfected cells (Fig. 6-b). The viability time course of the culture suggested that Bcl-2 and the caspase inhibitors suppressed apoptosis at early and late over-growth phase, respectively.

4. Acknowledgment

We thank Dr. A. Sugimoto (Univ. of Tokyo) for helpful advice. We thank Dr. S. Seshagiri (Univ. of Georgia), Dr. Y-P. Kuang (Univ. of Michigan Medical School), Dr. S. Hashimoto (Meiji Milk Products Co. Ltd.), Dr. Y. Tsujimoto (Osaka Univ.) and Dr. J. C. Reed (Burnham Institute) for the generous gifts of P35 gene, crmA gene, E1B19k gene, bcl-2 gene and bag-1 gene, respectively. We also appreciate Dr. M Haak-Frendscho (Promega Co.) for the generous gifts of anti-P35 antibody.

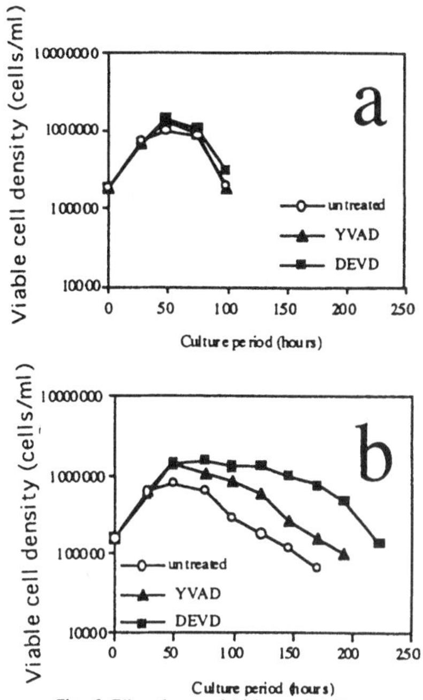

Fig. 6 Effect of caspase inhibitors on hybridoma : (a) wild type, (b) bcl-2 transfectant

5. References

[1] M. Pattersson, et.al. 1992. Blood 79 : 495-502.

[2] Y. Itoh, et.al. 1995. Biotechnol. Bioeng. 48 : 118-122

[3] S. Terada, et.al. 1997. Cytotechnology 24 : 135-141

[4] S. J. Martin, et.al. 1995. Cell : 349-352

[5] K. P. Wilson, et.al. 1994. Nature 370 : 270-275

[6] D. W. Nicholson, et.al. 1995. Nature 376 : 37-43

[7] N. A. Thornberry, et.al. 1992. Nature 356 : 768-774

[8] Y. Tsujimoto, et.al. 1989. Oncogene 4 : 1331-1336

[9] S. Takayama, et.al. 1995. Cell 80 : 279-284

[10] H. Karasuyama, et.al. 1990. J. Exp. Med. 172 : 969-972

IN VITRO IMMUNIZATION OF HUMAN PERIPHERAL BLOOD LYMPHOCYTES WITH 14-16 kDa RICE ALLERGEN PROTEINS

A. ICHIKAWA, Y. KATAKURA, K. TERUYA, S. HASHIZUME*
T. MATSUDA** AND S. SHIRAHATA
Graduate School of Genetic Resources Technology, Kyushu University, 6-10-1 Hakozaki, Higashi-ku, Fukuoka 812-0053, Japan
**Morinaga Institute of Biological Science, 2-2-1 Shimosueyoshi, Tsurumi-ku, Yokohama 230-0012, Japan*
***Department of Applied Biological Sciences, School of Agricultural Sciences, Nagoya University, Chikusa-ku, Nagoya 464-0814, Japan*

ABSTRACT Major rice allergen (RA) proteins which cause allergic diseases were isolated and characterized using IgE antibodies from patients allergic to rice recently. On the other hand, we had developed an *in vitro* immunization method to generate human monoclonal antibodies against soluble antigens. In this study, we tried to generate human monoclonal antibodies reactive to the 14-16 kDa RA proteins using *in vitro* immunization method. Human peripheral blood lymphocytes were isolated from a healthy donor and treated with L-leucyl-L-leucine methyl ester to remove suppressive cells such as killer and suppresser T cells and natural killer cells. The lymphocytes were cultured with muramyl dipeptide (10 mg/ml), IL-2 (10 units/ml), IL-4 (10 ng/ml) and the 14-16 kDa RA proteins mixture in 10% fetal bovine serum / ERDF medium for 7 days. Immunization were checked by measuring the concentrations and reactivities of antibodies in the culture supernatants. Lymphocytes immunized with the 14-16 kDa RA proteins were infected with Epstein-Barr virus to immortalize the RA specific B lymphocytes. We could obtain several EBV-transformed B cell lines secreting RA specific IgM antibodies.

1. Introduction

Human monoclonal antibodies are useful for diagnosis and treatment of various diseases. However, in order to obtain the human monoclonal antibodies, passive immunization, such as injection of dangerous antigens in human bodies, should not be permitted for ethical and moral reasons. One of the methods to solve this problem is the *in vitro* immunization method which is a primary activation of antigen-specific B lymphocytes. *In vitro* immunization techniques have great potential in the production of human monoclonal antibodies against various antigens. We have developed an *in vitro* immunization method against cultured lung adenocarcinoma cell line [1] and succeeded in generating several cancer specific human monoclonal antibodies [2), 3)]. However, this method could not be available for immunization against soluble antigens. In our previous

Y. Kitagawa et al. (eds.), Animal Cell Technology: Basic & Applied Aspects, Vol. 10, 191–197.

study, we established new *in vitro* immunization method for soluble antigens [4].

Allergic disease is one of the most serious diseases to be overcome for people today. It is important to establish the effective treatment of allergic diseases as well as to probe the mechanisms of its outbreak and progress. The 14-16 kDa rice allergen (RA) proteins which cause allergic diseases were isolated and characterized using IgE antibodies from patients allergic to rice [5), 6]. The purpose of this work is to generate human monoclonal antibodies against the 14-16 kDa RA proteins with new *in vitro* immunization method.

2. Materials and Methods

2.1. Antigens and reagent

The 14-16 kDa rice allergen proteins were prepared by the method reported previously [5), 6]. Recombinant human IL-2, was purchased from Genzyme corporation (Cambridge, MA, USA). Recombinant human IL-4 was purchased from Pepro Tech EC LTD. (London, England). Muramyl dipeptide (MDP) was purchased from Chemicon international Inc. (Temecula, CA, USA). L-Leucyl-L-leucin methyl ester was obtained from Boehringer GmbH (Mannheim, Germany).

2.2. Isolation of human peripheral blood lymphocytes

Human peripheral blood lymphocytes (PBL) were separated by density-gradient centrifugation from several healthy donors. In brief, 25 ml of peripheral blood was layered on 20 ml of lymphocyte separation medium (LSM; Organon Teknika, Durham, NC, USA) and was centrifuged at 400 x g for 30 min. The stratum of lymphocytes were harvested and washed three times with ERDF medium (Kyokuto Pharmaceutical Industrial Co., Japan). Lymphocytes from peripheral blood were treated with 0.25 mM Leu-Leu-OMe to remove the cytotoxic T cells, $CD8^+$ suppressor T cells, and natural killer cells before use.

2.3. *In vitro* immunization

In vitro immunization of the human PBL with the 14-16 kDa RA proteins was performed in a 24 well culture plate (Becton Dickinson). The Leu-Leu-OMe treated PBL were cultured for 7 days in ERDF medium containing 10% heat inactivated fetal bovine serum, MDP (10 μg/ml), IL-2 (10 units/ml), IL-4 (10 ng/ml), 2-mercaptoethanol (20 μM) and the 14-16 kDa RA proteins mixture (10 μg/ml).

2.4. Enzyme-linked immunosorbent assay

Ninety-six well microtiter plates were coated with the 14-16 kDa RA proteins or BSA in 0.1 M Na-carbonate buffer (pH 9.6) at concentration of 0.2 μg/well and incubated at 4 °C overnight. The plate were blocked with 1% BSA in PBS for 1 h at 37°C. Then supernatants of the *in vitro* immunized lymphocytes or the EBV transformed B cells were diluted 1:2 in 1 % BSA in PBS, added in 50 μl aliquots per well, and incubated at 4 °C overnight. The wells washed three times with PBS containing 0.05 % Tween 20 (PBS-

T) and incubated for 2 hours at 37 °C with 100 μl/well horseradish peroxidase-conjugated goat antibodies against human IgM. After washing three times with PBS-T, 100 μl/well of substrate solution [0.1 M citrate buffer (pH 4.0) containing 0.003 % H_2O_2 and 0.3 mg/ml p-2.2'-azino-di (3-ethylbenzothiazoline-6-sulfonic acid) diammonium salt] were added, and after 20 minutes, the absorbance at 405 nm was measured by using an ELISA reader.

2.5. EBV transformation and screening

After *in vitro* immunization, human B cells were Epstein-Barr virus (EBV) infected using supernatants from the B95-8 marmoset cell line. EBV-transformed cells were washed with ERDF medium containing 20 % FBS and seeded into 96 well culture plates at 1×10^5 cells/well. EBV-B cells producing antibodies to the 14-16 kDa RA proteins were screened by ELISA.

2.6. Immunoblotting

The 14-16 kDa RA proteins were electrophoresed in 15 % polyacrylamide gels using the Laemmli's buffer system. After electrophoresis, fractionated proteins in the gels were electrotransferred onto nitrocellulose membrane. The membrane was quenched with Block Ace (blocking reagent, Dainippon Pharmaceuticals, Japan) and then incubated with the EBV-B culture supernatants for 1 h at 37 °C. After washing with PBS-T, the membrane was incubated with horseradish peroxidase-conjugated goat antibody against human IgM diluted 1:500 in Block Ace for 1 h at 37 °C. After further washes, the membrane was incubated with a substrate solution containing 0.5 mg/ml 4-chloro-1-naphthol, 16 % methanol and 0.005 % H_2O_2.

3. Results and Discussion

3.1. *In vitro* immunization with the 14-16 kDa rice allergen (RA) proteins

We established the new protocol of *in vitro* immunization for soluble antigens [4] as shown in Fig. 1. Major difference of the new protocol for soluble antigens from that for cultured cancer cell antigens [1] was the use of IL-4 instead of IL-6 in the culture medium for *in vitro* immunization of lymphocytes. To immortalize the immunized B lymphocytes, we adopted Epstein-Barr virus (EBV) transformation instead of the somatic cell hybridization method because of low fusion efficiency of human fusion partner cells.

To illustrate the effect of the new protocol on the immunization with allergenic proteins, we actually performed *in vitro* immunization of human PBL with the 14-16 kDa RA proteins as an example of soluble allergenic protein antigens. We examined the effects of another protocol using the combination of MDP and IL-2, as well as MDP, IL-2 and IL-4. In the both protocols, IgM and IgG productions were approximately three times as much in the RA protein added culture as in the RA omitted culture (Fig. 2). This result indicates that the 14-16 kDa RA proteins have a high antigenicity. This high antigenicity may be associated with their allergenicity observed *in vivo*. Antibody reactive to the 14-16 kDa RA proteins was only detected in IgM antibodies (Fig. 3).

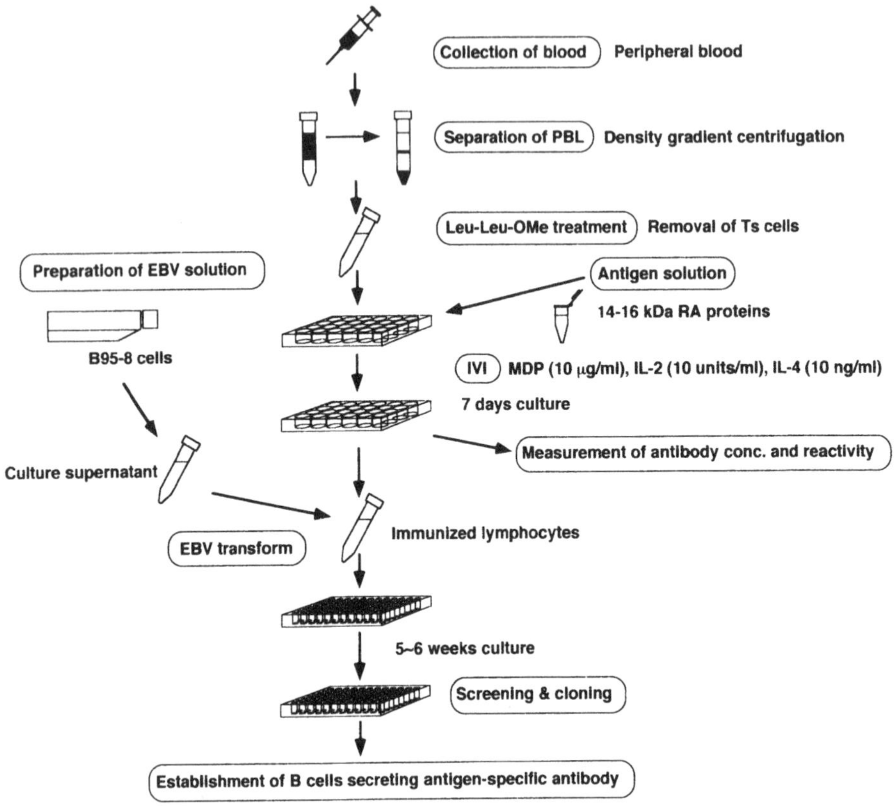

Fig. 1. Protocol of *in vitro* immunization for soluble antigens

Although IgM production was about twice times higher in the protocol using MDP and IL-2 than using MDP, IL-2 and IL-4, the highest reactivity to the 14-16 kDa RA proteins was observed in the protocol using MDP, IL-2 and IL-4. This result indicates that the new *in vitro* immunization protocol described previously is more effective to induce antigen specific human antibodies.

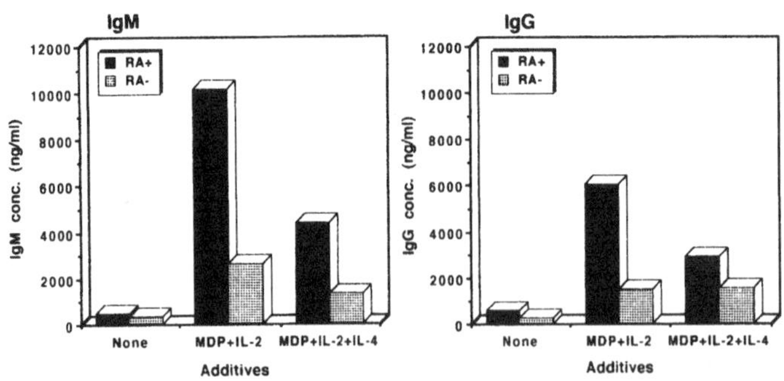

Fig. 2. Antibody productions of human peripheral blood lymphocytes (PBL) immunized with 14-16 kDa rice allergen (RA) proteins *in vitro*.

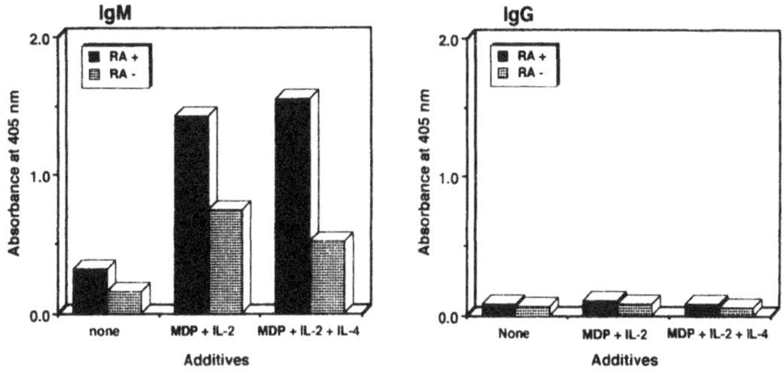

Fig. 3. Induction of specific antibodies by *in vitro* immunization against 14-16 kDa rice allergen (RA) proteins

Although a small amount of IgE production was detected in the culture of lymphocytes immunized with the protocol using MDP, IL-2 and IL-4, RA protein specific IgE was not detected (data not shown). This suggests that the PBL were derived from a healthy donor. Since one of our main purposes is a generation of allergen specific human monoclonal IgE antibodies, we have to lead to class switching from IgM to IgE. We are now searching for efficient *in vitro* class switching method.

3.2. Acquisition of EBV-B cell lines secreting rice allergen -specific antibodies

To immortalize the B lymphocytes after *in vitro* immunization with the 14-16 kDa RA proteins, we adopted the Epstein-Barr virus (EBV) transformation. When we screened EBV-B lymphocytes supernatants in ELISA, several EBV-B lymphocytes secreted antibodies reacting strongly with the 14-16 kDa RA proteins (Table 1).

Table 1. Production of rice allergen specific EBV transformed B cell lines from the lymphocytes immunized *in vitro*

Additives	No. of seeded wells [1]	No. of transformed wells [2]	No. of positive wells [3]	No. of established cell lines [4]	Class of positive antibodies [5]
None	384	291	0	0	—
MDP + IL-2	384	384	8	2	IgM
MDP + IL-2 + IL-4	384	384	22	4	IgM

[1] Number of wells seeded the cells after EBV infection.
[2] Number of wells containing EBV transformed B lymphocytes.
[3] Number of wells containing EBV transformed B lymphocytes secreting rice allergen specific antibodies.
[4] Number of established EBV transformed B cell lines secreting rice allergen specific antibodies.
[5] Class of antibodies reactive to the 14-16 kDa rice allergen proteins.

It was shown to be specific reactivity to the 14-16 kDa RA proteins in the six B cell lines, RA11C6, RA10B10, RA10C3, RA9G11, RA7C12 and RA7H3. These B cell lines were expanded and established as the 14-16 kDa RA specific B cell lines. The culture supernatants from these cell lines were subjected to immunoblotting analysis. Although all cell lines reactive to the 14-16 kDa RA proteins in ELISA, RA10B10 was only able to detect the 14-16 kDa RA proteins by immunoblotting under the condition with a reducing reagent (Fig. 4). These results suggested that every six cell lines recognizes the conformation of the 14-16 kDa RA proteins and only RA10B10 also recognizes the epitope which is not affected by reduction of intramolecular disulfide bonds. Matsuda *et al.* reported that the 14-16 kDa RA proteins contained 7 mol of cystine residue per mole of protein and no cysteine residues, and suggested that the cysteine residues would probably form intramolecular disulfide bonds [5]. They also reported that its immunoreactivity was quite stable to heating at 100 °C [5]. These reports suggest that the stability of the 14-16 kDa RA proteins was associated with its allergenic activity. Therefore various monoclonal antibodies recognizing nature and / or denatured allergenic proteins will be useful for development of hypoallergenic foods.

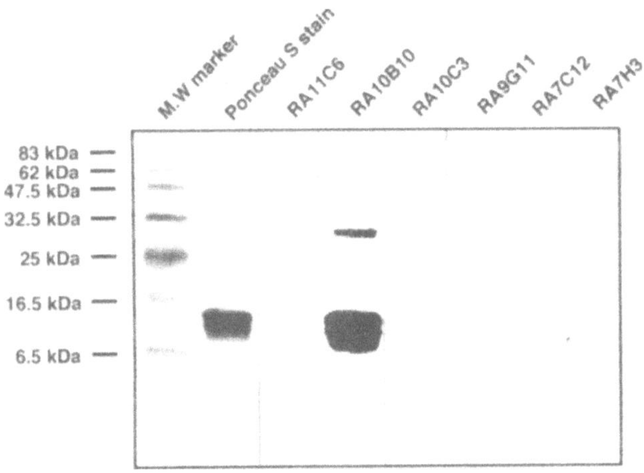

Fig. 4. Immunoblotting analysis of 14-16 kDa RA proteins using EBV-transformed B cell supernatants.

Because EBV transformed B cell is hard to clone by a conventional limiting dilution method, we are now generating hybridomas and cloning the 14-16 kDa RA protein specific human monoclonal antibody producing clones.

4. Acknowledgment

The authors are grateful to Dr. H. Shinmoto, National Food Research Institute, for a generous gift of the B95-8 marmoset cell line. This work is supported in part by a Grant-in-Aid for Scientific Research from the Ministry of Education, Science and Culture of Japan. A. I. is a research fellow of the Japan Society for the Promotion of Science.

5. References

1. Kawahara, H., Shirahata, S., Tachibana, H. and Murakami, H. (1992) *In vitro* immunization of human lymphocytes with human lung cancer cell line A549. *Hum. Antibod. Hybridomas* **3**, 8-13.
2. Shoji, M., Kawamoto, S., Sato, S., Kamei, M., Kato, M., Hashizume, S., Seki, K., Yasumoto, K., Nagashima, A., Nakanishi, H., Suzuki, T., Imai, T., Nomoto, K. and Murakami, H. (1994) Specific reactivity of human monoclonal antibody AE6F4 against cancer cells in tissues and sputa from lung cancer patients. *Hum. Antibod. Hybridomas* **5**, 116-122.
3. Shoji, M., Kawamoto, S., Setoguchi, Y., Mochizuki, K., Honjoh, T., Kato, M., Hashizume, S., Hanagiri, T., Yoshimatsu, T., Nakanishi, K., Yasumoto, K., Nagashima, A., Nakahashi, H., Suzuki, T., Imai, T., Nomoto, K. and Murakami, H. (1994) The 14-3-3 protein as the antigen for lung cancer-associated human monoclonal antibody AE6F4. *Hum. Antibod. Hybridomas* **5**, 123-130.
4. Ichikawa, A., Katakura, Y., Kawahara, T., Hashizume, S. and Shirahata, S. (1998) *In vitro* immunization of human peripheral blood lymphocytes with cholera toxin B subunit. *Animal Cell Technology: New developments - New applications*, Kluwer Academic Publishers, Dordrecht. *in press.*
5. Matsuda, T., Sugiyama, M., Nakamura, R. and Torii, S. (1988) Purification and properties of an allergenic protein in rice grain. *Agric. Biol. Chem.* **52** (6), 1465-1470.
6. Matsuda, T., Nomura, R., Sugiyama, M. and Nakamura, R. (1991) Immunochemical studies on rice allergenic proteins. *Agric. Biol. Chem.* **55** (2), 509-513.

OPTIONS TO ELIMINATE ANIMAL-DERIVED COMPONENTS OF CELL CULTURE MEDIA.

DAVID W. JAYME AND SHAWN R. SMITH
Life Technologies, Inc.
3175 Staley Road, Grand Island, NY 14072 USA

1. Introduction

For the past decade, nutrient medium development emphasized reduction or elimination of animal sera, largely for technical or economic reasons [1]. Supplementation of enriched basal nutrient formulations with insulin, transferrin, albumin and other protein materials, generally obtained from animal sources, resulted in various "serum-free" (but not necessarily "protein-free") culture media [2].

Recently, global concerns regarding adventitious contaminants (e.g., mycoplasma, virus, prion) and the evolution of therapeutic applications for cell culture-derived products have accelerated efforts to eliminate all constituents of animal origin from certain cell culture systems [3]. Depending upon the application, native protein components may still be acceptable, given demonstrated integrity of the collection and purification processes, origin of source material (e.g., geographic location and endemicity, animal species, and tissue/organ of interest), and validation of the treatment or inactivation method against an adventitious agent challenge panel.

Recombinant proteins may be first-line substitutes for animal-sourced materials, but various options also exist for re-engineering of traditional serum-free formulations (containing insulin, transferrin and other protein factors) with non-protein substitutes. Caution must also be exercised with sourcing of non-protein additives, particularly amino acids and lipids, to avoid introduction of adventitious contaminants.

2. Methods

Our initial effort was to identify those components in our catalog serum-free formulations which were derived from animal materials [4]. As noted in Table 1, some protein additives were obvious candidates for substitution. However, we were surprised to discover that several amino acids and lipids often used in serum-free formulations were derived from animal sources.

For example, cyst(e)ine has been historically produced from human hair, tyrosine is most readily available derived from either human hair or poultry feathers, and hydroxyproline is traditionally obtained from porcine skin. Some amino acids are obtained by controlled

Y. Kitagawa et al. (eds.), Animal Cell Technology: Basic & Applied Aspects, Vol. 10, 199–203.

acid hydrolysis of bovine and porcine bone gelatins. The cholesterol used to supplement culture media for sterol-requiring cell types is commonly derived from sheep "wool grease" (lanolin). Lipids may be derived from fish oils or various plant sources.

Table 1. Media Components Commonly Derived from Animal Sources.

Medium Component	Common Animal Source	Non-Animal Source
Insulin	Bovine pancreas	Bovine or human recombinant
Transferrin	Bovine, porcine or human plasma fraction	Inorganic iron carriers/chelates
Serum protein fractions (e.g., albumin, fetuin, etc.)	Bovine or other animal serum	Lipid delivery alternatives
Protein hydrolysates	Lactalbumin, Peptones, Casein	Plant-derived hydrolysates
Lipids/sterols	Ovine cholesterol, marine lipids, porcine liver	Plant-derived sterols
Growth/attachment factors	Murine organ digests	Recombinant factors
Amino acids (e.g., tyrosine, cyst(e)ine, hydroxyproline)	Human hair, avian feathers, bovine collagen, bovine/porcine bone gelatin	Recombinant amino acids

Having targeted these animal-derived constituents, our next task was to identify and qualify alternative suppliers of these raw materials which used recombinant production techniques or purified the target molecules from non-animal starting materials. We formulated custom serum-free media which deleted the target component. Complete media were reconstituted by aseptic addition of equimolar amounts of either the original animal-derived component (control) or a non-animal sourced substitute (test).

Parallel cultures were maintained for several passages to ensure adequate adaptation to novel media and to minimize residual contribution from the former formulation prior to quantitative assay of biological performance. Culture adaptation and preliminary screening were performed in 125 mL shake flasks. Nutrient optimization and culture performance equivalence were demonstrated in pilot-scale bioreactors prior to commercial release. Relative performance was monitored by conventional assays using relevant cell types for each formulation and examining both cellular proliferation and production of the desired biological product.

3. Results

3.1. Transferrin Substitution

For certain *in vitro* diagnostic applications not requiring elimination of animal-derived components, replacement of human origin transferrin by transferrin derived from plasma of other animal species, particularly sourced from biosecure geographies, may be suitable. Studies in our laboratory and elsewhere [5] note that bovine-derived transferrin may be an inferior substitute to porcine-derived transferrin, while porcine transferrin appeared to yield comparable performance data to human transferrin at equivalent supplementation levels.

Multiple laboratories [6-8] have examined various inorganic iron salts for direct delivery of iron to cells, rather than bound to transferrin. Diminished cytotoxicity was reported [8] when the inorganic iron salts were chelated with citrate or EDTA.

We have demonstrated that transferrin may be eliminated from protein-free, chemically-defined formulations capable of supporting hybridoma, recombinant Chinese hamster ovary (CHO) cell and virus production applications [9-11]. However, we have observed, unfortunately, that a single iron carrier substitute is not always superior and that the most effective compound must be titered to optimize performance for the specific cell system. For this purpose, we prepare a transferrin-free formulation and supplement with a panel of iron carrier concentrates, supplemented at a target concentration of approximately 50 uM. Following selection of the preferred iron compound, a titration experiment over a range of 20 - 100 uM will yield the optimized transferrin-free formulation.

3.2. Insulin Substitution

The most straightforward replacement for animal-derived insulin is recombinant insulin, which is commercially available and can substitute for most cell culture applications. We have demonstrated biological equivalence in multiple mammalian cell culture systems.

Recent development of insulin-free formulations for recombinant CHO cells (e.g., CD CHO and CHO III) and for virus production applications (e.g., VP-SFM) confirm that absolute insulin requirement may be eliminated by appropriate adjustment of non-protein medium constituents [11]. While insulin addition typically has a beneficial culture effect, we observed a surprising inhibition of transient expression of beta galactosidase activity in transfected CHO cells cultured in insulin-supplemented medium, compared with cells grown in identical protein-free medium under insulin-free conditions.

3.3. Albumin Replacement

Albumin, as a primary protein component of animal sera, performs various functions within the cell culture environment, including binding, inactivation or sequestration of toxic materials (e.g., organic compounds, trace metals, cell-secreted products) and protection against proteolysis, shear damage and other biophysical phenomena [1]. Its most widely-reported cell culture function, however, is its ability to bind lipids and deliver them to the cell. Effective, stable preparation of lipid micellar suspensions may be accomplished through appropriate addition of amphoteric molecules (detergents) to lipid suspensions and mechanical emulsification processes. Synthetic and plant-derived materials have also successfully replaced albumin for specific applications.

3.4. Fetuin Replacement

Fetuin is an operational definition for a crude purified alpha globulin, glycoprotein fraction of fetal bovine serum. Fetuin has been frequently used as a serum-free additive to support CHO cell applications. However, its exorbitant cost, limited availability and animal origin have encouraged efforts to identify more suitable substitutes for large-scale biotechnology applications. Recently, protein-free formulations for CHO cell culture have been commercialized which have eliminated all constituents of animal origin. Some of these media contain plant-derived hydrolysates, while other formulations are completely biochemically-defined.

4. Discussion and Application

Animal sera (particularly fetal bovine serum) have been widely used as culture additives because they support proliferation of a broad range of cell types and require minimal adaptation. Motivations to reduce or eliminate animal sera are also diverse. Historically, these concerns have focused upon either business (e.g., cost, availability) or technical (e.g., lot-to-lot variation; yield reduction due to requisite downstream purification steps, enzymatic degradation or neutralizing antibodies; culture differentiation, senescence or apoptosis). More recently, however, regulatory concerns regarding antigenicity of residual animal proteins and contamination by adventitious agents (e.g., mycoplasma, virus, prion) have emerged as the primary motivator for eliminating serum. Product recalls relating to donor-associated Creutzfeld-Jakob disease and general anxiety over adventitious contaminants in human plasma have escalated industry concerns regarding introduction of any materials of animal origin into the cell culture environment.

Re-engineering of existing formulations and *de novo* development have resulted in commercialization of chemically-defined media which contain no constituents of animal origin. These formulations were designed for critical biotechnology applications (e.g., expression of recombinant therapeutic proteins, vaccine production, therapeutic and diagnostic monoclonal antibody production, cell/gene therapy, and tissue engineering).

Table 2. Impact of Component Substitution on Cell Culture Parameters.

Cell Culture Parameter	Reduced Serum Medium	Serum-Free Medium	Protein-Free Medium	Chemically-Defined Medium
Potential for adventitious agents	+++	++	+	--
Ease of downstream product purification	+	++	+++	+++
Improved biological performance	+/-	++	++	++
Ease of cell culture adaptation	+++	+	+/-	+/-

Substitution of non-animal components impacts a variety of cell culture parameters (Table 2). Elimination of animal-derived protein elements reduces the potential for adventitious agent contamination and may facilitate downstream product purification. Elimination of serum-associated inhibitory factors may also result in improved biological performance. Cellular adaptation from serum-supplemented medium or from protein-supplemented serum-free formulations to protein-free, biochemically-defined formulations may require patience and careful weaning over multiple subcultures and proportionate mixtures of fresh:conditioned media [2]. Working cell banks should be evaluated post-weaning to ensure persistent expression and to guard against inadvertent selection of undesirable culture subpopulations.

5. Conclusion

This paper has focused upon potential replacements for animal origin medium components which exhibit comparable biological performance in cell culture systems but eliminate concerns regarding introduction of adventitious contaminants. Nutrient media devoid of components derived from animals are commercially available. Similar strategies may be exploited to modify customized formulations to generate protein-free and even biochemically-defined second generation media to support cellular proliferation and biological production.

Practical considerations for development and quality assurance of nutrient media demand comparable emphasis on process controls for media manufacture. Careful maintenance of facilities and equipment, environmental controls and monitoring, design and control of manufacturing processes, and validated equipment sanitization are required to ensure final product integrity.

6. References

1. Jayme DW and Blackman KE (1985) *Culture media for propagation of mammalian cells, viruses and other biologicals,* In: Advances in Biotechnological Processes, 5: 1-30 (Liss) New York.
2. Ham RG and McKeehan WG (1979) *Media and growth requirements,* In: Methods in Enzymology 58: 44-93.
3. Bader F, Davis G, Dinowitz M, Garfinkle B, Harvey J, Kozak R, Lubiniecki A, Rubino M, Schubert D, Wiebe M and Woollett G (1998) *Assessment of risk of bovine spongiform encephalopathy in pharmaceutical products, part I,* BioPharm 11: 20-32.
4. GIBCOBRL Products & Reference Guide (1997/1998).
5. Young SP and Garner C (1990) *Delivery of iron to human cells by bovine transferrin,* Biochem J., 265: 587-591.
6. Keenan J and Clynes M (1996) *Replacement of transferrin by simple iron compounds for MDCK cells grown and subcultured in serum-free medium,* In Vitro Cell. Dev. Biol. - Animal, 32: 451-453.
7. Kovar J and Franek F (1987) *Iron compounds at high concentrations enable hybridoma growth in a protein-free medium,* Biotechnol. Lett. 9: 259-264.
8. Sauer P, Burky J, Wesson M and Sternard H (1997) *Investigating alternatives to transferrin,* In: Process development and production issues in monoclonal antibody manufacturing, Waterside Monoclonal Conference, April 1997, Norfolk, VA.
9. Fike RM, Pfohl JL, Epstein DA, Jayme DW and Weiss SA (1991) *Hybridoma growth and monoclonal antibody production in protein-free hybridoma medium.* BioPharmocology 4: 26-29.
10. Gorfien SF, Dzimian JL, Tilkins ML, Godwin GP and Fike RM (in press) *Recombinant protein production by CHO cells cultured in a chemically defined medium,* In: Animal Cell Technology: Basic & Applied Aspects.
11. Price PJ and Evege EK (1997) *Serum-free medium without animal components for virus production,* Focus, 19: 67-69.

CHARACTERIZATION OF THE SUBPOPULATIONS OF RAT ASCITES HEPATOMA AH109A CELLS WITH DIFFERENT INVASIVE AND METASTATIC ACTIVITIES

Y. MIURA, M. MIYAUCHI, K. KUBOTA, Q. LI,
H. TACHIKAWA, D. FUJIMOTO AND K. YAGASAKI
Department of Applied Biological Science,
Tokyo Noko University, Fuchu, Tokyo 183, Japan

1. Introduction

It is well known that many tumor cells metastasize actively and that resulting metastatic nodules become fatal to the host in many cases. This makes it important to identify the precise mechanisms of tumor metastasis and to prevent them. We have been using the rat ascites hepatoma cell line of AH109A as a model to study the molecular mechanisms of tumor metastasis and to screen the effective compounds that have the ability to inhibit them. We reported earlier that the green tea extract and coffee have anti-proliferative and anti-invasive activities to AH109A cells, using the *in vitro* and *ex vivo* assay systems we developped (Miura *et al.*, 1997a and Miura *et al.*, 1997b). In the course of these studies, we found that there are two populations of AH109A cells which have different invasive and metastatic activities. In the present study, we report the separation and characterization of these two populations and the molecular cloning of the genes which express at the different levels between them.

2. Separation of two populations in AH109A cells

Rat ascites hepatoma cell line of AH109A was supplied by SRL (Tokyo, Japan). AH109A cells were maintained in the peritoneal cavity of a male Donryu rat (NRC Haruna, Gunma, Japan) and cultured as described in Miura *et al.* (1997a). AH109A cells are usually cultured in the presence of 10% calf serum (CS) like many other animal cells. They grow well in the floating state in the presence of 10% calf serum. However, drastic morphological changes occur when they are cultured in the presence of 10% rat serum. That is, some AH109A cells adhere to the culture dish, actively extending the pseudopodia while others remain in the floating state.

Y. Kitagawa et al. (eds.), Animal Cell Technology: Basic & Applied Aspects, Vol. 10, 205–209.

Interestingly, to the extent we have examined, only rat and mouse sera have the ability to cause these morphological changes to AH109A cells. At present, we do not know why only rodent sera have such an ability. Nevertheless, this result clearly shows that there are two populations in AH109A cells which respond differently to rat serum.

These two populations can be easily separated, because in the presence of rat serum one population adheres to the culture dish and the other does not. After repeating the separation procedure three time, we designate the population that continues to adhere to the culture dish as adhesive AH109A (AH109A-Ad) and that which remains in the floating state as floating AH109A (AH109A-Fl) (Figure 1).

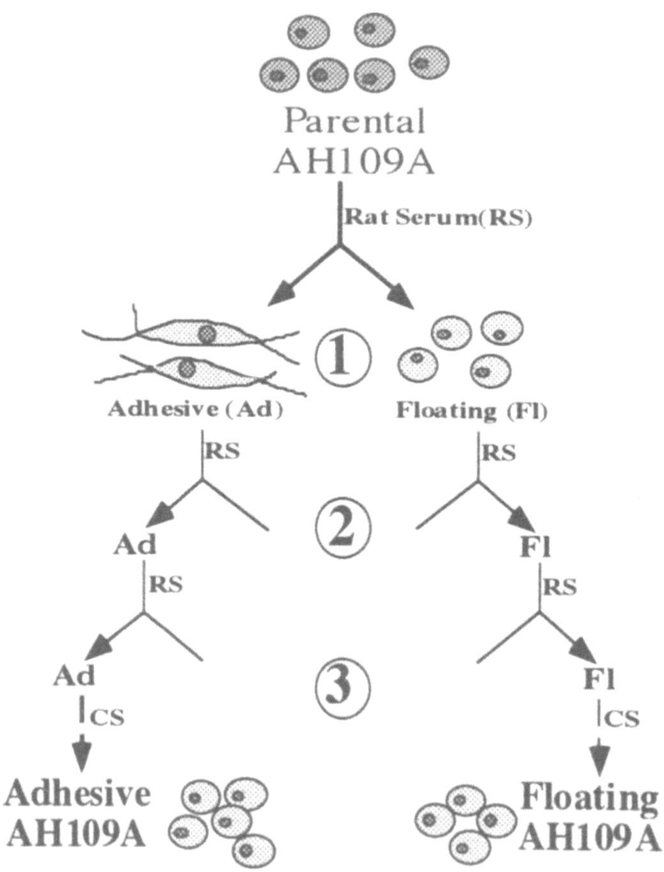

Figure 1. Separation procedure of AH109A-Ad and AH109A-Fl

3. Proliferative, invasive and metastatic activities of AH109A-Ad and AH109A-Fl

The *in vitro* proliferative and invasive activities of these two populations of AH109A cells were compared. The *in vitro* proliferative activities were assessed by the MTT method (Mosmann, 1983) and AH109A-Ad showed about two times higher proliferative activity than AH109A-Fl. The *in vitro* invasive activities were assessed by the *in vitro* invasion assay we have reported (Miura et al., 1997a). The invasive activity of AH109A-Ad was also two or three times higher than that of AH109A-Fl (Table 1). These results shows that AH109A-Ad proliferate more rapidly and invade more actively than AH109A-Fl when they are cultured *in vitro*.

Table 1. The *in vitro* proliferative and invasive activities of AH109A-Ad and AH109A-Fl

Proliferative Activity (%)	
AH109A-Ad	119±1.4
AH109A-Fl	69±1.0
Invasive Activity (No. of invaded cells & colonies)	
AH109A-Ad	1095±95
AH109A-Fl	478±87

The *in vivo* metastatic activities of AH109A-Ad and AH109A-Fl were next examined. These two populations were injected via the tail vein and rats were dissected after three weeks. Surprisingly, AH109A-Ad metastasized specifically to lung and AH109-Fl specifically to mesentery. This result suggests that these two populations metastasize with different tissue-specificities.

4. Molecular cloning of genes that express at different levels in AH109A-Ad and AH109A-Fl and analyses of their functions

We used the differential display method to clone the genes that express at the different levels between these two populations of AH109A to further characterize them (Liang and Pardee, 1992). Total RNA was prepared from AH109A-Ad and AH109A-Fl by AGPC method (Chomczynski and Sacchi, 1987) and reverse-transcribed by Superscript^{TM} reverse transcriptase (GIBCO BRL, MD, USA) using T_{12}MN primers (GenHunter, MA, USA). PCR was performed using T_{12}MN and AP primers (AP-1~AP-10) (GenHunter, MA, USA).

Thirty cDNAs were obtained by this method and 2 of them were positive clones; other clones were revealed to be pseudopositive, when we measured the exact amount of their mRNA by slot blot analysis.

One of the positive clones, #C5-1, which expressed at the higher level in AH109A-Ad than in AH109A-Fl, is a short fragment of 90bp. Its sequence completely coincides with the sequence of rat stathmin cDNA. Stathmin is reported to be expressed ubiquitously and to be involved in the intracellular signal transduction (Doye *et al.*, 1989, Maucuer *et al.* 1995). To examine whether stathmin is one of the genes that determine the differences between AH109A-Ad and AH109A-Fl and whether it is invloved in cell adhesion and pseudopodia extension induced by rat serum, the effect of antisense oligo deoxynucleotide (ODN) of stathmin was assessed. Antisense, sense and nonsense ODNs were all S-oligo synthesized by Pharmacia Biotech. (Uppsala, Sweden) and transfected by cationic liposome (Tfx-20™, Promega, WI, USA). The sequences of these ODNs were followed as described by Di Paolo *et al.* (1996). After transfection, cells were cultured for 20 hr in the presence of 10% CS. Transfected cells were collected and cultured for a further 24 hr in the presence of 10% rat serum. AH109A cells that were transfected by liposome only, sense ODN and nonsense ODN adhered to culture dishes, actively extending pseudopodia in the presence of 10% rat serum. On the contrary, AH109A cells transfected by antisense ODN adhered to the culture dish, but their extension of pseudopodia was significantly inhibited. This result shows that stathmin is involved in cell extension induced by rat serum.

Clone #T3-3, the other positive clone we obtained, which expressed at the higher level in AH109A-Fl than in AH109A-Ad, was sequenced and shown to be a novel gene, but it has significantly high homology with *YHR74*, the hypothetical ORF of *Saccharomyces cerevisiae*. The nucleotide sequence homology between these two genes is 58% and their putative amino acid sequences are highly homologous. This high homology between them suggests that clone #T3-3 may be a rat homologue of yeast *YHR74*. We have no data as yet about the function of clone T3-3 because *YHR74* is a hypothetical ORF obtained by the yeast genome project and its expression is not confirmed nor is its function known. We are now constructing the disruptant strain of *YHR74* and analyzing its phenotype.

Clone #T3-3 gene is expressed in brain and thymus as 1.4kb mRNA. In AH109A-Fl, the expression is induced by rat serum. We are now planning to screen rat thymus or brain cDNA library to obtain its full length cDNA.

5. Summary

We have found that there are two populations in rat ascites hepatoma cell line of AH109A which have different invasive and metastatic activities. We have also obtained two candidate genes that may determine the differences between these two populations. We plan to identify the precise functions of these two genes and continue to screen other candidates. We will also examine in more detail the biological characteristics of the two populations, such as the regulation of adhesion molecules and cytoskeletons. Such studies will help our understanding of the molecular mechanisms involved in tumor invasion and metastasis.

6. References

Chomczynski, P. and Sacchi, N. (1987) Single-step method of RNA isolation by acid guanidium thiocyanate-phenol-chloroform extraction. *Anal. Biochem.*, **162**, 156-159.

Di Paolo, G., Pellier, V., Catsicas, M., Antonsson, B., Catsicas, S. and Grenningloh, G. (1996) The phosphoprotein stathmin is essential for nerve growth factor-stimulated differentiation. *J. Cell Biol.*, **133**, 1383-1390.

Doye, V., Soubrier, F., Bauw, G., Boutterin, M. -C., Beretta, L., Koppel, J., Vandekerckhove, J. S. and Sobel, A. (1989) A single cDNA encodes two isoforms of stathmin, a developmentally regulated neuron-enriched phospho- protein. *J. Biol. Chem.*, **264**, 12134-12137.

Liang, P. and Pardee, A. B. (1992) Differential display of eukaryotic messenger RNA by means of the polymerase chain reaction. *Science*, **257**, 967-971.

Maucuer, A., Camonis, J. H. and Sobel, A. (1995) Stathmin interaction with a putative kinase and coiled-coil-forming protein domains. *Proc. Natl. Acad. Sci. USA*, **92**, 3100-3104.

Miura, Y., Shiomi, H., Sakai, F. and Yagasaki, K. (1997a) Assay systems for screening food components that have anti-proliferative and anti-invasive activity to rat ascites hepatoma cells: *In vitro* and *ex vivo* effects of green tea extract. *Cytotechnology*, **23**, 127-132.

Miura, Y., Furuse, T. and Yagasaki, K. (1997b) Inhibitory effect of serum from rats administered with coffee on the proliferation and invasion of rat ascites hepatoma cells. *Cytotechnology*, **25**, 219-223.

Mosmann, T. (1983) Rapid colorimetric assay for cellular growth and survival: Application to proliferation and cytotoxicity assays. *J. Immuno. Methods*, **65**, 55-63.

FUNCTIONAL EXPRESSION OF NMDA RECEPTOR CHANNELS IN DIFFERENTIATING NEURAL PRECURSOR CELLS

K. Yamada, T. Hisatsune, S. Uchino[#], T. Nakamura[*], Y. Kudo[*], and S. Kaminogawa
Dept. Appl. Biol. Chem., Univ. Tokyo, [#]Yokohama Res. Center, Mitsubishi Chem., [*]Sch. Life Sci., Tokyo Univ. Pharm. & Life Sci. 1-1-1 Yayoi, Bunkyo-ku, Tokyo 113, Japan

Abstract

We have established neural precursor cell lines isolated from various regions of mouse embryonic brain to study a mechanism of neural development. MSP-1 cells, derived from basal ganglia, has the potencial to differentiate into neurons and grial cells. In this study, we examined the expression of *N*-methyl-D-aspartate (NMDA) receptor channels in MSP-1 cells. RT-PCR analysis using the NMDA receptor channel subunit specific primers suggested that the ε2 subunit mRNA was expressed in the precursor stage and during differentiation, however, the ζ1 subunit mRNA was expressed in the later stages of differentiation. Furthermore, MSP-1 cells in the late stage of differentiation apparently showed the NMDA-induced intracellular Ca^{2+} elevation, suggesting that the differentiated MSP-1 cells express the functional NMDA receptor channels. These results indicate that MSP-1 cells will provide a useful system to investigate the molecular basis of neural differentiation.

1. Introduction

The mature nervous system is composed of a large number of terminally differentiated neuronal and glial cell types, which develop from precursor cells in the embryonic nervous system (Mckay et al., 1997). To reduce cellular and molecular complexity of the brain, we are studying particular developmental stages and steps in neural differentiation using cell lines in an in vitro culture system. One of the neural precursor cell lines, MSP-1 cells, established from p53-deficient mouse embryonic brain by introducing c-Myc estrogen receptor (ER) -fusion gene, can proliferate and differentiate into neuronal and glial cell types (Hisatsune et al., 1998).

Maturation of terminally differentiated neuron can be assessed by several criteria: production of neurotransmitters, expression of their receptors, outgrowth and branching of axons, and synapse formation. Recently it has been reported that neurotransmitters and their

Y. Kitagawa et al. (eds.), Animal Cell Technology: Basic & Applied Aspects, Vol. 10, 211–215.

receptors play an important role in the development of central nervous sysytem, neurogenesis and synapse formation as well as in adult neural functions. In this study, we focused on the NMDA receptor channel.

The NMDA receptor channel is highly permeable to Ca^{2+} and is essential for synaptic transmission and synaptic plasticity underlying memory, learning and development. Recent studies have elucidated the molecular diversity of the NMDA receptor channel (Mori and Mishina, 1995). Highly active NMDA receptor channels are formed in vitro through the co-expression of two members of glutamate receptor (GluR) channel subunit families, the GluRε (ε1-ε4 or NR2A-2D) and GluRζ (ζ1 or NR1) subunit families.The four ε subunits are distinct in distribution, properties, regulation and physiological roles, thus providing the molecular basis of the functional diversity of the NMDA receptor channel. Here we report the expression of the functional NMDA receptor channels in the differentiated MSP-1 cells.

2. Materials and Methods

2.1. Culture of MSP-1

MSP-1 cells were cultured at 37 °C in DMEM/F12 medium containing 10% FCS, 10 ng/ml bFGF and 1μM β-estradiol. For RNA preparation, MSP-1 cells were plated on an ornitine and fibronectine-coated dishes at a density of 10^5 cells/100-mm dishes. For measurement of intracelllular Ca^{2+} concentration ($[Ca^{2+}]_i$), MSP-1 cells were plated on an ornitine and fibronectine-coated glass coverslip with a silicone rubber wall at a density of 5 x 10^4 cells/well. MSP-1 cells were induced to differentiate by withdrawal of bFGF from the medium after an incubation for 4 days with DMEM/F-12 medium containing N2-supplement and 10 ng/ml bFGF.

2.2. RT-PCR

Total RNA was extracted from MSP-1 cells using ISOGEN reagent (Nippon Gene). RT-PCR proceeded as described (Okabe et al., 1996). The following NMDA receptor channel subunit specific primers were used; NMDA-R ζ1 forward primer ACCCTGTCCTCTGCCATGTGGTTTTC; reverse primer ACATTCTTGATACCGAA CCCATGTC; NMDA-R ε1 forward primer ACCCCAAGGACTGTAGTGAGGTTG; reverse primer ATGTCATAGAGGTTCCCCATCCGCAG; NMDA-R ε2 forward primer ATGACTGTGACAACCCACCCTTT; reverse primer ACTGACCGAATCTCGCTTG AAGT; NMDA-R ε3 forward primer ACGTGGCTCAAACACAGTCGTTG; reverse primer TGACCTCTTCTAGCACCCCACTG; NMDA-R ε4 forward primer AACCACTTCTTCTGCCAGGAGG; reverse primer TGAAGGAGTAGTCTCGGTTA

TC; β-actin forward primer TGGAATCCTGTGGCATCCATGAAAC; reverse primer TAAAACGCAGCTCAGTAACAGTCCG.

2.3. Measurement of $[Ca^{2+}]_i$

MSP-1 cells were exposed for 45 min to 7.5 μM fura-2 acetoxymethyl ester dissolved in a balanced salt solution consisting of 130 mM NaCl, 5.4 mM KCl, 2.0 mM $CaCl_2$, 5.5 mM glucose and 10 mM HEPES-5.0 mM NaOH-HCl (pH 7.3). NMDA-induced increase in $[Ca^{2+}]_i$ was measured using an image processor (Argus 50; Hamamatsu Photonics).

3. Results and Discussion

3.1. Expression of the NMDA receptor channel subunit mRNAs in MSP-1 cells

We extracted total RNA from MSP-1 cells at 0, 2, 4, 6, 9, 12 and 15 days after depletion of bFGF and investigated the expression of the NMDA receptor channel subunit mRNAs by means of RT-PCR analysis (Fig. 1). The band of the predicted size (494 bp) of the ε2 subunit mRNA was observed at all days. However, the band of the predicted size (591 bp) of the ζ1 subunit mRNA was detected at 12 and 15 days. The band of the ε1 and ε4 subunit mRNAs were slightly detected and that of the ε3 subunit mRNA was hardly (data not shown).

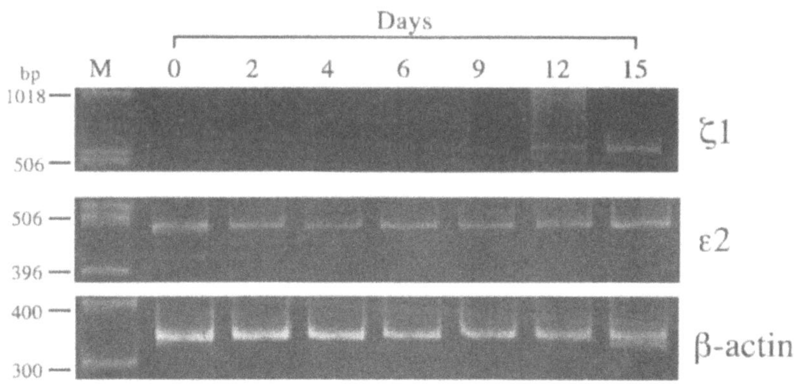

Figure 1. RT-PCR analysis of NMDA receptor channel subunit mRNAs from differentiating MSP-1 cells. Total RNA was extracted from MSP-1 cells at 0, 2, 4, 6, 9, 12 and 15 days after depletion of bFGF. 35 cycles of PCR reactions were performed for NMDA receptor subunits and 22 cycles for β-actin.

3.2. Functional expression of the NMDA receptor channel in MSP-1 cells

To examine whether or not MSP-1 cells express the functional NMDA receptor channel, we measured an NMDA-induced increase in $[Ca^{2+}]_i$ by means of fura-2 fluorometry at 0 and 12 days after depletion of bFGF. No cells exhibited the increase in $[Ca^{2+}]_i$ by the administration of 100 μM NMDA and 100 μM glycine at 0 day (data not shown). On the other hand, several cells at 12 day apparently showed the NMDA-induced increse in $[Ca^{2+}]_i$ which was maintained during the administration of agonists (Fig. 2A, 2B). These responses were completely inhibited by 100 μM MK-801 (a channel blocker of the NMDA receptor) (data not shown). Together with considerations of RT-PCR analysis mentioned above, these results indicate that the differentiated MSP-1 cells can express the ε, especially ε2 subunit mRNA, and ζ1 subunit mRNAs and produce the functional NMDA receptor channel, in accordance with the notion that highly active NMDA receptor channels are formed when ε and ζ subunit families are expressed together (Mori and Mishina, 1995).

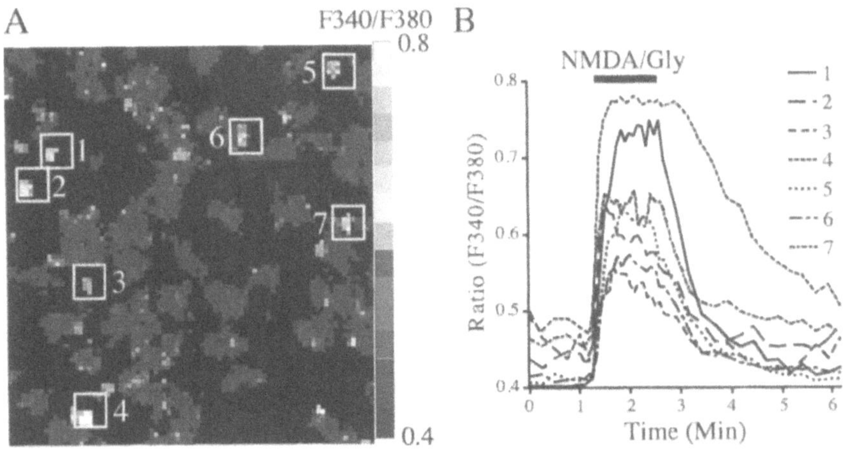

Figure 2. NMDA-induced intracellular Ca^{2+} increase in differentiating MSP-1 cells.
(A, B) MSP-1 cells were cultured in the FGF-depleted medium for 12 days. 100 μM NMDA and 100 μM glicine were applied for 60 sec at day 12 (A). Cells exhibitting $[Ca^{2+}]_i$ increase are marked with a square. (B) Time courses of the $[Ca^{2+}]_i$ increase in the marked cells in (A). NMDA and glicine were applied for 60 sec as indicated by a bar.

3.3. Concluding Remarks

The present findings demonstrate that the established neuronal cell line, MSP-1 cells, can differentiate into neuronal and glial cells and express the functional NMDA receptor channel. To gain further insight into the subtypes of the NMDA receptor channel expressed in MSP-1 cells, we are carrying out the pharmacological investigation using subunit specific antagonist, such as ifenprodil (Williams et al., 1993). Moreoveer, we attempt to examine the effects of neurotrophic factors, like as NGF, BDNF and NT-3, to regulate the neuronal survival, differentiation. Finally, we have established not only MSP-1 cells but also other neural precursor cell lines, MHP series isolated from hippocampus and MTP series isolated from forebrain. These cell lines will provide a simple and reproducible system to elucidate the mechanism of neural development and search for a novel factor promoting neural maturation. Furthermore, these precursor cells may be capable of implantation into the particular regions of diseased or damaged brain.

4. References

Hisatsune, T., Yamada, K., McKay, R.D.G., and Kaminogawa, S. (1998) Immortalization of mammalian central nervous system stem cells from p53 knock-out forebrain, in Nagai, K. and Wachi, M. (eds.), *Animal Cell Technology: Basic & Applied Aspects*, Kluwer Academic Publishers, Dordrecht, Vol. 9, pp. 17-22.

McKay, R.D.G. (1997) Stem cells in the central nervous system, *Science* **276**, 66-71,

Mori, H. and Mishina, M. (1995) Structure and function of the NMDA receptor channel, *Neuropharmacology* **34**, 1219-1237.

Okabe, S., Forsberg, N. K., Spiro, A. C., Segal, M., and McKay, R. D. (1996) Development of neuronal precursor cells and functional postmitotic neurons from embryonic stem cells in vitro, *Mech. Dev.* **59**, 89-102.

Williams, K. (1993) Ifenprodil discriminates subtypes of the N-methyl-D-aspartate receptor: selectivity and mechanisms at recombinant heteromeric receptors, *Mol. Pharmacol.* **44**, 851-859.

TGF-β REVERSIBLY INDUCES CELL SENESCENCE IN ADENOCARCINOMA CELL LINE

YOSHINORI KATAKURA, ERIKO NAKATA, TAKUMI MIURA AND SANETAKA SHIRAHATA
Graduate School of Genetic Resources Technology, Kyushu University
Hakozaki 6-10-1, Higashi-ku, Fukuoka 812-81, Japan

1. Introduction

It has been reported that TGF-β exhibits an anti-tumor activity via inhibiting its growth in several cancer cell lines derived from breast carcinoma, melanoma, mammary epithelium and so on. However, several cancer cell lines have not been affected for its growth by TGF-β. This insensitivity to TGF-β have been explained by the functional deletion of molecules, such as cycline dependent kinase inhibitor p15, protein phosphatase cdc25A, Smads and so on, which lie in the TGF-β signaling cascade. In this study, we investigated the effect of TGF-β on the growth of human lung adenocarcinoma cell line, A549 cells. In A549 cells, p15 gene is demonstrated to be homozygously deleted and cdc25A was functionally deleted.

2. Materials and Methods

2.1. CELL CULTURE

Human lung adenocarcinoma cell line (A549 cells) was cultured in ERDF medium (Kyokuto Pharmaceuticals, Tokyo, Japan) supplemented with 5% fetal bovine serum at 37°C in a 95%/5% CO_2 atmosphere. Recombinant human transforming growth factor β1 (TGF-β) (AUSTRAL Biologicals, San Ramon, CA) was added every other days to the culture at the final concentration of 10 ng/ml where indicated.

2.2 TELOMERASE ASSAY

We used the PCR-based TRAP (Telomeric Repeat Amplification Protocol) assay for detecting telomerase activity with some modification (Kim *et al.*, 1994, Piatyszek *et al.*, 1995). To measure telomerase activity semi-quantitatively, cell lysates prepared from

Y. Kitagawa et al. (eds.), Animal Cell Technology: Basic & Applied Aspects, Vol. 10, 217–221.

10^6 cells were titrated between 10 and 10^4 cells equivalent per assay, and a linear range of telomerase activity versus cell number was determined to estimate the appropriate cell number to be used in each assay. In our case, TRAP assay was quantitative when lysate equivalent to a range from 50 to 1000 cells per reaction were used (Katakura, *et al.*, 1997). TRAP assay products were visualized by SYBR Green I (TAKARA, Shiga, Japan) staining and quantified by using Kodak EDAS system (Eastman Kodak Co., Rochester, NY).

2.3 DNA EXTRACTION AND ANALYSIS

Genomic DNA was prepared using the DNA Extractor WB Kit (Wako, Kyoto, Japan) according to the manufacturer's protocol. Telomere length distributions were analyzed at a series of time points during serial passaging. Length of the terminal restriction fragments (TRF) was determined by Southern blot analyses with a non-radioisotopic telomeric sequence $(TTAGGG)_4$ probe, the 3'-end of which was labeled with fluorescein-11-dUTP using the ECL 3'-oligolabelling system (Amersham, Buckinghamshire, UK) and used to measure the length of the telomere by detecting the telomeric band with the Gene Images CDP-Star detection module (Amersham) as described previously (Katakura *et al.*, 1997).

2.4 β-GALACTOSIDASE ACTIVITY

β-galactosidase staining was performed according to the method described by Dimri *et al.* (1995). Briefly, cells were fixed in 3% formaldehyde and then incubated with fresh senescence-associated β-Gal stain solution. Staining was carried out at 37°C for 12 h.

2.5 NORTHERN BLOT ANALYSIS

Ten μg of total RNA was resolved by electrophoresis, transferred onto Hybond N^+ membrane, and probed with the 3'-end labeled human plasminogen activator inhibitor-1 (PAI-1) DNA (CALBIOCHEM, Cambridge, MA) according to the standard protocol.

2.6 RT-PCR ANALYSIS

Human telomerase reverse transcriptase (hTRT) and telomerase-associated protein related to Tetrahymena p80 (TP1) mRNAs were detected by RT-PCR methods as described by Nakamura *et al* (1997). Total RNA was prepared from each subconfluent culture by using TRIzol reagent (GIBCO BRL, Grand Island, NY) and further purified with ice-cold 4 M LiCl. First strand cDNA was synthesized by using SuperScript II Rnase H^- reverse transcriptase (GIBCO BRL). hTRT mRNA was amplified using oligonucleotide primers LT5 (CGGAAGAGTGTCTGGAGCAA) and LT6 (GGATGAAGCGGAGTCTGGGA) for 26 cycles (94°C for 45 s, 60°C for 45 s, 72°C for 90 s). TP1 mRNA was amplified using oligonucleotide primers TP1.1 (TCAAGCCAAACCTGAATCTGAG) and TP1.2 (CCCGAGTGAATCTTTCTACGC) for 26 cycles (94°C for 45 s, 60°C for 45 s, 72°C for 90 s). Reaction products were resolved on an 8% polyacrylamide gel and stained with SYBR Green I.

3. Results and Discussion

3.1 TGF-β DOWNREGULATE TELOMERASE ACTIVITY IN A549 CELLS

In TGF-β responsive cancer cells, cells can not proliferate in the presence of TGF-β, but in A549 cells, the antiproliferative response against TGF-β is found to be impaired (data not shown), which can be explained by the functional deletions of cycline dependent kinase inhibitor p15 and protein tyrosine phosphatase cdc25A molecules. Although weak but detectable antiproliferative response can be observed in A549 cells, which suggests a novel signaling molecules other than p15 and cdc25A which participates in the TGF-β signaling cascade. We expected that this novel signaling molecules and pathway might elicit other anti-tumor response against TGF-β.

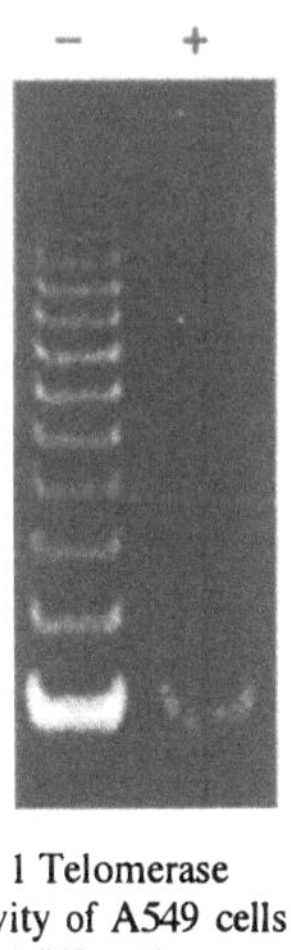

Fig. 1 Telomerase activity of A549 cells (-) and A549 cells treated with TGF-β (+)

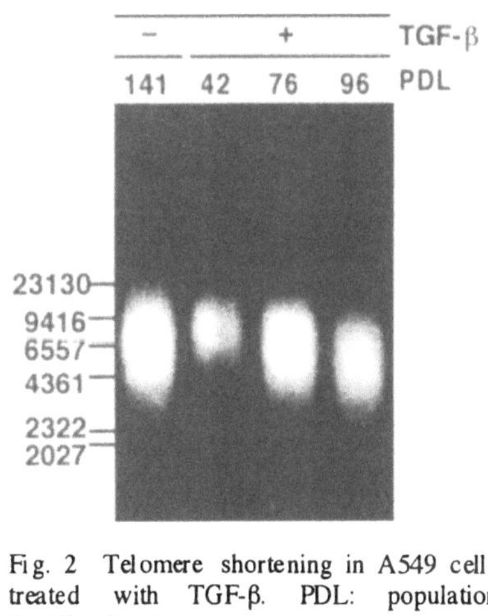

Fig. 2 Telomere shortening in A549 cells treated with TGF-β. PDL: population doubling level.

In order to elucidate the novel TGF-β signaling cascade, we measured the telomerase activity by the TRAP method and telomere length of A549 cells treated with TGF-β by Southern blot analysis using telomere sequence as probe. Fig 1 shows that telomerase activity greatly decreased by the addition of TGF-β, and the corresponding telomere shortening was observed during the long-time culture (Fig. 2).

These results suggest that A549 cells downregulates the telomerase activity by the addition of TGF-β and would enter into the senescent state during the culture accompanied by the telomere shortening.

3.2 TGF-β INDUCED CELLULAR SENESCENCE IN A549 CELLS

Fig. 3 clearly showed that A549 cells treated with TGF-β shows a flat and large morphology and expressed senescence marker of β-galactosidase, indicating that TGF-β induced cellular senescence in A549 cells.

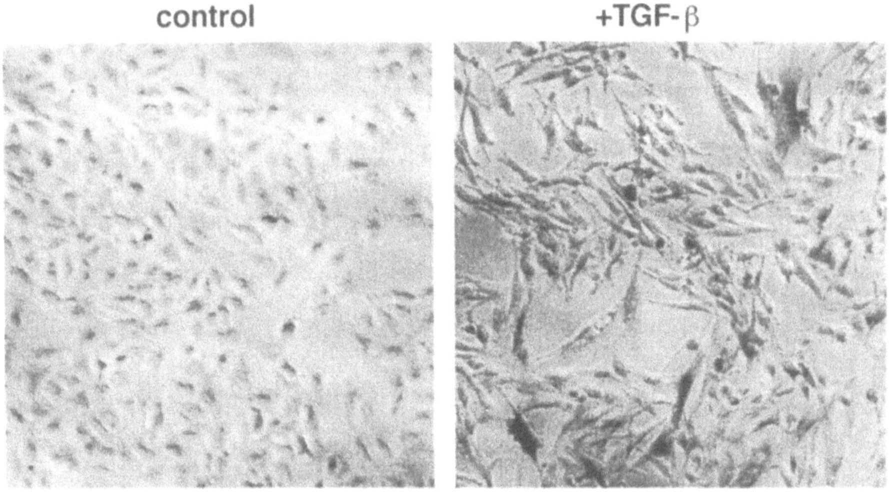

Fig. 3 β-galactosidase activity in A549 cells and that treated with TGF-β

3.3 RT-PCR ANALYSIS FOR TELOMERASE COMPONENT

Table 1 Signaling molecules participating in telomerase repression

TGF-β	PAI-1 induction	hTRT	TP1
–	–	+	+
+	+	–	+

PAI-1 induction was assessed by Northern blot analysis as described in Materials and Methods. hTRT and TP-1 expression in A549 cells treated with TGF-β was analyzed by RT-PCR analysis.

Next we analyzed the signaling pathway downstream from the TGF-β receptor in A549 cells. Transcriptional regulation of plasminogen activator inhibitor-1, PAI-1, in A549 cells treated with TGF-β was analyzed by Northern blot, indicating that PAI-1 transcription in A549 cells treated with TGF-β is upregulated (Table 1). Considering together with the fact that transcription of PAI-1 is promoter by functional Smads proteins, this result indicates that Smads are activated in A549 cells treated with TGF-β and that the signaling pathway from the TGF-β receptor to PAI-1 which does not mediate p15 and cdc25A do exist in A549 cells. Second, we analyzed the transcriptional regulation of human telomerase components, human telomerase reverse transcriptase (hTRT) and human telomerase-associated protein (TP1). hTRT is only telomerase component whose expression was completely correlated with telomerase activity in cells. Table 1 shows that hTRT transcription in A549 cells is downregulated by the addition of TGF-β. This down-regulation would explain the reduction of telomerase activity in A549 cells treated with TGF-β.

3.4 REVERSIBILITY OF TGF-β SIGNALS IN A549 CELLS

TGF-β signal was demonstrated to be reversible in A549 cells, which was evidenced by repression of PAI-1 and β-galactosidase activity and reactivation of telomerase activity and hTRT expression with the removal of TGF-β (data not shown). This indicates that senescence induction by TGF-β in A549 cells is not fixed alteration and is caused by the stimulation of the specific signaling cascades.

4 Conclusion

Hypothetical strategies to vest a limited replicative lifespan in cancer cells are as follows, one is to exogenously downregulate the hTRT transcription, which certainly leads to the reduction of telomerase activity, because the telomerase activity has been demonstrated to completely correlated with the hTRT transcription. Second is to inhibit the binding of telomerase complex to telomere. When these strategies are successfully pursued, cancer cells would shorten the telomere and senesce. In this study, we clarified that TGF-β reversibly downregulates the hTRT transcription and induces the cellular senescence in A549 cells.

5. References

Dimri, G. P., Lee, X., Basile, G., Acosta, M., Scott, G., Roskelley, C., Medrano, E. E., Linskens, M., Rubelj, I., Pereira-Smith, O., Peacocke, M., and Campisi, J. (1995) A biomarker that identifies senescent human cells in culture and in aging skin in vivo, *Proc. Natl. Acad. Sci. USA*, **92**, 9363-9367

Katakura, Y., Yamamoto, K., Miyake, O., Yasuda, T., Uehara, N., Nakata, E., Kawamoto, S., and Shirahata, S. (1997) Bidirectional regulation of telomerase activity in a subline derived from human lung adenocarcinoma, *Biochem. Biophys. Res. Commun.* **237**, 313-317

Kim, N. W., Piatyszek, M. A., Prowse, K. R., Harley, C. B., West, M. D., Ho, P. L. C., Coviello, G. M., Wright, W. E., Weinrich, S. L., and Shay, J. W. (1994) Specific association of human telomerase activity with immortal cells and cancer, *Science* **266**, 2011-2015

Nakamura, T. M., Morin, G. B., Chapman, K. B., Weinrich, S. L., Andrews, W. H., Lingner, J., Harley, C. B., Cech, T. R. (1997) Telomerase catalytic subunit homologs from fission yeast and human, *Science* **277**, 955-959

Piatyszek, M. A., Kim, N. W., Weinrich, S. L., Hiyama, K., Hiyama, E., Wright, W. E., and Shay, J. W. (1995) Detection of telomerase activity in human cells and tumors by a telomeric repeat amplification protocol (TRAP), *Meth. Cell Sci.* **17**, 1-15

EVALUATION OF A FIXED-BED TYPE BIOREACTOR SYSTEM USING RECOMBINANT CHO CELLS PRODUCING HUMAN TRANSFERRIN

A. NEMOTO, Y. HIROTA-IWAI, Y. MIYAZAKI, N. NAKAMICHI, M. ITO, T. TOKIWA*, M. KODAMA* AND T. MATSUMURA

*RITE Meinyu Branch Laboratory and Meiji Milk Products Co. Cell Technology Center, Odawara 250, and *Bionic Design Research Group, National Inst. for Advanced Interdisciplinary Research, 1-1-4 Higashi, Tsukuba, Ibaraki 305, Japan*

ABSTRACT To improve production efficiency of recombinant animal proteins, we evaluated a fixed bed type bioreactor, which we call circulatory flow bioreactor, using human transferrin-producing recombinant CHO cells. In the circulatory flow bioreactor, the cells are immobilized on the surface of vertically arrayed substrate sheets, between which culture medium circulate in the vessel. Cell immobilization technology, therefore, plays a key role in the successful operation of this bioreactor, and probably of all fixed-bed type bioreactors. We paid special attention to immobilize CHO cells in serum-free condition, and found that expanded porous polytetrafluoroethylene (ePTFE) membrane and woven glass fiber cloth are suitable for immobilizing the cells. It is rather striking to note that ePTFE is hydrophobic and chemically inert, and still can immobilize the cells, suggesting that the porous surface structure of the substrate allows cells to anchor stably. Using a serum-free culture medium containing fructose and mannose in place of glucose as sugar source, we obtained more than $2x10^{10}$ cells producing recombinant human transferrin in the spent culture fluid for up to a month.

1. Introduction

For economical production of recombinant mammalian proteins, technologies are needed for highly efficient gene expression, a bioreactor for high-density cell culture, and for a long and stable operation of such a bioreactor in a serum-free condition. Previously, we developed a bioreactor of fixed bed type, intending to maintain adhesive cells for months, i.e., the circulatory flow bioreactor (1). To improve and evaluate the efficiency of recombinant production using the circulatory flow bioreactor, we chose a recombinant CHO cell line producing human transferrin (hTfr) as a model cell line (2). Transferrin (Tfr) is a glycosylated metal-binding protein with industrial usefulness. So far, successful production of functional recombinant hTfr (rhTfr) has been reported using BHK cells (3,

Y. Kitagawa et al. (eds.), Animal Cell Technology: Basic & Applied Aspects, Vol. 10, 223–227.

4). During the improvement and evaluation, we paid special attention to immobilize CHO cells, since they often detach from the substrate in a serum-free condition. Previously, we noted that cells adhere to expanded polytetrafluoroethylene (ePTFE) membrane in a plate culture (5, 6). Here, we found conditions for successful cell immobilization and operation of the bioreactor in a serum-free condition, using ePTFE as well as glass fiber cloth as immobilizing substrates.

2. Materials and Methods

2.1. RECOMBINANT CHO CELLS

Recombinant CHO clone #1-23 (2), which produces human transferrin (hTfr), has been used throughout this study (2). This CHO clone was obtained using SRα expression vector, which is characteristic with its SV40 promoter-enhancer element and HTLV LTR sequence (7).

2.2. BIOREACTOR

The structure and mode of function of this reactor was described in details previously (1). A reactor of mini-scale model used in this study consisted of a one liter glass vessel, a stainless steel core structure with a cell-immobilizing sheet material stored in the core, a stirrer bar, and regulatory equipment (Fig. 1).

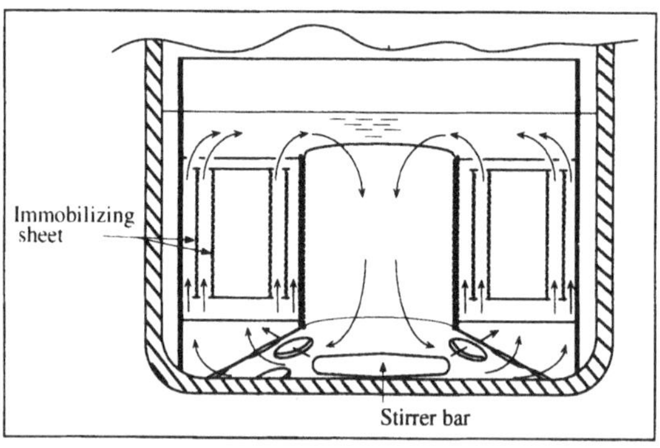

Fig. 1. Schematic illustration of circulatory flow bioreactor
In the one liter scale model reactor, the core structure consists of a cylindrical bobbin and a long strip sheet (ca. 2200cm^2) of culture substrate which is rolled up with a constant space (1.5mm or 2mm) between sheets in a physical support. The effective volume of the core structure in which cells are immobilized is 0.24 liter. The bottom end of the bobbin is attached to a corn-shaped frame on which small holes are open to permit medium flow. The bobbin-sheet-corn complex is vertically set in a vessel. A magnetic stirrer bar, or an impeller in a large reactor, is set at the bottom of the cylindrical space.

As cell-immobilizing materials, we used ePTFE membrane sheet (5, 6), and glass fiber cloth (1), as described in details below. When the vessel is filled with medium of ca. 800ml, and the stirrer bar driven in it, medium circulates in the vessel so that it flows through the space of immobilizing sheets in a laminar constant flow. The gas, containing 5% CO_2 in air supplemented with O_2, was supplied on the surface of culture medium, where medium flows rapidly. When cells are introduced, they are immobilized on the sheet material and grown.

2.3. REACTOR OPERATION AND TRANSFERRIN DETERMINATION

Before inoculation into the reactor, the cells are grown in E-MEM (GIBCO BRL, Maryland) supplemented with non-essential amino acids and 10% fetal bovine serum (FBS, JRH Bioscience, Lenexa, Kansas) (growth medium), and trypsinized for passage cultures in plastic containers. Separately, the reactor vessel was filled with the growth medium. For inoculation, a cell suspension containing $2x10^8$ cells was obtained and introduced into the reactor, and the vessel stirred at a low rotational speed (100-200 rpm) for two days, which was then fed constantly with the growth medium for the period of two to three weeks with constant monitoring of sugar consumption, lactate and ammonia production. During this period feeding rate was gradually enhanced up to ca. 1 litter per day, and the stirrer rotation up to 600 rpm.

After the above growth phase, the medium was replaced with the serum-free medium, UC212 (Nissui, Tokyo), which contains fructose and mannose (4mg/ml and 1mg/ml respectively) but no glucose. Catalase and insulin (50μg/ml and 1μg/ml, respectively) were supplemented, but no fibronectin nor transferrin.

Transferrin concentration in the medium was quantitatively determined by a standard EIA technique using rabbit anti-hTfr IgG fraction (DAKO), and POD-conjugated rabbit anti-goat IgG (DAKO). We confirmed that bovine Tfr minimally interferes with hTfr in the immunological detection system.

3. Results and Discussion

3.1. MERITS AND PROBLEMS OF A FIXED-BED TYPE BIOREACTOR

CHO cell line is the most popular recombinant host for mammalian protein production. They adhere in monolayer when they are grown in serum-supplemented medium. When they are grown in a serum-free condition, they tend to grow in suspension, or adhere only weakly to substrate. CHO cells, therefore, are suitable targets to improve the technology of cell immobilization.

In a series of trial operation of the circulatory flow bioreactor, we accumulated experiences for successful serum-free cultivation of the rCHO cells in the reactor. Although purely empirical, factors influencing successful cultivation are summarized in Table 1. The circulatory flow bioreactor was designed originally for cells which strongly adhere to culture substrate. For such cells, the reactor could extend its maximum capability, providing a prolonged operation for more than 6 months at a high cell density of more than $4x10^7/cm^3$ (Details to be described elsewhere). We believe that such high performance of the reactor is attained because medium flows in the reactor at a relatively high velocity, i.e., ca. 2-3cm/sec on the surface of the substrate, allowing efficient exchange of gas and medium nutrients on cell surface, as well as on the air-medium interface. However, when cells adhere weakly, they partly adhere to the substrate, while partly remain in suspension, and the advantage of the reactor system is reduced greatly. Therefore, technology for strong cell immobilization on substrate surface, without interfering cellular metabolic activity, is a key element for a successful operation of the reactor. This is probably true to all fixed bed type bioreactors.

Table 1. Factors influencing successful reactor operation

Conditions	Our choice
Inoculation cell density	**1-2x10^{8}cell per 1 liter reactor**
With a low inoculation, cells tends to form colonies during growth in the reactor, which often dislodge from the substrate and cause cell loss. 10^8cells provide ca. $3x10^4/cm^2$ surface cell density in the reactor.	
Rotation of stirrer bar	**600 rpm**
A higher rotation will provide more supply of gas and nutrient, but tends to lead dislodging and loss of cells from the reactor. 600rpm causes ca. 2cm/sec medium flow on the cell surface.	
Culture medium	**UC212 with insulin and catalase**
UC212 (Nissui) is originally designed for surface culture in combination with the addition of fibronectin. The use of culture substrates with large surface area and complexity, as in this study, permits cell immobilization without fibronectin.	
Culture substrate	**ePTFE or glass fiber cloth**
ePTFE has much more surface complexity and surface area than glass fiber cloth. Since ePTFE has no charge nor hydrophilic residues, cell attachment on the membrane is considered to be purely physical.	

3.2. EXAMPLES OF SUCCESSFUL OPERATION

After a series of trial operation of the reactor, CHO cells were successfully immobilized and maintained for one month in the circulatory cell bioreactor on ePTFE membrane sheet: (FP1000, effective pore size=10µm, Sumitomo Electric Industry, Osaka, Japan), and on glass fiber cloth (LT13. 10µm diameter fiber, Nittobo, Osaka, Japan), in a synthetic serum-free medium, UC212, without supplementation of fibronectin. Cell density of more than 10^7/ml reactor volume, and hrTfr production of more than 10mg/l effective volume of the reactor per day were attained in the reactor either equipped with glass fiber cloth or with ePTFE. It should be noted here that the hrTfr production rate was obtained without using gene amplification technology.

Morphological observation suggests that cells attach and grow more on ePTFE membrane than on glass fiber cloth, which may be due to the fact that the surface structure of ePTFE membrane used in this study is more fine and complex than that of available glass fiber cloth materials (Fig. 2). Since ePTFE is highly inert and inactive to chemicals and heat, and also highly insoluble, it will be a useful candidate material for cell immobilization for production purposes. Moreover, it may provide bioreactor with new applications including artificial hybrid organs. We are currently improving expression vectors with intention of obtaining a more efficient recombinant production system as a goal.

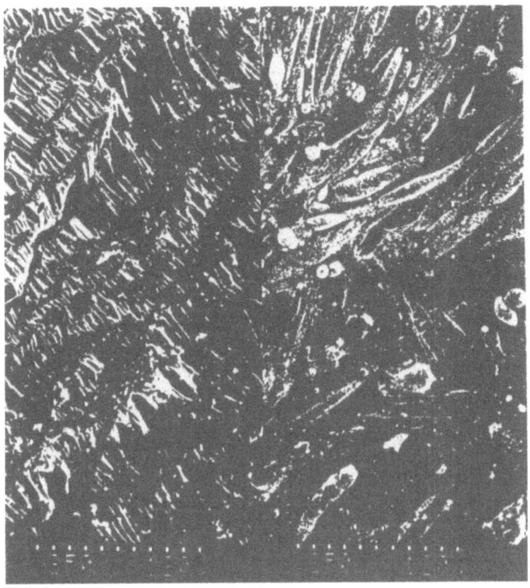

Fig . 2. Expanded polytetrafluoroethylene (ePTFE) membrane cultivated with (right) and without (left) CHO cells. Scanning electron micrographs (x250).

4. Acknowledgements

This work was supported in part by the New Energy and Industrial Technology Development Organization (NEDO).

5. References

1. Matsumura, T., Sawai, Y., Suzuki, J., Fujimori, T., (1993) Circulatory culture equipment. U.S. Patent 5270207.
2. Nakamichi, N., Ito, M., Hirota, Y., Matsumura, T., (1997) Establishment of recombinant human transferrin-producing CHO cell lines, *Animal Cell Technology: Basic & Applied Aspects*, Kluwer Academic Publishers, Dordrecht. **8**, 373-379.
3. Mason, A.B. Miller, M.K., Funk, W.D., Banfield, D.K., Savage, K.J., Oliver, R.W.A., Green, B.N., MacGillivray, R.T.A. & Woodworth, R.C., (1993) Expression of glycosylated and nonglycosylated human transferrin in mammalian cells: Characterization of recombinant proteins with comparison to three commercially available transferrins. *Biochemistry* **32**, 5472-5479.
4. Mason, A.B., Tam, B.M., Woodworth, R.C., Oliver, R.W.A., Green, B.N., Lin, L-N, Brandts, J.F., Savage, K.J., Lineback, J.A., & MacGillivray, R.T.A., (1997) Receptor recognition sites reside in both lobes of human serum transferrin. *Biochem. J.*, **326**, 77-85
5. Tokiwa, T., Kano, J., Kodama, M. & Matsumura, T., (1997) Multilayer rat hepatocyte aggregates formed on expanded polytetrafluoroethylene surface, *Cytotechnology* **25**, 135-142.
6. Kano, J., Tokiwa, T., Meng, X., Kodama, M., & Matsumura, T., (1997) Immobilization of CHO-K1 cells in serum-free medium using glass fiber fabrics or expanded polytetrafluoroethylene membranes. *Seitai-Zairyo* **15**, 55-62 (in Japanese).
7. Takebe, Y., Seiki, M., Fujisawa, J., Hoy, P., Yokota, K., Arai, K., Yoshida, M., and Arai, N. (1988). Srα promoter: an efficient and versatile mammalian cDNA expression system composed of the simian virus 40 early promoter and the R-U5 segment of human T-cell leukemia virus type 1 long terminal repeat. *Mol. Cell. Biol.* **8**, 466-472.

FORMATION OF CYLINDRICAL MULTICELLULAR AGGREGATE (CYLINDROID) OF RAT HEPATOCYTES ON PRESSED SHEET OF POLYURETHANE FOAM

Hiroshi Mizumoto, Mitsuru Kaneko, Kohji Nakazawa, Hiroyuki Ijima, Taku Matsushita and Kazumori Funatsu

Department of Chemical Systems and Engineering,
Graduate School of Engineering, Kyushu University
6-10-1 Hakozaki, Higashi-ku, Fukuoka 812-81, Japan

ABSTRACT

Primary rat hepatocytes formed cylindrical multicellular aggregates on a pressed sheet of polyurethane foam (pressed-PUF) as a culture substratum. The sheet was made up by pressing PUF during the process of foaming. The diameter and length of the hepatocyte cylindroid is approximately 200 ~ 500 μm and 500 μm ~ 2 mm, respectively. Its activities of liver specific functions (albumin secretion and ammonia metabolism) were equal to those of hepatocyte spheroids. These results indicated that hepatocytes cylindroid are a kind of organoid, and will provide a useful culture technique for high cell density culture.

INTRODUCTION

It is very important to maintain differentiated functions of isolated hepatocytes and to immobilize the cells in an artificial matrix when making a hybrid type artificial liver as a bioreactor. However, full maintenance of the differentiated functions of hepatocytes *in vitro* is still difficult. For example, in most cases, albumin secretion activity gradually decreased when the cells were cultured in a monolayer[1]. Recently, it has become clear that multicellular aggregates formed of hepatocytes express highly differentiated functions[2]. So, a technology for reorganization of tissue or organ *in vitro* using cultured cell (Tissue Engineering) is being tested[3]. We have already reported that primary rat hepatocytes formed spherical multicellular aggregate (spheroid) in the pores

Y. Kitagawa et al. (eds.), Animal Cell Technology: Basic & Applied Aspects, Vol. 10, 229–233.

of polyurethane foam (PUF) as a culture substratum, and that liver specific functions can be maintained for more than 2 weeks[4].

In this manuscript, studying the effect of the structure of PUF pores on spheroid formation, we used pressed-PUF (no pore structure) as a culture substratum. To characterize the hepatocytes inoculated on pressed-PUF, morphological analysis and measurement of the activities of liver specific functions were performed.

MATERIALS AND METHODS

PUF and pressed-PUF as a culture substratum

PUF has a sponge-like macroporous structure with each pore made up of smooth thin films and thick skeletons, as reported previously[5]. For preparation of stationary culture, the block of PUF was cut into flat plate (1.0 × 25 × 25 mm). It was put into a tissue culture dish of 35 mm diameter.

Pressed-PUF was film-like PUF with a smooth surface. The raw materials of PUF were mixed, and the PUF was pressed in the process of foaming. For preparation of stationary culture, the pressed-PUF was cut into discs of 35 mm diameter.

Hepatocytes culture

Hepatocytes were isolated from the liver of an adult male Wistar rat (7-8 weeks, 200-215g) by the liver perfusion method with 0.05 % collagenase. The isolated hepatocytes were inoculated on each pressed-PUF plate at a density of 1.02×10^5cells/cm^2 (1×10^6cells/ml-medium). The cell culture dishes were incubated in a CO_2 incubator (5% CO_2, 95% air) at 37℃. Medium exchange was performed 4 and 24 hours after cell inoculation, then every 48hours. As a control experiment, PUF/spheroid culture[4] (inoculum density: 1×10^6cells/ml) was performed in parallel studies.

Morphological analysis

Morphology of hepatocytes inoculated on pressed-PUF was observed by a phase contrast microscope with video camera and time lapse video. The culture dish was put on the microscope stage and maintained under 5% CO_2, 95% air, at 37℃.

Liver specific functions (albumin secretion and ammonia metabolism)

Albumin concentration in the culture medium was measured by the ELISA method. When we evaluated the ammonia metabolic activity, culture medium was changed to fresh medium containing 1mM-NH_4Cl, and the ammonia concentration was measured at

6 hours after incubation. For each normalized albumin secretion and ammonia metabolic activity, the rates were divided by the nuclei number which was obtained by the nucleus counting method of van Wezel[6]. To compare with the functional activity of PUF/cylindroid culture, PUF/spheroid culture was performed as a control experiment.

RESULT

Cylindroid formation of adult rat hepatocytes

When freshly isolated hepatocytes were inoculated on pressed-PUF, they initially attached to the pressed-PUF, and spread out to form monolayer. Subsequently, the hepatocytes peeled off from the pressed-PUF surface, and formed small cell aggregates or large strand like aggregates. The small aggregates grew to form spheroids after 1 ~ 3 days of culture. On the other hand, the large aggregates grew, sometimes incorporating previously formed spheroid, form into cylindroids at 5 ~ 7 days. The diameter and length of cylindroids was approximately 200 ~ 500 μm and 500 μm ~ 2mm, respectively.

Liver specific functions of hepatocytes cylindroid

The activity of albumin secretion and ammonia metabolism of cultured hepatocytes in PUF/cylindroid culture as compared with that of PUF/spheroid culture is shown in Table 1 and 2. The albumin secretion activity in PUF/cylindroid culture maintained for 7 days of culture and the level of the activity increased after 3 days of culture. The activity of albumin secretion in PUF/cylindroid and PUF/spheroid culture was almost the same during the culture period (Table 1). On the other hand, the ammonia metabolic activity in PUF/cylindroid culture decreased after 3 days of culture, but it maintained for at least 7 days of culture. The hepatocytes in PUF/cylindroid culture exhibit a similar ammonia metabolic rate as that in PUF/spheroid culture just as those in albumin secretion (Table 2).

Table 1 Comparison between PUF/cylindroid and PUF/spheroid culture on albumin secretion rate

Culture time [day]	Albumin secretion rate of PUF/cylindroid culture [μg/10^6nuclei/day]	Albumin secretion rate of PUF/spheroid culture [μg/10^6nuclei/day]
1	26.56±1.47	35.19±6.26
3	21.59±3.97	20.46±7.22
5	34.81±4.83	53.69±11.75
7	45.49±4.36	37.35±1.90

Table 2 Comparison between PUF/cylindroid and PUF/spheroid culture on ammonia metabolic rate

Culture time [day]	Ammonia metabolic rate of PUF/cylindroid culture [μmol/10^6nuclei/h]	Ammonia metabolic rate of PUF/spheroid culture [μmol/10^6nuclei/h]
1	0.208±0.005	0.241±0.004
3	0.223±0.042	0.128±0.042
5	0.122±0.007	0.116±0.010
7	0.069±0.011	0.044±0.008

DISCUSSION

In previous studies about hepatocytes reorganization, the formation of spheroids on various substratum have been reported[2) 4)]. In many cases, hepatocytes spheroid are formed by the following steps, attachment, peeling and aggregation. In this study, we demonstrated that the hepatocytes cylindroid formed following the same steps as spheroid formation. Moreover, the activity of liver specific functions (albumin secretion and ammonia metabolism) of hepatocytes organized as cylindroids was equal to the activity of hepatocytes spheroid. These results indicate that cylindroid is also a kind of organoid. For development of a hybrid artificial liver, high cell density is required since the volume of bioreactor in an artificial liver needs to be compact. At this point, hepatocytes cylindroid may be useful. When a huge spheroid is formed, necrosis will occur in the center part of the spheroid. Therefore, the size of spheroid is limited. In contrast, hepatocytes cylindroid formed with a diameter of 200 ~ 500 μm and a length of 500 μm ~ 2 mm in this study. Hepatocytes cylindroid can in theory grow without limitation of length . Therefore, if it is possible to control the diameter of cylindroids, the cell density in PUF/cylindroid culture may be higher than that in PUF/spheroid culture.

Consequently, hepatocytes inoculated on pressed-PUF formed cylindroid, and the hepatocytes cylindroid show highly differentiated functions and high cell density culture. Hepatocytes cylindroid may be useful for development of hybrid artificial liver in the future.

Reference

1) Crane LJ, Miller DL (1977) Plasma protein synthesis by isolated rat hepatocytes. *J. Cell. Biol.* **72**: 11-25.

2) Koide N, Sakaguchi K, Koide Y, Asano K, Kawaguchi M, Matsushima H, Takenami T, Shinji T, Mori M and Tsuji T (1990) *Exp. Cell Res.* **186**: 227-235.

3) Langer R, Vacanti JP (1993) Tissue Engineering. *Science* **260**: 920-926

4) Matsushita T, Ijima H, Koide N and Funatsu K (1991) High albumin production by multicellular spheroids of adult rat hepatocytes formed in the pores of polyurethane foam. *Appl. Microbiol. Biotechnol.* **36**: 324-326.

5) Matsushita T, Ketayama M, Kamihara K and Funatsu K (1990) Anchorage-dependent mammalian cell culture using polyurethane foam as a new substratum for cell attachment. *Appl. Microbiol. Biotechnol.* **33**: 287-290.

6) Wezel AL van (1967) Growth of cell-strains and primary cells on micro-carriers in homogeneous culture. *Nature* **216**: 64-65.

PRODUCTION OF A NOVEL RECOMBINANT TGFα-LIKE FUSION PROTEIN FOR USE IN SERUM-FREE MEDIA.

A.P. SIMULA, C. BARONE*, I.K. PRIEBE*, G. FRANCIS, F.J. BALLARD, J. WHELDRAKE*, C. GODDARD.
GroPep Pty. Ltd. and Cooperative Research Centre for Tissue Growth & Repair, P.O. Box 10065, Gouger St., Adelaide S.A. 5000, Australia.

1. Abstract

A bacterial expression system has been developed which allows for high level expression of recombinant proteins in the form of fusion proteins. This system was used to manufacture a transforming growth factor (TGF)α-like fusion protein (LongTGFα) for use as a supplement in serum-free culture of eukaryotic cells. The expressed fusion protein, recovered in the form of inclusion bodies, was solubilised using chaotropic agents, correctly refolded and then purified, prior to determination of its biological activity. The purified molecule, LongTGFα, displayed mitogenic activity in mammalian cells. Moreover, it demonstrated a 14-fold greater EGF receptor affinity than recombinant EGF when assessed in cultured chick embryo fibroblasts. This differential affinity for EGF receptors in avian cells indicates a potential use for LongTGFα in the development of serum-free media for the production of viral vaccines in such cells.

2. Introduction

The need for product safety has lead the biotechnology industry to move away from the use of animal- and human-derived raw materials during the production of biopharmaceuticals in cell culture. This trend allied to pressure from all the national and international regulatory authorities, has led to a growing demand for completely defined, serum-free media formulations containing only recombinant proteins. To answer this challenge GroPep has developed and patented an *E. coli* expression system for the production of recombinant proteins fused to the first 11 amino acids of methionyl porcine growth hormone (Wells et al. 1989; King et al. 1992).

This system has been used for the manufacture of growth factor analogues as supplements for the development of serum-free media for the cell culture industry. The company currently manufactures analogues of insulin-like growth factor-I (LongR^3IGF-I) and EGF (Long EGF), which are used in commercial serum-free culture media. In this study we describe the expression, purification and characterisation of a TGFα-like fusion

Y. Kitagawa et al. (eds.), Animal Cell Technology: Basic & Applied Aspects, Vol. 10, 235–239.

protein with potential applications in the biopharmaceutical, gene therapy vector and vaccine production industry.

3. Materials and Methods

3.1. CLONING OF LongTGFα cDNA AND FERMENTATION

Using polymerase chain reaction technology the TGFα-like cDNA was engineered into GroPep's expression vector. JM101 cells were transformed with the recombinant expression vector and a fusion protein-expressing clone was selected for fermentation. This was performed in two 1 litre Applicon fermenters, with each litre composed of buffer solution (30mM NH_4Cl, 7mM K_2SO_4, 12mM KH_2PO_4 and 19mM Na_2HPO_4), 100ml of glucose solution (1.4M glucose, 225mM $MgSO_4.7H_2O$), 2ml of trace elements (68mM $FeSO_4.7H_2O$, 10mM $MnSO_4.H_2O$, 15mM $ZnSO_4.7H_2O$, 1.5mM $CuSO_4.5H_2O$, 150mM trisodium citrate), 50mg ampicillin and 2ml of 4% thiamine solution. The medium was inoculated with a 10ml log-phase culture and when an A600 absorbance of 4-12 OD units was attained, the cells were induced with isopropylthiogalactoside and the fermentation terminated once the glucose was depleted.

3.2. PURIFICATION OF LongTGFα FROM INCLUSION BODIES

Cells collected from the fermentation broth were disrupted with an APV30 high-pressure homogeniser and the inclusion bodies (IBs) recovered by centrifugation after two washes in 30mM NaCl/10mM KH_2PO_4. The IB paste was reduced and dissolved in a solubilisation buffer (8M urea, 0.1M Tris.HCl, pH9.0, 40mM glycine, 0.5mM $ZnCl_2$, 40mM dithiothrietol) and then passed through a 0.1μm filter prior to downstream processing.

The dissolved fusion protein was purified in a refolded form using the following combination of steps:

1. Size-exclusion chromatography; Cellufine GCL-1000 matrix, 100 x 5cm (diameter).
2. Refolding of protein by reduction of urea from 8M to 3M, addition of 2-hydroxyethyldisulphide, and room temperature incubation for 3hr.
3. Reverse phase (RP)-HPLC; C18 silica LC matrix, 25 x 5cm (diameter), 0.08% TFA/80% acetonitrile.
4. RP-HPLC; Waters Delta Prep 3000, 10 x 2.5cm (diameter), 0.08% TFA/20-52% acetonitrile gradient.
5. Cation exchange chromatography; S-Sepharose (Fast-flow), 10 x 5cm (diameter), 0-1M NaCl gradient in 8M urea/50mM ammonium acetate, pH4.8.
6. RP-HPLC; Waters Delta Prep 3000, 10 x 2.5cm (diameter), 0.13% HFBA/25-55% n-propanol gradient.

Protein elution was monitored by absorbance at 280nm and fractions assessed for presence of recombinant protein by Coomassie Blue-stained SDS-polyacrylamide gels and RP-HPLC on a C4 Aquapore butyl column.

3.3. CHARACTERISATION OF RECOMBINANT LongTGFα

3.3.1. *Methylene blue mitogenic assay*

Balb/c 3T3 cells were seeded at $2x10^4$ cells/well in 96-well plates in DMEM/10%FBS. Following a 24hr incubation, medium was replaced with fresh medium containing varying concentrations of LongTGFα or recombinant EGF, in triplicate, and incubated for a further 45hrs. After the final incubation cells were washed and fixed prior to staining the cellular protein with methylene blue. The bound dye was then solubilised in acid/ethanol and the absorbance at 655nm was recorded.

3.3.2. *Radioreceptor assay*

Chick embryo fibroblasts were isolated and maintained in DMEM/10%FBS. At a passage number of 7, the cells were trypsinised and plated out at $1x10^5$cells/well in 24-well plates. After 48hrs of culture, triplicate wells were incubated with ^{125}I-TGFα and varying concentrations of LongTGFα or recombinant EGF for 24hrs. After washing cells they were harvested, lysed and transferred to a gamma counter to determine the level of bound ^{125}I-TGFα tracer.

4. Results

Fermentation of JM101 cells carrying the LongTGFα expression vector yielded 12g of IBs, of which 76mg was the recombinant fusion protein. Upon refolding and purification, a final yield of 30mg (40%) of LongTGFα was obtained with a correct mass of 7,113Da as determined by mass spectrometry. Figure 1 shows an EGF receptor-binding profile for fractions recovered from the second RP-HPLC step.

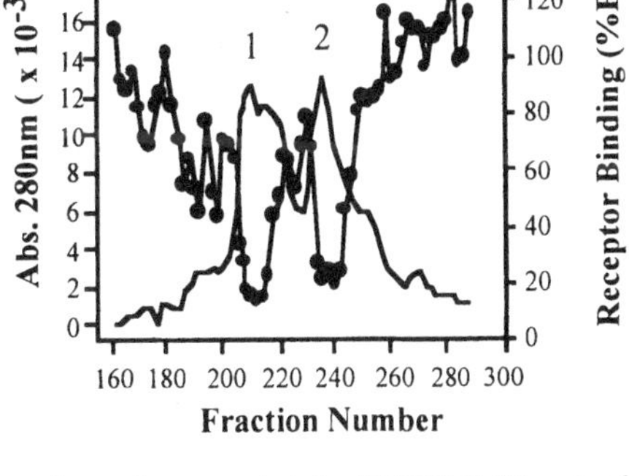

Figure 1. The protein peaks separated by RP-HPLC (Step 4 of purification protocol) demonstrating EGF receptor-binding activity in Ag2804 cells. Mass of peaks 1 and 2 are 7,113 and 7,141Da, respectively.

The first peak had the correct mass for LongTGFα, while the second was the formyl derivative of the N-terminal methionine of LongTGFα. The RP-HPLC analysis of the starting material (inclusion body paste) and positive fractions from the chromatography steps are illustrated in Figure 2.

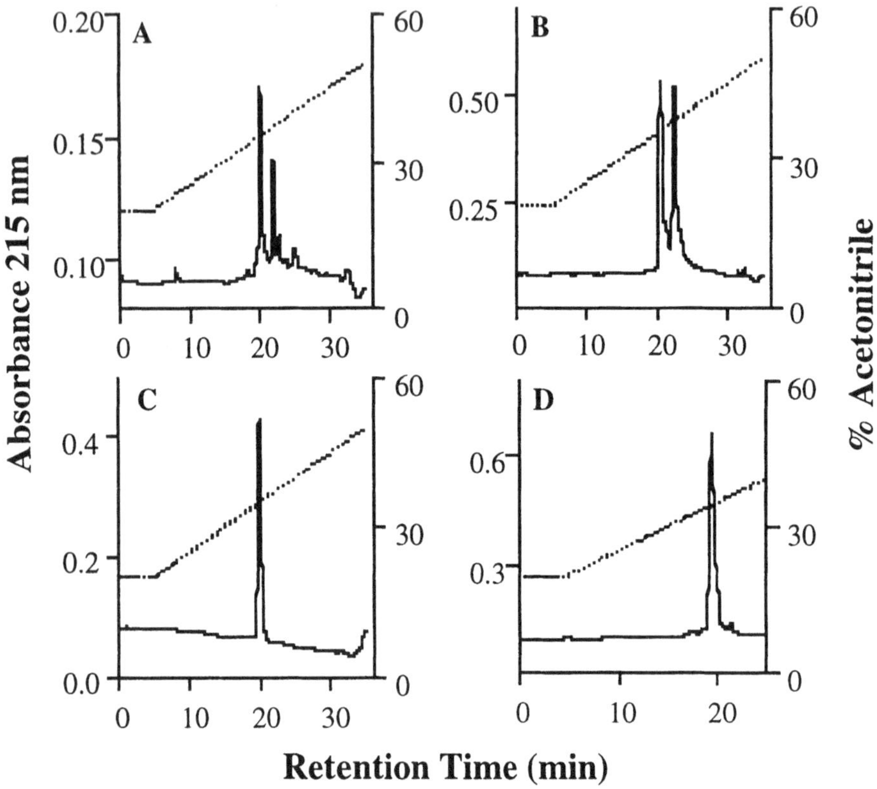

Figure 2. Analysis of protein purification on analytical RP-HPLC. Chromatogram A respresents the dissolved IBs and B-D represent the product obtained after steps 1, 4 and 6 of the purification method (Section 3.2.).

Following the purification of the recombinant protein its biological activity was determined. In Balb/c 3T3 cells LongTGFα demonstrated mitogenic activity (Figure 3A), although lower than for recombinant EGF. When compared to EGF in the EGF receptor-binding assay the TGFα-like fusion analogue had a similar affinity in human Ag2804 cells (data not shown), but in chick embryo fibroblasts (Figure 3B), it had a 14-fold greater affinity for the EGF receptor.

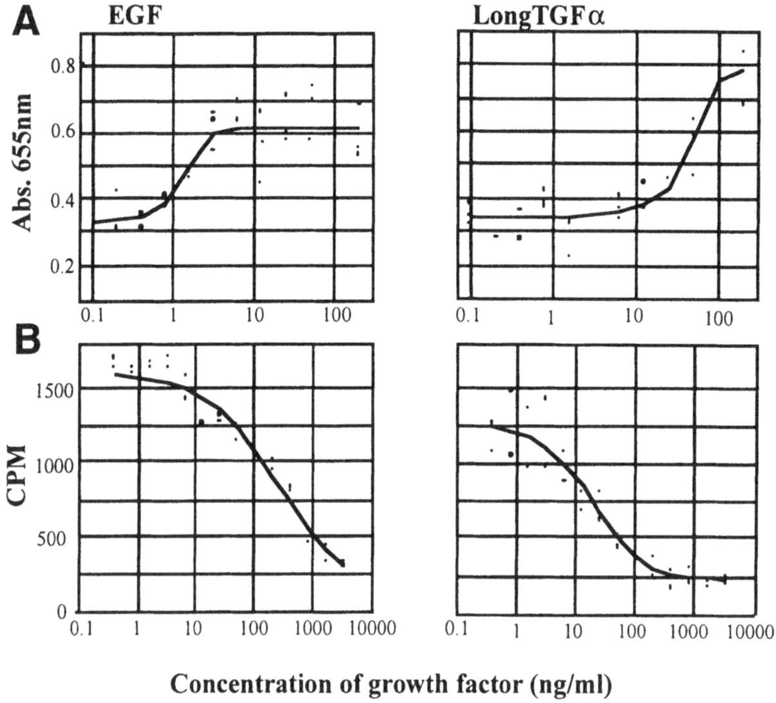

Figure 3. Balb/c 3T3 mitogenic assay (A) and chick embryo fibroblast EGF-radioreceptor assay (B) for LongTGFα and recombinant EGF.

5. Discussion

GroPep has developed a cost-effective manufacturing process for a new recombinant TGFα-like fusion protein, LongTGFα, for the cell culture market. While this molecule has a biological potency in mammalian cells which is lower than EGF, it has been shown to have a 14-fold greater affinity for the EGF receptor then recombinant EGF in primary cultured avian fibroblasts. The majority of viral vaccine manufacturers rely on the use of serum-containing media for their production using avian cell lines. Hence, there is a need for the development of fully-defined media for these cells. The data presented for LongTGFα would suggest that it has great potential as a supplement in the formulation of new serum-free media for the culture of primary avian fibroblasts.

6. References

King, R., Wells, J.R.E., Krieg, P., Snoswell, M., Brazier, J., Bagley, C.J., Wallace, J.C., Ballard F.J., Ross, M., and Francis G.L. (1992) Production and characterisation of recombinant insulin-like growth factor-I (IGF-I) and potent analogues of IGF-I, with Gly or Arg substituted for Glu^3, following their expression in *Escherichia coli* as fusion proteins, *J. Molecular Endocrinology* **8**, 29-41.

Wells, J.R.E., King, R.M., and Francis G.L. (1989) Growth hormone fusion proteins, methods of production and methods of treatment, United States patent 5,330,971.

EFFECT OF THE MACROMOLECULAR METABOLISM OF MURINE HYBRIDOMA CELLS ON ANTIBODY PRODUCTIVITY IN BATCH AND FED-BATCH CULTURE

JAE DEOG JANG, JOHN P. BARFORD*

** Department of Chemical Engineering, Sydney University, Sydney, NSW 2006, Australia; telephone:+61-2-9351-2455; fax:+61-2-9351-2854; email: johnb@chem.eng.usyd.edu.au*

Abstract

Batch and fed-batch cultures of a murine hybridoma cell line (AFP-27) have been performed in a stirred tank reactor to estimate effect of macromolecular composition on antibody production. Macromolecular composition was found to change dynamically during batch culture of hybridoma cells possibly due to active production of RNA and protein and high metabolic rate of total cellular carbohydrates during exponential phase. Antibody synthesis from the amino acid pool is expected to compete with production of cellular proteins. Therefore, it is necessary to examine the relationship between cell growth in terms of cellular macromolecules and antibody production. A set of fed-batch cultures of a murine hybridoma cell line (AFP-27) have been performed in a stirred tank reactor to elucidate this relationship. In this study, we searched for an optimum feeding strategy by changing the target specific growth rate in fed-batch culture to give higher antibody productivity while examining the macromolecular composition. From this study, the specific production rate of cellular protein and RNA were found to have maximum rates in the fed-batch culture with a target specific growth rate of 0.035 hr^{-1} and to decrease with target specific growth rate below 0.035 hr^{-1}. Specific antibody production rate was found to be significantly increased in the fed-batch cultures with target specific growth rate of 0.01 hr^{-1} in which production of cellular RNA and protein were strictly limited by slow feeding of glucose and other nutrients.

Y. Kitagawa et al. (eds.), Animal Cell Technology: Basic & Applied Aspects, Vol. 10, 241–245.

1. Introduction

Antibodies have two important features to be considered for large scale production. Since they are proteins, their production require all the cellular machinery for protein synthesis including transcription, translation, posttranslational modification, and secretion. Since they are not an essential product for cell growth, their production is expected to compete with the production of cellular proteins. Therefore, to achieve mass production of antibodies, the relationship between cell growth in terms of macromolecules and antibody production should be elucidated.

2. Materials and Methods

2.1. CELL LINE AND MEDIUM

The cell line used in this study is a murine hybridoma AFT-27, producing an IgG1 antibody which reacts with human a-fetoprotein. The medium used in this study was based on DR medium (1:1 mixture of DMEM and RPMI) or eDR medium (DR with additional amino acids) with various concentration of glucose and glutamine. 20 mM glucose and 4 mM glutamine in DR medium was used for batch culture in 4 liter scale. For the inoculum of fed-batch culture, subculture volume was doubled daily up to 1 liter with eDR medium of 5 mM glucose and 2 mM glutamine (eDR5/2) to inoculate 1.5 liter of fresh eDR5/2 medium . 60 mM glucose and 20 mM glutamine in eDR medium was used for feed in fed-batch cultures. Foetal calf serum (2 %, v/v) was added to the main culture medium and feed medium.

2.2 CULTIVATION CONDITIONS

Batch and fed-batch cultures for 2-4 L running volume were performed in a stirred tank reactor in which pH and temperature were automatically controlled to 7.1 and 37°C ,respectively. DO was controlled automatically to be maintained between 10 and 50 % by direct sparging of compressed air.
In fed-batch operation, Feed rate (F,liter/hr) was calculated from target specific growth rate (μ), cell density (X, cell/liter), running volume (V, liter), cell yield from glucose consumed (Y_{xs}, cell/mmole), and glucose concentration in feed medium (F_{glc} , mM) according to following equation:

$$F = \frac{\mu X V}{Y_{xs}\, F_{glc}} \qquad (1)$$

Feed rate was recalculated after every sampling with updated cell numbers and running volume.

2.3. ANALYTICAL METHODS

Glucose, lactate, and ammonia were assayed using an enzymatic technique. Total amount of antibody was measured using enzyme-linked immunosorbent assay

(ELISA). Pre-column derivatization technique using FDNDEA(N,N. diethyl-2,4-dinitro-5-fluoro-aniline, Fluka, Buchs, Switzerland) was used to assay the concentration of amino acids in the cell culture supernatant (Fermo *et al.*, 1988).
DNA concentration in the sample was determined by the diphenylamine method (Dische,1955). RNA concentration was assayed by the orcinol method (Brown, 1946). Total cellular protein was assayed by using commercial Coomasie Blue reagent (Biorad) after extraction with sodium hydroxide. Total cellular carbohydrates were determined by the phenol-sulfuric method (Dubois *et al.*, 1956) after lysis of cell pellets with lysis buffer(0.5% Triton X-100, 1 mM EDTA, 0.2 mM PMSF). The extraction of total cellular lipids was carried out using the methods of Folch *et al.* (1956). The final lower phase contained lipids and was dried in a weighing pan at 40°C in a vacuum desiccator until a constant weight.

3. Results and Discussion

3.1. CELL GROWTH AND CULTURE PERFORMANCE

Hybridoma cell growth was successfully controlled during fed-batch operation by limiting glucose concentration. Viable and total cell density were increased about 3-fold in the fed-batch with target specific growth rate of 0.025 hr^{-1} compared to the batch level. Antibody titre in the fed-batch with target growth rate of 0.01 hr^{-1} was 5-fold that of the batch case (Table 1).

Table 1. Culture performance in batch and fed-batch cultures of AFP-27.

Target specific growth rate (hr^{-1})	0.01	0.0175	0.025	0.035	0.06	batch
Maximum viable cell density ($x10^4$ cells/ml)	82.9	207.2	312.8	244.8	97.8	104.4
Maximum total cell density ($x10^4$ cells/ml)	193.1	351.2	382.8	280.2	118.5	113.2
Maximum Mab titre (mg/L)	180.0	56.5	73.3	84.7	43.9	34.2

3.2. METABOLITES

Glucose has been limited during fed-batch operation with different limitation span according to the target specific growth rate used. Glutamine has been maintained low concentration less than 1.5 mM but not exhausted during fed-batch operations. Specific lactate production rate was decreased with decreasing target specific growth rate. This implies that by limiting glucose in the hybridoma cell culture, pyruvate in glycolysis pool has been more efficiently metabolized to secrete less lactate, which otherwise will be accumulated in the culture medium by overflow metabolism (Ljunggren *et al.*, 1994). Ammonia was accumulated up to 4.5 mM in the fed-batch culture with much higher cell growth than batch case in which final ammonia level was 2.0 mM. This can be explained by adaptation of hybridoma cells to higher ammonia level with slow increase rate (Nielsen, 1992, Newland *et al.*, 1994).

3.3. MACROMOLECULAR METABOLISM AND ANTIBODY PRODUCTION

DNA proportion to cell mass was maintained nearly constant through all the fed-batch cultures, while RNA proportion was dynamically changed in response to the target specific growth rate. RNA proportion was highest in the fed-batch with target specific growth rate of 0.035 hr^{-1}. Lipid proportion was slightly increased with decreasing target specific growth rate. Total cellular carbohydrate content was nearly constant through all the fed-batch cultures while it was notably high in the batch case (Fig. 1).

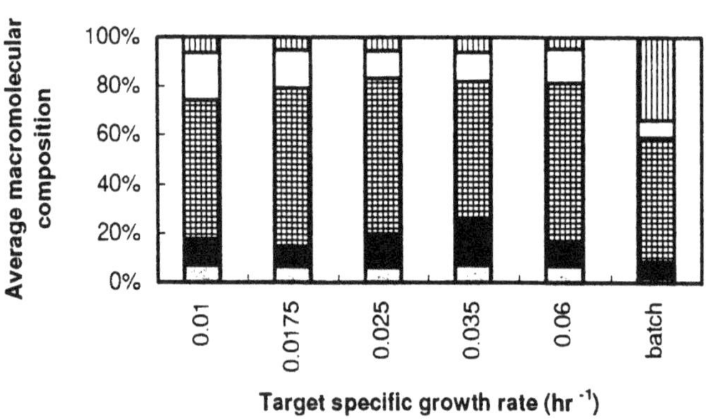

Fig. 1. Macromolecular composition of murine hybridoma cells , AFP-27 during fed-batch culture. Grey: DNA, Black: RNA, Meshed: Protein, White: Lipid, Vertical-striped: Polysaccharide

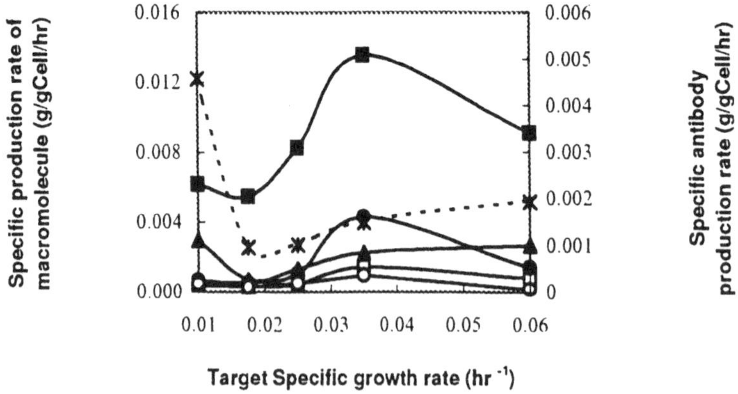

Fig. 2. Comparison of specific production rate of cellular macromolecules (DNA (○), RNA (●), protein (■), lipid (▲), polysaccharide (□)) and monoclonal antibody (∗) during fed-batch culture of AFP-27.

Specific production rate of cellular RNA and cellular protein were changed most dynamically as target specific growth rate was changed, in contrast to specific production rate of other cellular macromolecules which have been maintained relatively constant (Fig. 2). Specific production rate of protein in the fed-batch with target specific growth rate of 0.035 hr^{-1} was 2.2-fold and specific production rate of RNA in this fed-batch was nearly 6.8 fold compared to those of lowest cases in the fed-batch cultures. On the other hand, in the fed-batch with target specific growth rate of 0.01 hr^{-1}, specific production rate of antibody scored 4.6 mg/gcell/hr which was much higher than any other fed-batch cases, while maximum cell density was only a quarter of another fed-batch case with target specific growth rate of 0.035 hr^{-1}.

4. Conclusions

This work suggests that the specific growth rates favoring high specific antibody production rate and high antibody concentration were different from those which favored high viable cell concentration. Fed-batch culture is a compromise with high specific antibody production rate and high antibody concentration at high but not maximal viable cell concentration. It offers simplicity, stability and reproducibility of operation, economy of medium compared to perfusion culture which may attempt to achieve both high specific antibody production rate and maximal viable cell density.

5. References

Brown, A.H. (1946) Determination of pentose in the presence of large quantities of glucose, *Arch. Biochem.* **11**, 269-278.

Dische, Z. (1955) Color reactions of nucleic acid components, *Nucleic Acids*, ed. by E. Chargaff and J. N. Davison, vol. 1, p.287, Academic Press, New York

Dubois,M., Gilles, K.A., Hamilton, J.K., Rebers, P.A.and Smith , F., (1956) Colorimetric method for determination of sugars and related substances, *Anal. Chem.* **28**, 350-356.

Fermo,I. and Vecchi ,E.D. (1988) Pre-column derivatization of amino acids with N,N,-diethyl-2,4-dinitro-5-fluoroaniline and reversed-phase liquid chromatographic separation, *J. Chromatography*, **433**, 53-62.

Folch J. L and Sloane ,S.G. (1957) A simple method for the isolation and purification of total lipids from animal tissues, *J. Biol. Chem.* **226**, 497-509.

Halliburton, I.W. and Thompson, R. Y. (1965) Chemical aspects of compensatory renal hypertrophy, *Cancer Research* **25**, 1882-1887.

Ljunggren, J. and Haggstrom, L. (1994) Catabolic control of hybridoma cells by glucose and glutamine limited fed batch culture, *Biotech. Bioeng.*,**44**, 808-818

Nelson, N. (1994) Energising porters by proton-motive force, *J. of Exp. Biol.* **196**, 7-13.

Newland, M., Kamal, M. N., Grrenfield, P. F., and Nielsen, L. K. (1994) Ammonia inhibition of hybridomas propagated in batch, fed-batch, and continuous culture, *Biotech. Bioeng.* **43**, 434-438.

EFFECT OF GLYCINE BETAINE AS OSMOPROTECTANT ON THE PRODUCTION OF ERYTHROPOIETIN BY CHO CELLS IN HYPEROSMOTIC SERUM FREE MEDIA CULTURE

YONG-HO AHN, KYUBEOM HAN, SUNG KWAN YOON AND JI-YONG SONG

Biotech Research Institute II, Research Park, LG Chemical Ltd., Munji-dong 104-1, Yusung-ku, Taejon, Korea 305-380

Abstract. We investigated effects of glycine betaine on erythropoietin(EPO) production by CHO cells in hyperosmotic serum free media. It has been known that the physiological osmolality of media in animal cell culture is 320 to 340 mOsm/kg. Most of researchers usually ignore osmolality change of media in cell culture. In this report we showed that the osmolality of media might be important factor in CHO cell culture for protein production.

In these experiments, specific EPO productivity(q) was high and longevity of cells was longer but cell growth was low in hyperosmotic media. The cell growth seemed to be inhibited by high osmotic stress of media. We added 15mM of glycine betaine, one of osmoprotectants, to hyperosmotic media in order to protect cells from that inhibition. Addition of glycine betaine increased cell growth and specific productivity. The maximal quantity of EPO produced, however, was smaller even in hyperosmotic media containing glycine betaine than in physiological media. So, we applied two-step culture method that hyperosmotic stress and addition of glycine betaine were done after enough cell growth. That method gave rise to the elevated level (2~3 fold) of maximal EPO production.

All experiments were done under both of static and suspension culture conditions. The results were about the same irrespective of culture condition.

1. INTRODUCTION

Osmolality is an important process variable during the cultivation of animal cells, but its importance seems to be ignored in many cases. The effect of osmolality on hybridoma growth and its product has been reported in a lot of literature. However, few reports about CHO (Chinese Hamster Ovary) cell culture have been published.

Kurano et al (12) reported that in the range between 320 and 390 mOsm/kg specific growth rate of CHO cell was not greatly influenced. Meanwhile, in hybridoma culture following results have been reported. Oh et al (1-2), Ozturk et al (3), and Øyaas et al (5-7) showed that q_{Mab} (specific antibody production rate) was higher in hyperosmotic medium. Especially Øyaas et al reported that growth inhibition under hyperosmotic condition was prevented using osmoprotectants (glycine betaine etc.).

Y. Kitagawa et al. (eds.), Animal Cell Technology: Basic & Applied Aspects, Vol. 10, 247–250.

The osmoprotectants have been known to protect cells from external high osmotic pressure. Rudulier et al (8) explained that osmoprotectants allowed cells to balance the osmotic strength of its cytoplasm with that of its surroundings to prevent a net loss of water. Most osmoprotectants showing those effects have structures of zwitterionic QACs (quaternary ammonium compounds). (9) The most effective osmoprotectant is known as glycine betaine (N-trimethylated glycine). The application of that material can be found in many reports. (5-7,10-11)

In this study we investigated EPO production by CHO cells under NaCl-stressed hyperosmotic medium and compared results from presence or absence of glycine betaine.

2. MATERIAL AND METHOD

2.1. CELL LINE AND MEDIUM

The cell line was created by transfecting a dihydrofolate reductase(DHFR) deficient CHO cell line (ATCC CRL 9096) with a vector containing the DHFR and human EPO genes. Culture medium was adopted according to culture purposes, i.e., IMDM (Gibco) containing 10%(v/v) of serum (fetal bovine, Gibco) was used to attach cells on the surface of culture vessel and SF-2 (serum free medium) was used to undertake hyperosmolality experiment. The SF-2 medium is based on IMDM(Gibco) and modified by ourselves. Osmolality of medium was altered with addition of NaCl. In addition, 15mM of glycine betaine (Sigma) was used as osmoprotectant.

2.2. ANALYSIS

EPO was quantified with ELISA method. Viable cells were estimated using the trypan blue dye exclusion method. Specific production rate (q_{epo}) was calculated by plotting the concentration of EPO at each time versus the integral over the time of viable cells curve.

Specific growth rate (μ) was calculated in the initial growth phase. For indirect cell viability, we measured LDH activity, which is the degree of cell disruption. Low LDH activity indicates high cell viability.

3. RESULTS

As shown in Fig.1 and Fig.2, cell growth was markedly inhibited in hyperosmotic medium under both static and suspension culture condition. We observed that cell size under hyperosmolality was much larger than under normal condition. The higher medium osmolality was, the less cells were grown while the cell longevity was enhanced in hyperosmotic medium.

Addition of glycine betaine increased the specific growth rate of CHO cells in hyperosmotic medium. In addition, q_{epo} was higher in hyperosmotic medium than in normal one and the addition of glycine betaine to hyperosmotic medium increased q_{epo} additionally. (Table1) It seems to be more effective in higher osmotic conditions.

Large-scale static culture using roller bottle was performed with hyperosmotic medium containing 15mM of glycine betaine. EPO concentration increased about three-fold compared with conventional process which was undertaken in normal osmotic

medium. (Fig.3, A) LDH activity was lower in hyperosmotic medium. (Fig.3, B)
In suspension culture , NaCl and glycine betaine were added to the culture medium after cell number reached plateau. About three-fold increase of EPO concentration was observed compared with conventional process. (Fig.4)

TABLE 1. Specific growth rate and specific EPO production rate in various osmotic SF-2 media under static or suspension culture

	Static Culture					*Suspension culture*				
Osmolality	316	408	422	508	522	314	425	443	528	544
Glycine betaine(15mM)	-	-	+	-	+	-	-	+	-	+
μ (hr^{-1})	0.018	0.015	0.024	0.012	0.025					
q_{epo}($\mu g/10^8$cells ·day)	1.393	0.911	1.655	2.205	3.006	0.023	0.097	0.136	0.082	0.101

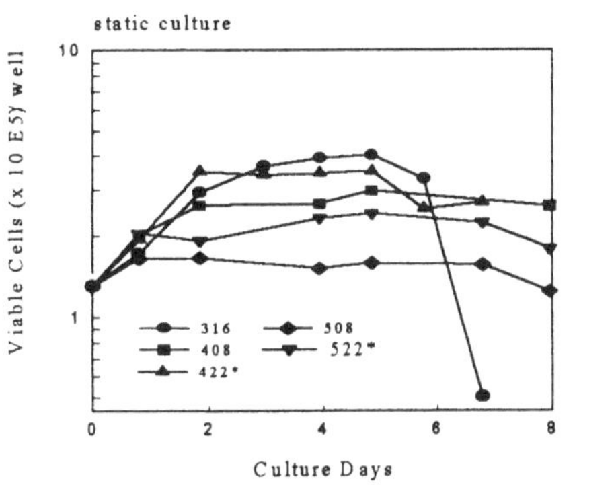

Fig 1. Growth curves of CHO cells in various osmolaities of SF-2 medium. * indicates the presence of glycine betaine (15mM) in culture medium.

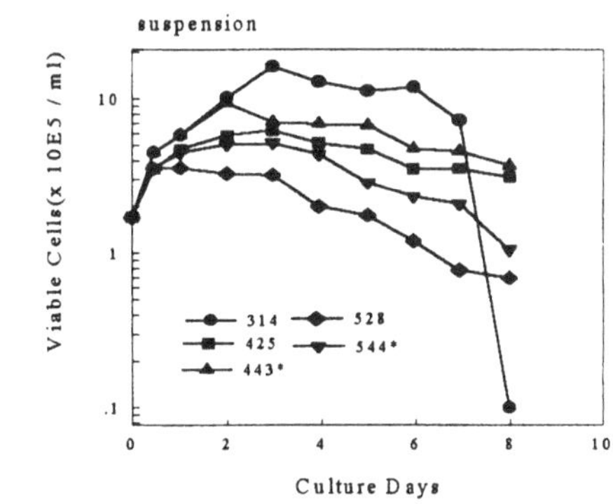

Fig.2 Growth curves of CHO cells in various osmolalities of SF-2 medium. * indicates the presence of glycine betaine (15mM) in culture medium.

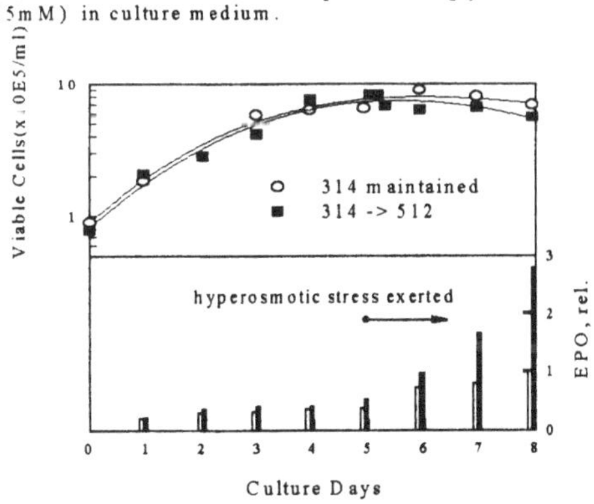

Fig.3 EPO production using spinner flask(125ml). At production phase 120mM of NaCl and 15mM of glycine betaine were added to SF-2 culture medium. Numbers indicate osmolalities.

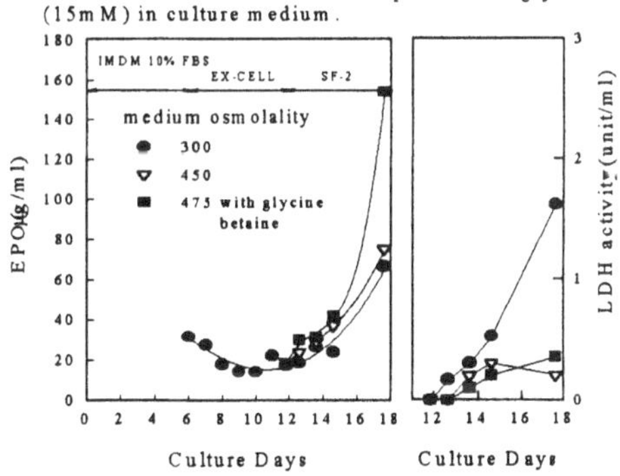

Fig.4 A: EPO production using roller bottles(1750cm^2, 500ml); First stage: Culture step in serum(10%) containing IMDM medium ; Second stage: Medium of each bottle was replaced with fresh serum free medium(EX-CELL, JRH) daily.;Third stage: Long term culture in hyperosmotic NaCl-stressed medium
B: LDH(Lactate DeHydrogenase) activity in third stage; It indicates the degree of cell disruption.

4. DISCUSSION

Growth depression in hyperosmotic medium has been reported for hybridoma cells. We have observed similar effects in CHO cells culture. In CHO cells as well as hybridoma cells, the specific production rate of EPO produced by cells increased under hyperosmotic environment compared with normal one. Reasons for it have been suggested as followings.

Oh et al (1-2,4) reported that hybridomas increased their metabolic activities and amino acids uptake via the Na^+ -dependent symports to compensate for the osmotically elevated external environment. In addition, they explained that these effects contributed directly and indirectly to the increased cell mass, RNA, cellular proteins, and secreted antibody product. In our study, we observed intracellular EPO was higher in hyperosmotic culture than in normal condition (data not shown). This result seems to be attributed to increased metabolic activity in hyperosmotic condition.

As mentioned in previous results we observed that cell size was larger in hyperosmotic medium. According to several works, cell size may be related with cell metabolism. Lager cells may contain a greater amount of enzymes that are requred for protein synthesis and secretion. Ozturk et al (3) reported that a part of the increase in the metabolic rates and the antibody production rate might be attributed to changes in cell volume.

Addition of glycine betaine to hyperosmotic medium increased the specific growth rate in this study. This application was undertaken according to following suggestions. Øyaas et al (5-7) reported that the osmoprotectants were accumulated inside the hybridoma cells to balance an increased extracellular osmolality and they enabled cells to grow even in hyperosmotic medium. The results presented here suggest that in CHO cells as well as hybridoma cells the specific production rate increases in hyperosmotic condition and glycine betaine is effective during growth phase of CHO cells via protecting them from hyperosmotic depression.

In addition, the process we suggested is available in large-scale production of EPO because it reduces labor and cost due to high q_{epo}. LDH activity was lower in hyperosmotic medium and this shows the low probability to deactivate EPO as a biopharmaceutical because some intracelluar enzymes, for example sialidase, liberated by cell lysis have been known to cleave the glycosylated structures of EPO , which are important in EPO activity.

5. REFERENCES

1.Oh,S.K.W., P.Vig, F.Chua, W.K.Teo, and M.G.S.Yap; 1993, Biotech. Bioeng., **42**, 601-610
2.Oh,S.K.W., F.K.F.Chua, and A.B.H.Choo; 1995, Biotech. Bioeng., **46**, 525-535
3.Ozturk,S.S and B.O.Palsson; 1991, Biotech. Bioeng., **37**, 989-993
4.Yang,X., G.W.Oehlert, and M.C.Flickinger; 1996, Biotech. Bioeng., **50**, 184-196
5.Øyaas,K., T.E.Ellingsen, N.Dyrset, and D.W.Levine; 1994, Biotech. Bioeng., **43**, 77-89
6.Øyaas,K., T.E.Ellingsen, N.Dyrset, and D.W.Levine; 1994, Biotech. Bioeng., **44**, 991-998
7.Øyaas,K., T.E.Ellingsen, N.Dyrset, and D.W.Levine; 1994, Cytotechnology, **17**, 143-151
8.Le Rudulier,D.,A.R.Strom,A.M.Dandekar,L.T.Smith and R.C.Valentine; Science, **224**, 1064-
9.Hanson,A.D., B.Rathinasabapathi, J.Rivoal,M.Burnet, M.O.Dillon, and D.A.Gage; 1994, Proc.Natl.Acad.Sci.USA, **91**, 306-310
10.Y.Kawahara, Y.Yoshihara,S.Ikeda, and Y.Hirose; 1990, Appl.Microbiol.Biotechnol., **33**, 574-
11.Y.Kawahara, Y.Yoshihara, S.Ikeda, and Y.Hirose; 1990, Appl.Microbiol.Biotechnol., **34**, 87-90
12.Kurano,N, C.Leist, F.Messi, S.Kurano, and A.Fiechter; 1990, J.of Biotechnology, **15**, 101-112
13. M.J.Keen and N.T.Rapson; 1995, Cytotechnology , **17**, 153-163

ESTABLISHMENT OF RECOMBINANT CHO CELL LINES PRODUCING WILD-TYPE AND MUTANT HUMAN α2-MACROGLOBULINS

NOBORU NAKAMICHI, MASARU SHIMIZU*, MAMIKO ITO, ATSUSHI IKAI* AND TOSHIHARU MATSUMURA

Meiji Milk Products Co. Cell Technology Center/Meiji Institute of Health Sciences. Odawara, Japan 250.
**Laboratory of Biodynamics, Tokyo Institute of Technology, 4259 Nagatsuta, Yokohama, Japan 227*

ABSTRACT Human alpha 2-macroglobulin (hα2M) cDNA was obtained from HepG2 human hepatoma cell line, expression vectors constructed, with which a recombinant hα2M (rhα2M)-producing CHO cell line, rCHO/hα2M ,was obtained. By N-terminal amino acid sequence determination, purified rhα2M protein obtained was confirmed to be of human type. The profile of conformational change due to chemical and enzymatic reactions, as studied electrophoretically, support that thioester-bonds were properly formed in the rhα2M, as in plasma derived hα2M. In addition, we obtained cDNAs with modified DNA sequence in their 'bait region', the region where the specificity for α2M to inhibit a proteinase is determined in accordance with the substrate specificity of the proteinase, and constructed expression vectors, and a recombinant CHO cell line thereof. The potential usefulness of the recombinant wild type-, as well as mutant-, hα2Ms is discussed.

1. Introduction

Alpha 2-Macroglobulin (α2M) is one of the glycosylated serum proteins produced mainly by the liver (1). It shows inhibitory activities against a wide variety of proteinases, and plays homeostatic roles against proteolytic reactions in the body (1, 2). Its versatile inhibitory activity against proteinases is explained by the trapping theory by which a peptide region called 'bait region' is specifically cleaved with a variety of proteinases, leading to a conformational change of the protein, trapping the proteinases, and finally inhibition of the proteinase activity (3, Fig. 1). If the theory is true, then the inhibitory specificity of α2M can be modulated by altering the bait region amino acid sequences. Furthermore, α2M could

Y. Kitagawa et al. (eds.), Animal Cell Technology: Basic & Applied Aspects, Vol. 10, 251–257.

be one of the major components of artificial plasma, when a large scale production for recombinant human α2M (rhα2M) is in hand. So far, a successful production of functional rhα2M was reported using baby hamster kidney cells (4). To accomplish our objectives, we obtained a complete hα2M cDNA sequence and a rα2M production system using CHO cells. Furthermore, we modified the bait region sequence, and cloned mutant rhα2M producing CHO cell lines, and characterized the proteins in part. Biochemical characteristics in details of these engineered proteins will be described elsewhere (Ikai et al. submitted).

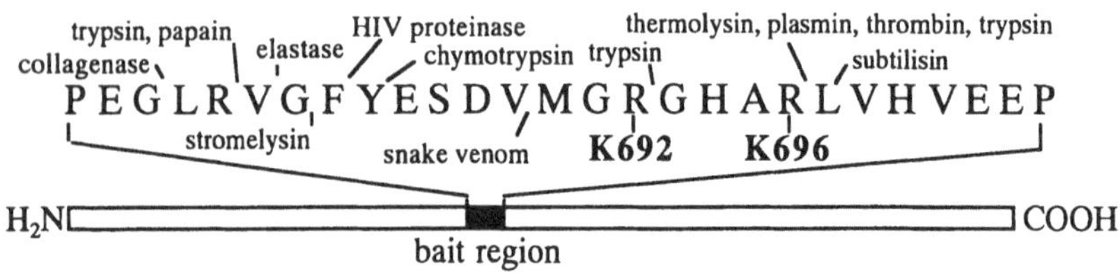

Figure 1. The location and sequences of proteolytic cleavage sites within the bait region of the α2M subunit (modified from Ref. 2).

2. Materials and Methods

2.1. CELL CULTURES, PLASMID AND SYNTHETIC DNA

Chinese hamster ovary (CHO) K1 (ATCC CCL61) cells and HepG2 (ATCC HB8065) cells were grown as described (5). HepG2 cells were used as a source for total RNA isolation. Eukaryotic expression vector, pcDSRαΔ-neoΔ11, was provided by Dr. Noda (6). Oligomers were synthesized by an automated DNA synthesizer (Applied Biosystems DNA synthesizer, Perkin Elmer, Chiba, Japan), and used as hybridization probe, primers for PCR and mutagenesis.

2.2. PREPARATION OF PROBE DNAs BY RT-PCR

α2M 5'- and 3'-end fragment probes were obtained by the reverse transcriptase-coupled polymerase chain reaction (RT-PCR) with primer combinations, 5'caacgatctgcaggctagCATG GGGGAAGAACAAACTCCTTC3' (lower case letters indicate Pst I site containing the tail sequence and upper case letters correspond to the nucleotide numbers 43 to 65 of the reported sequence (7)) and 5'ACAGATGCAGTGAGGCGCTTGTTCC3' (Hae II site containing the primer sequences corresponding to the nucleotide numbers 2566 to 2590 of the reported sequence (7)) for 5'-end fragment; or 5'TGGAGAAGGAACAAGCGCCTCACTG3' (Hae II site containing primer sequences correspond to the nucleotide numbers 2559 to 2583 of the reported sequence (7)) and 5'gttgctaggtaccctcgagTGTGGTCTTCATGCATTTCCAAG 3' (lower case letters indicate Kpn I site containing the tail sequence and upper case letters correspond to the nucleotide numbers 4454 to 4476 of the reported sequence (7)) for 3'-end fragment. These fragments were separately subcloned into pUC19 plasmid and their sequences were confirmed as the segments of α2M gene by sequencing (Sequenase Ver. 2, United states Biochemical, Co., Cleveland, Ohaio, USA). Plasmid DNAs were labeled using [α-^{32}P]dCTP (3000 Ci/mmol, Amersham, England) and a Random Primer Labeling kit (Takara, Shiga, Japan), and used as probes to isolate full length α2M cDNAs from HepG2 cDNA library.

2.3. ISOLATIONS OF hα2M AND ITS MUTANT cDNAS

Total RNA was isolated using RNA purification reagent (RNAzol B, Biotex Laboratories, Houston Texas, USA) and poly $(A)^+$ RNA was selected from total RNA using oligodT cellulose column. Double stranded cDNA was synthesized using HepG2 poly $(A)^+$ RNA and cloned into λgt10 phage vector. Candidate clones obtained by ^{32}P-labeled α2M 5'- and 3'-end fragment probes were further screened to select full length cDNAs using PCR method. The full length α2M cDNA was subcloned into pUC19 and nucleotide sequence was determined.

Using present hα2M cDNA as a template and in vitro mutagenesis method, we have isolated two bait region mutant cDNAs, which we will refer to as K692 and K696 (Fig. 1). They were designed such that Arg residues at 692 and at 696 are respectively substituted with Lys residues when translated.

2.4. CULTIVATION OF rCHO CELLS AND PURIFICATION OF rα2M

Recombinant CHO cell clones producing either wild-type or mutant rhα2M, ca. 2 x 10^7 cells, was inoculated into each 1750 cm^2 roller bottle containing 200 ml of the serum-supplemented complete medium and grown for two days. The complete medium was replaced by a serum free medium (UC212, Nissui, Tokyo, Japan) and incubated for two days. Fresh serum free medium was again added for second round cultivation for additional two days. Spent medium of 4-5 liters was collected during the serum free culture.

For the purification of wild type rhα2M and mutant type rhα2M (rhα2M/K692), spent media were treated with hydrochloric acid to adjust its pH to 6.0 and passed through a Chelating Sepharose Fast Flow column saturated with Zinc ion (Pharmacia Biotech, Japan). Bound α2M was eluted and the elution buffer was replaced by PBS buffer. α2Ms from human and bovine plasma were purified by the published method (8).

3. Results and Discussion

3.1. CONSTRUCTION AND EXPRESSION OF RECOMBINANT α2M VECTOR

We took several steps to isolate the α2M cDNA sequence. First to evaluate a quality of poly $(A)^+$ RNA prepared from HepG2 total RNA, synthetic DNA was used to probe a Northern blot. The probe DNA detected α2M RNA in both total and poly $(A)^+$ RNAs (Fig. 2). The α2M mRNA was shown to be efficiently concentrated in the poly $(A)^+$ fraction (Fig. 2, lane

b) compared to that of total RNA (Fig. 2, lane a). This poly (A)$^+$ fraction was used to construct cDNA library.

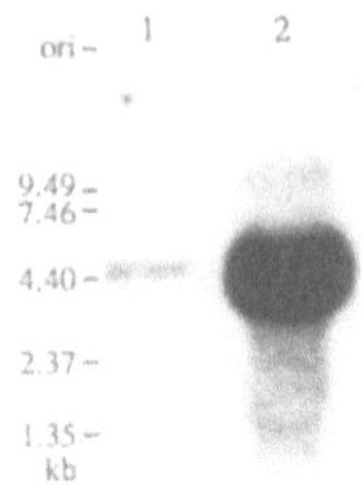

Figure 2. Detection of α2M mRNA by Northern blot analysis.
Total (20μg) and poly (A)$^+$ (6μg) RNAs were resolved in a 1% formaldehyde-agarose gel and α2M mRNA was detected using ^{32}P-end labeled oligo-probe (5'TGGTCTTCAAGCATTTCAAG3'). lane a: total RNA, Lane b: poly (A)$^+$ RNA.

RT-PCR amplified α2M cDNA fragments were used as probes for full length cDNA screening as shown in the Materials and Methods. Determined α2M cDNA sequence was compared with published sequences (Table 1). Nine silent changes were identified between the liver sequence and the present cDNA (Table 1, liver[a] and HepG2[c]). Among these changes, a new silent codon change not identified previously was found at codon 125 in the present cDNA (Table 1, indicated by a dashed underline). Codon 413 and 796 in the present cDNA were identical to those of the liver sequence (Table 1, liver[a] and HepG2[c]).

TABLE 1. Silent codon changes in human α2M cDNA

codon	liver[a]	HepG2[b]	HepG2[c]
125(Glu)	GAG	GAG	GAA
413(Asn)	AAC	AAT	AAC
495(Phe)	TTT	TTC	TTC
750(Gly)	GGG	GGT	GGT
796(Leu)	CTT	CTC	CTT
835(Leu)	CTT	CTA	CTA
1266(Ala)	GCC	GCA	GCA
1296(Asn)	AAT	AAC	AAC
1326(Thr)	ACC	ACA	ACA
1442(Leu)	CTC	CTG	CTG
1460(Ile)	ATC	ATT	ATT

a: (7), b: (4), c: (Present work)

Codon 1000 was ATC encoding isoleucine in the present cDNA and reported HepG2 cDNA by others (4) while the published liver sequence is GTC encoding Valine. These sequence analyses indicate that the cDNA isolated here is a representative cDNA clone for hα2M expressed in HepG2 cells. Therefore this cDNA was ligated to pcDSRαΔ-neoΔ11 vector and the recombinant plasmid was designated as pcDSRαΔ-neoΔ11/hα2M. This recombinant plasmid was transfected to CHO cells by calcium phosphate coprecipitation method followed by neomycin selection. After 10 days of drug selection, one colony (#8) out of 96 primary neomycin resistant colonies was selected on the basis of cell staining method and used for further purification by limiting dilution method. A monoclone, 8d (Fig. 3), was used to purify rhα2M by using roller bottles as described in the Materials and Methods. This cultivation method yielded 5.2mg of purified rα2M protein per liter of spent medium.

As to the two mutant cDNAs, we determined their bait region sequences, and confirmed that they were as designed. Using K692 cDNA, a recombinant CHO cell clone, rCHO/K692, was obtained. We obtained 1.2mg of purified rhα2M/L692 protein per liter of spent medium.

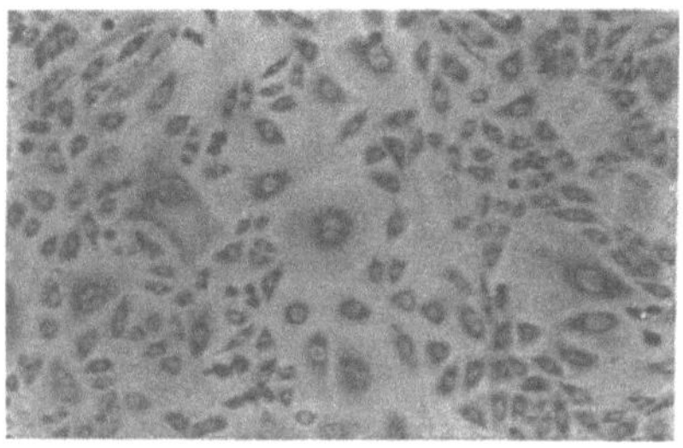

Figure 3. Immunostaining and phase-contrast microscopy of α2M-producing CHO cell line. α2M-producing cells were immunologically stained using an improved immunostaining method presented at the JAACT meeting (July 1995, Yamagata, Japan).

3.2. AUTHENTICITY OF RECOMBINANT HUMAN α2M

Although we used serum free medium to obtain rhα2M preparations, we could not exclude the possibility that the purified rα2M is contaminated by bovine α2M (bα2M).

To detect contamination of bα2M, if any, we treated an rhα2M protein preparation with methylamie, and submitted to electrophoresis, since it has been known that hα2M changes its conformation by a methylamie treatment of cleave its thioester bonds, and runs fast in electrophoreses (fast-form), while bα2M remains slow (8). Results show that, after methylamine treatment, the rhα2M preparation gave a band in the position of the fast-form, without any trace band at the slow-form position (Fig. 4 lane 9), while plasma derived bα2M (Fig. 4 lane 3) remained mostly as the slow-form (Fig. 4 lane 5), and plasma derived hα2M (Fig. 4 lanes 1 and 6) were converted to the fast-form (Fig. 4 lanes 2 and 7), as reported (8). Also as expected, plasma derived bα2M was converted to the fast-form by trypsin treatment (Fig. 4 lane 4).

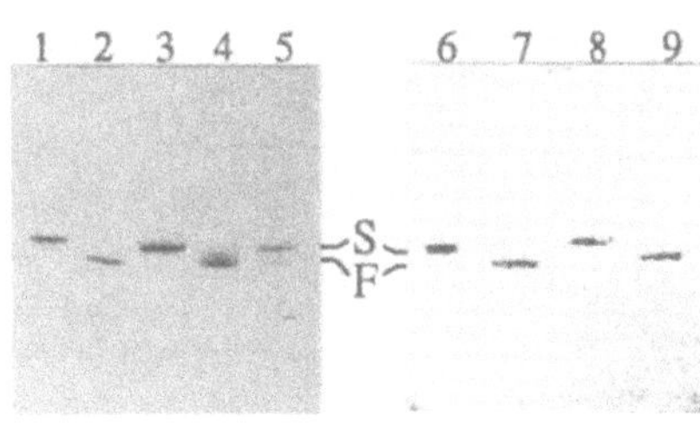

Figure 4. Comparison of rhα2M with human and bovine serum α2Ms.
α2Ms purified from different sources were treated with methylamine or trypsin, and analyzed by native-polyacrylamide gel electrophoresis. Lanes 1, 2, 6 and 7: plasma derived hα2M; Lanes 3-5: plasma derived bα2M; Lanes 8 and 9: rhα2M purified from rCHO culture medium. Lanes 2, 5, 7 and 9: α2Ms treated with 100mM methylamine for 15 hours at 37 °C. Lane 4: treated with trypsin.

Another characteristics of human serum α2M is the difference in two amino acid residues at codons 1 and 3 from those of bovine α2M (4). To confirm this, we determined N-terminal amino acid sequence for purified rhα2M. Comparison of N-terminal sequences of present and published human and bovine α2Ms showed that rα2M purified here composed of the same amino acid sequence as the plasma derived hα2M (Table 2).

TABLE 2. Comparison of N-terminal amino acid (A.A.) sequences

A.A.number	1	2	3	4	5	6	7	8	9	10	11	12
rhα2M	Ser-	Val-	Ser-	Gly-	Lys-	Pro-	Gln-	Tyr-	Met-	Val-	Lew-	Val[a]
human serum α2M	Ser-	Val-	Ser-	Gly-	Lys-	Pro-	Gln-	Tyr-	Met-	Val-	Lew-	Val[b]
bovine α2M	Ala-	Val-	Asp-	Gly-	Lys-	Pro-	Gln-	Tyr-	Met-	Val-	Lew-	Val[b]

a: present work, b: (4)

These results demonstrated that the rhα2M obtained in this study is the human type and contaminated minimally with bα2M.

3.3. CHARACTERIZATION OF NATURAL TYPE AND K692 MUTANT α2Ms

It has been shown that thioester-bonds within the plasma derived ha2M break when heated the nearby peptide bonds, and yeald 130kDa and 50kDa peptides (9). In a heat treatment experiment, plasma derived hα2M (Fig. 5A lanes 4) was converted to 130kDa and 50kDa peptides as expected (Fig. 5A lane 3). Under the same reaction condition, rhα2M and rhα2M/K692 also produced 130kDa and 50kDa peptides (Fig. 5A lanes 1-2). Next, we compared reactivity with trypsin. As expected, the plasma derived hα2M without trypsin treatment migrated as the slow-form (Fig. 5B lane 1), while trypsin treated hα2M migrated as the fast-form (Fig. 5B lane 4). Present rhα2M (Fig. 5B lane 2) and rhα2M/K692 (Fig. 5B lane 3) also migrated as the fast-form (Fig. 5B lanes 5 and 6, respectively) after trypsin treatment.

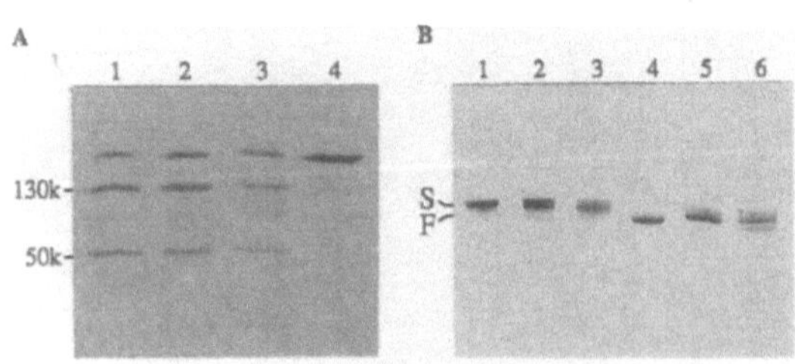

Figure 5. Characterization of rhα2M and rhα2M/K692 with human serum α2M.

α2Ms purified from different sources were treated with heat or trypsin and analysed by polyacrylamide gel electrophoresis. A; α2Ms were treated for 15 min. at 95 °C and analyzed on SDS-PAGE. lane 1: human serum α2M, lane 2: rhα2M, lane 3: rhα2M/K692 lane 4: untreated human serum α2M. B; α2Ms were treated with trypsin and analyzed on native SDS-PAGE. lanes 1and 4: human serum α2M, lanes 2 and 5: rhα2M, lanes 3 and 6: rhα2M/K692, lanes 1-3: untreated, lanes 4-6 : trypsin treated.

Results of heat treatment and trypsin digestion experiments together with the methylamine experiment support that thioester- bonds were properly formed in the rhα2M and rhα2M/K692 molecules, and that their conformation, as studied electrophoretically, is indistinguishable from the plasma derived hα2M. Further details of rhα2M/K692 and rhα2M/K696 mutants will be reported elsewhere.

4. References

1. Borth, W. (1992) α2-macroglobulin, a multifunctional binding protein with targeting characteristics, *FASEB* **10**, 3345-3353.
2. Chu, C. T., and Pizzo, S. V. (1994) Biology of disease; amacroglobulin, complement, and biologic defence: antigens, growth factors, microbial proteiases, and receptor ligation, *Lab. Invest.* **71**, 792-812.
3. Sottrup-Jensen, L., Sand, O., Kristensen, L., and Fey, G. H. (1989) The α-macroglobulin bait region: Sequence diversity and localization of cleavage sites for proteinases in five mammalian a-macroglobulins, *J. Biol. Chem.* **264**, 15781-15789.
4. Boel, E., Kristensen, T., Peterson, C. M., Mortensen, S. B., Gliemann, J., and Sottrup-Jensen, L. (1990) Expression of human α2-macroglobulin cDNA in baby hamster kidney fibroblasts: Secretion of high levels of active α2-macroglobulin, *Biochemistry* **29**, 4081-4087.
5. Nakamichi, N., Ito, M., Hirota, Y., and Matsumura, T. (1997) Establishment of recombinant human transferrin -producing CHO cell lines, in K. Funatsu et al. (eds.), *Animal Cell Technology: Basic &Applied Aspects* **8**, 373 -379.
6. Noda, A., Ning, Y., Venable, S. F., Pereira-Smith, O. M., and Smith, J. R. (1994) Cloning of senescent cell-derived inhibitors of DNA synthesis using an expression screen, *Exp. Cell Res.* **211**, 90-98.
7. Kan, C.-C., Solomon, E., Belt, K. T., Chain, A. C., Hiorns, L. R., and Fey, G. (1985) Nucleotide sequence of cDNA encoding human α2-macroglobulin and assignment of the chromosomal locus, *Proc. Natl. Acad. Sci. USA* **82**, 2282-2286.
8. Dangott, L. J. and Cunningham, L. W. (1982) Residual alpha 2-macroglobulin in fetal calf serum and properties of its complex with thrombin, *Biochem. Biophys. Res. Commun.* **107**, 1243-1251.
9. Howard, J. B. (1980) Reactive site in human alpha 2-macroglobulin: circumstantial evidence for a thioester, *Proc. Natl. Acad. Sci. USA* **78**, 2235-2239.

EVALUATION OF STABILITY IN THE DHFR GENE AMPLIFICATION SYSTEM USING FLUORESCENCE *IN SITU* HYBRIDIZATION

Fumi Nakanishi, Tomohiro Yoshikawa, Seima Itami, Takeshi Omasa, Yoshio Katakura, Michimasa Kishimoto and Ken-ichi Suga
Department of Biotechnology, Graduate School of Engineering, Osaka University,
2-1 Yamada-oka, Suita, 565 JAPAN
Telephone: +81-6-879-7437, Facsimile: +81-6-879-7439
e-mail: omasa@bce.bio.eng.osaka-u.ac.jp

ABSTRACT: In order to obtain a stable and highly productive gene-amplified recombinant Chinese Hamster Ovary (CHO) cell line, five kinds of stepwise methotrexate (MTX) selection were carried out. The specific growth and production rates of the cells were compared with each other, and the distribution of the location of amplified genes was determined using fluorescence *in situ* hybridization (FISH). The specific growth and production rates of the cell line obtained under the selection condition in which the stepwise increase in the MTX concentration was most gradual reached the highest levels and under this condition about 80% of the amplified genes were observed near the telomeric site. During long-term cultivation without MTX, the percentage of amplified genes near the telomeric site hardly changed, but that of amplified genes at other sites decreased. The specific production rate gradually decreased during cultivation without MTX. To clarify the relationship between the specific production rate and the location of amplified genes, a cloned cell line, DR1000L4N, was obtained. This cell line showed higher productivity, and the amplified genes were all situated near the telomeric site.

1. Introduction

Gene amplification techniques using recombinant mammalian cells are frequently employed for the production of glycoproteins, among which the dihydroforate reductase (DHFR) gene amplification system in the CHO cell line is the most common [1]. In this gene amplification system, highly productive CHO cells can be selected by increasing the MTX concentration in the culture medium stepwise. In the industrial production of recombinant proteins, it is most important to be able to construct a recombinant CHO

Y. Kitagawa et al. (eds.), Animal Cell Technology: Basic & Applied Aspects, Vol. 10, 259–263.

cell line that stably produces the desired recombinant protein. We selected the human granulocyte-macrophage colony-stimulating factor (hGM-CSF) as a model glycoprotein and investigated the relationship among the specific production rate, the location distribution and stability of the amplified genes, and the pattern of increase in the MTX concentration, for the selection of a stable and highly productive cell line.

2. Materials and Methods

2.1 CELL LINE AND CULTURE CONDITIONS

The CHO DG44 cell line (*dhfr*$^-$), which was kindly provided by Dr. L. Chasin of Columbia University, was used as the host. The host cells were grown in Iscove's modified DMEM (Sigma) with 10% fetal bovine serum (Gibco), hypoxanthine (13.6 mg/L), and thymidine (2.42 mg/L). Cells were incubated in 5% CO_2 at 37°C.

2.2 TRANSFECTION AND SELECTION

The vector pSV2-dhfr/hGM-CSF was constructed from pSV2-dhfr (ATCC 37146) and pcD-hGM-CSF (ATCC 57594). CHO DG44 cells were transformed with pSV2-dhfr/hGM-CSF using the calcium phosphate co-precipitation method. For dhfr gene amplification, cells were cultivated in a selection medium (Iscove's modified DMEM with 10% dialyzed fetal bovine serum) in which the MTX concentration was increased under one of five patterns (Figure 1).

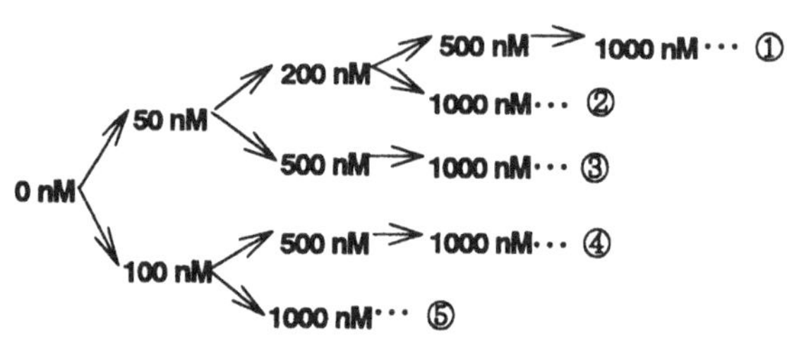

Figure 1. Stepwise selection in the dhfr gene amplification system.

2.3 ELISA ANALYSIS

The hGM-CSF concentration was measured by sandwich ELISA using 96-well plates; 100μL of culture supernatant was added to each well previously coated with sheep anti-hGM-CSF antibody (Endrogen p-521). Polyclonal rabbit anti-hGM-CSF (Genzyme LP-714) was added as the first step antibody. Alkaline phosphatase-conjugated anti-rabbit IgG antibody (Tago ALI3405) was used as the enzyme-linked second-step antibody. Purified recombinant hGM-CSF (Genzyme RH-CSF-C) was used as a standard.

2.4 FLUORESCENCE *IN SITU* HYBRIDIZATION

Chromosomes were fixed onto slides by the standard technique [2]. Fixed chromosomal DNA was denatured in 70% formamide and 2×standard saline citrate (SSC) at 70°C for 2 min. The DHFR probe was biotinylated by nick-translation with biotin-16-dUTP (Boehringer Mannheim 1093070). The probe was denatured for 10 min at 80°C and then mounted on slides shielded by a paper bond. Hybridization was carried out overnight at 37°C. The probe was detected using fluorescein isothiocyanate (FITC)-labeled streptoavidine. The chromosomes were stained with 4,6-diamidino-2-phenylindole (DAPI) and observed under a Zeiss Axioplan microscope. On the basis of the observations, the amplified genes were classified into three groups (telomere type, other types, and those with no signal).

3. Results and Discussion

3.1 EFFECT OF STEPWISE INCREASE OF MTX CONCENTRATION ON SPECIFIC GROWTH AND PRODUCTION RATES AND AMPLIFIED GENE LOCATION

Recombinant cells were cultivated in the medium with MTX, which inhibits DHFR. As a result of the stepwise increases in the MTX concentration, most cells died, but a few cells were able to survive because of the dhfr gene amplification. Since a suitable pattern for the increase in the MTX concentration has not hitherto been clarified, we carried out stepwise MTX selection under five different selection conditions (patterns① to ⑤ in Figure 1), from which five cell lines (named DR1000L1 to DR1000L5) were obtained. The specific growth (μ) and hGM-CSF production (ρ) rates of these cell lines were calculated and the amplified gene locations were determined using FISH. The distribution of amplified gene locations was found to depend on the pattern of MTX increase. The specific hGM-CSF growth and production rates reached the highest levels when the stepwise increase in the MTX concentration was most gradual, i.e., under pattern ① (Figure 2a), with about 80% of the amplified genes being observed near the telomeric site (Figure 2b).

(a) (b)

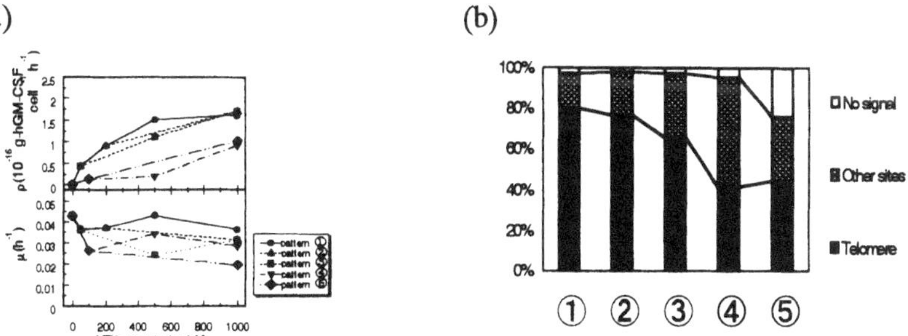

Figure 2. Relationship between pattern of increase in MTX concentration and specific growth and production rates (*a*) and the amplified gene locations (*b*).

3.2 STABILITY OF SPECIFIC GROWTH AND PRODUCTION RATES AND AMPLIFIED GENE LOCAITON IN LONG-TERM CULTIVATION WITH OR WITHOUT MTX

MTX is used as an anticancer compound. However, because liver and kidney functional lesions have been induced as a result of administering MTX to patients, in industrial pharmaceutical production, MTX has to be removed from the cultivation medium. In order to evaluate the relationship between stability and the locations of amplified genes, we carried out long-term cultivations with or without MTX. In these experiments, the DR1000L1 cell line, which had the highest specific production rate, was used. During the cultivation with MTX, the specific growth and production rates remained constant. However, in the cultivation without MTX, the specific hGM-CSF production rate had decreased by approximately 30% compared with the initial rate after 45 days (Figure 3a). Without MTX, the percntage of amplified genes near the telomeric site hardly changed, while the percentage of genes giving no signal gradually increased, along with a decrease in that of amplified genes situated at other sites (Figure 3b).

(a) (b)

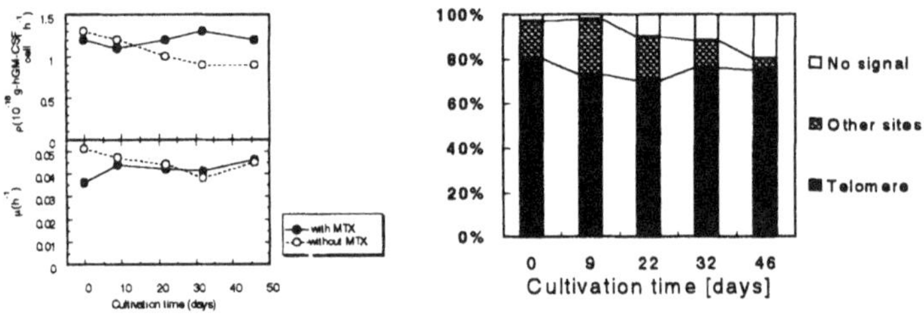

Figure 3. Changes in the specific growth and production rates (*a*) and in the amplified gene locations (*b*) during long-term cultivation without MTX using DR1000L1 cell line.

3.3 RELATIONSHIP BETWEEN SPECIFIC PRODUCTION RATE AND AMPLIFIED GENE LOCATION

In order to clarify the relationship between the specific production rate and the amplified gene location, two cloned cell lines (named DR1000L4N and DR1000L4F) were obtained from the pattern ④ cell line. In DR1000L4N, amplified genes were all situated near the telomeric site, whereas in DR1000L4F they were located at other sites (Figure 4).

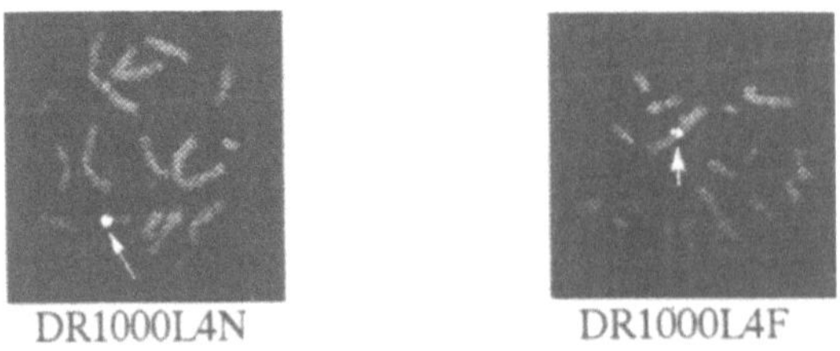

DR1000L4N DR1000L4F

Figure 4. Results of FISH in cloned cell lines.

Long-term cultivations with or without MTX were carried out using the two cloned cell lines. The results showed that the cell line in which the amplified genes were situated near the telomeric site had higher productivity (Figure 5). The amplified gene location thus affects the specific production rate.

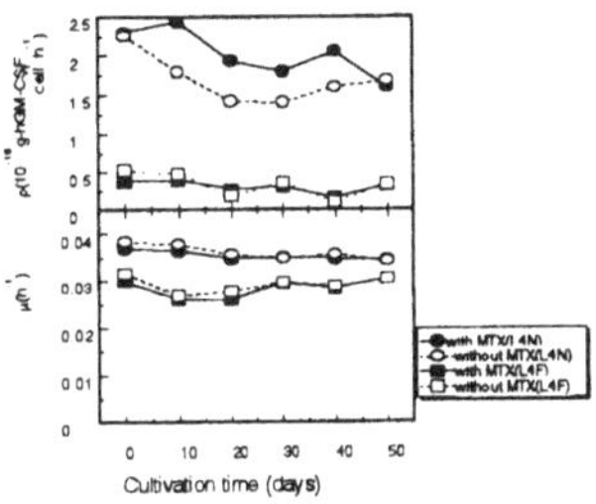

Figure 5. Changes in the specific growth rate (μ) and specific production rate (ρ) during long-term cultivation using the DR1000L4N and DR1000L4F cell lines.

The results of the study showed that the cell line with amplified genes near the telomeric site had higher productivity. In other words, it is desirable to select the cells that have amplified genes in the telomeric region for the construction of a stable and highly productive cell line. Such a cell line can be obtained by gradual stepwise MTX selection.

We are now plannning to investigate a method for the selection of cell lines with amplified genes situated near the telomeric site from heterogeneous cell lines using a cell sorter, in order to establish a rapid means of constructing stable and highly productive cell lines.

4. Acknowledgments

We thank Dr. L. Chasin of Columbia University for providing the CHO DG44 cell lines and Dr. K. Okumura of Mie University for directing us in FISH hybridization techniques.

5. References

[1] Kaufman R. J. (1985) 'Coamplification and coexpression of human tissue-type plasminogen activator and murine dihydrofolate reductase sequences in Chinese hamster ovary cells.' Mol. Cell. Biol, 5, 1750-1759.

[2] Abe T., Inazawa J. (1997) *Rinsyou FISH protocol,* Shujunsha, Tokyo (in Japanese).

EXPRESSION AND CHARACTERIZATION OF HUMAN CALPAIN AND CALPASTATIN USING BACULOVIRUS SYSTEM

KIYOTAKA HITOMI and MASATOSHI MAKI

Department of Applied Biological Sciences, School of Agricultural Sciences, Nagoya University, Nagoya 464-8601 Japan

ABSTRACT. Calpain, the Ca^{2+}-dependent cysteine endopeptidase, and its endogenous inhibitory protein calpastatin are distributed in animal cells and are involved in Ca^{2+} signalling pathway. Using baculovirus expression system, human μ-calpain with active-site-mutation and calpastatin are expressed and purified. The large subunit of the mutant μ-calpain was processed by the wild type calpain, supporting the intermolecular cleavage mechanism of procalpain during activation. The mutant μ-calpain had no proteolytic activity, but retained the high ordered structure of wild type calpain. The recombinant calpastatin had inhibitory activity for μ- and m-calpain.

INTRODUCTION A series of stimulators such as peptide hormones, growth factors, and biogenic amines elevate the concentration of intracellular Ca^{2+} and induce various biological reactions. Calpain (E.C.3.4.22.17), a Ca^{2+}-dependent cysteine protease, is suggested to be involved in limited proteolysis of certain enzymes, receptors, cytoskeletal proteins, transcription factors, etc. coupled with calcium mobilization[1-3]. Based on the Ca^{2+} requirement for their protease activity, two major isoforms of calpain have been identified: μ-calpain (μM-requiring form) and m-calpain (mM-requiring form). Calpastatin is an endogenous inhibitor protein that acts specifically on both types of calpain. Both calpain and calpastatin are known to be widely distributed in animal tissues and cells[4].

The calpain-calpastatin system has been suggested to be involved in various physiological and pathological phenomena as platelet activation, development, and Alzheimer disease[5-7]. Purifications of calpain and calpastatins from various tissues have been reported[8]. Analysis of interaction has been also studied[9], however, the amounts of native protein from human sources are so limiting that basic biochemical studies are restricted.

In this paper, we will describe expression, purification and characterization of active site-mutated human μ-calpain and calpastatin in baculovirus system. This system have

Y. Kitagawa et al. (eds.), Animal Cell Technology: Basic & Applied Aspects, Vol. 10, 265–269.

been frequently used to produce and characterize high-molecular weight proteins expressed in eukaryotic cells[10].

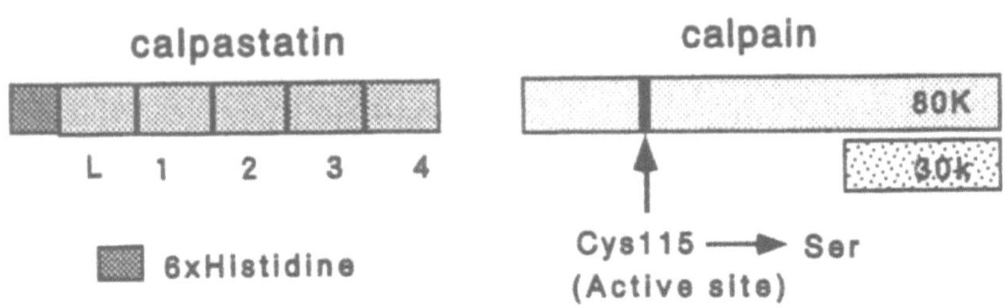

Figure 1. **Schematic structure of calpastatin and calpain for expression of recombinant protein in baculovirus system.** (left) Calpastatin. L indicates unique L (leading) domain. Four repeats of mutually homologous domains following L domain. Hexahisitidine residues were tagged at the N-terminal side for purification. (right) Calpain consists of large (80k) and small (30k) subunit. Active site was mutated from Cys to Ser.

RESULTS AND DISCUSSION

1 *Expression, purification and characterization of human calpastatin* [11]

A calpastatin molecule contains four mutually similar regions of 140 amino acid residues (domains 1-4) and a unique domain on the N-terminal side (domain L)(Fig.1, left). For detection and purification of recombinant calpastatin in Sf9 cells, cDNAs were modified to attach tags of hexahistidine (His). For expression in *E.coli*, pET24d-derived plasmids that bear the cDNAs under the control of the T7 promoter were used to transform into BL21(DE3)LysE. As in the baculovirus expression system, cDNAs were modified to produce recombinant proteins tagged with His.

Both in baculovirus and *E.coli* systems, recombinant calpastatins were successfully expressed. As shown in Fig.2, both recombinant protein migrated aberrantly at 110 kDa in SDS-PAGE as previously reported for calpastatins isolated from animal tissues[8]. This abnormal mobility is assumed to be due to the deviant amino acid composition:rich in Asp, Glu, Ser, and Pro.

Since calpastatin is known to be resistant to denaturing treatments, recombinant calpastatins from both systems were purified by nearly identical methods including denaturation steps. The initial extract of 6 M guanidine-HCl soluble protein was affinity purified by Ni-NTA chromatography. The eluted His-tagged calpastatin preparation appeared almost homogeneous. The fraction was further boiled to inactivate potentially contaminating proteases. To obtain finally purified calpastatin, contaminated proteins with low molecular weight in the soluble fraction were excluded by size fractionation. The SDS-PAGE profile of proteins at each step of the purification is shown in Fig.2.

After these purification steps, approximately 100 μg of His-tagged calpastatin was obtained from 100 ml of suspension culture (2x10^7) of Sf9 cells. In the case of E.coli, approximately 1 mg of His-tagged calpastatin was obtained from 100 ml of culture.

Purified recombinant calpastatin were investigated its inhibitory activity for calpain. Proteolysis of bovine casein , which were commonly used substrates of calpain[2], was monitored in the presence or absence of recombinant calpastatin. The enzymatic activity of m-calpain and μ-calpain was inhibited depending on the content of recombinant calpastatin from both systems.

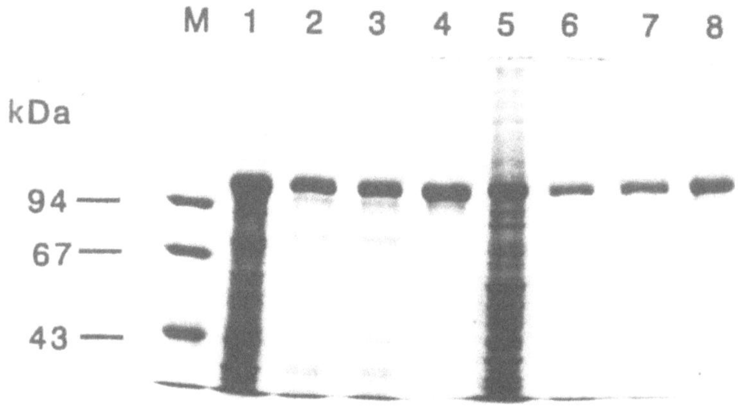

Figure 2. **Purification of the calpastatins expressed in Sf9 cells and in *E.coli*.**
Proteins (approximately 1-10μg) of each steps from E.coli (lanes1-4) and baculovirus-infected Sf9 cells (lanes5-8) were separated by 7.5% SDS-PAGE and the gel was stained with CBB staining. Lanes 1, 5, total cellular extract;lanes 2, 6, eluted fraction of affinity chromatography;lanes 3, 7, supernatant by heat treatment;lanes 4, 8, peak fraction of size separation.

2 *Expression, purification , and characterization of mutant calpain[12]*

For recombinant expression of calpain, baculovirus system were used as in the case of calpastatin. Since calpain consisted of two subunits, simultaneous expression of both subunits are required using dual promoter vector. Recombinant expression virus for active calpain have been reported recently [13]. To analyze the interaction with calpastatin, we decided to express calpain at an inactive form. The mutant calpain, in which the active site was mutated (Cys115 to Ser), was attempted to express(Fig.1, right). The calpain large(80kDa) subunit, however, was found mostly in the insoluble fraction. On the other hand, the expression level of small(30kDa) subunit was low, but the protein was recovered mostly in the soluble fraction.(data not shown). From the suspension-cultured Sf9 cells infected, mutant μ-calpain were partially purified by DEAE-Sephacel and Phenyl-Sepharose chromatography (data not shown).

Analysis by the Fluorescence Polarization revealed that the mutant calpain could interact with fluoroscein-labeled functional domain of calpastatin in a calcium-dependent manner(data not shown).

To confirm that recombinant mutant calpain had same structural conformation as the wild type, the autolysis by active calpain was investigated (Fig.3). Both native μ-and m-calpain are zymogen and changed to activated form in the presence of calcium ion. Upon activation, the large subunit are autolysed from 80kDa to 76kDa by limited proteolysis[1]. When incubating 1 μg of mutant calpain with smaller amounts of wild type pig erythrocyte μ-calpain, large subunit were processed to 76kDa. Casein were simultaneously digested in the reaction showing the calpain activity. The N-terminal sequence of processed large subunit (76kDa) of mutant calpain identical with that of wild type (data not shown). These results indicated that mutant recombinant calpain had maintained the native structure except for the active site.

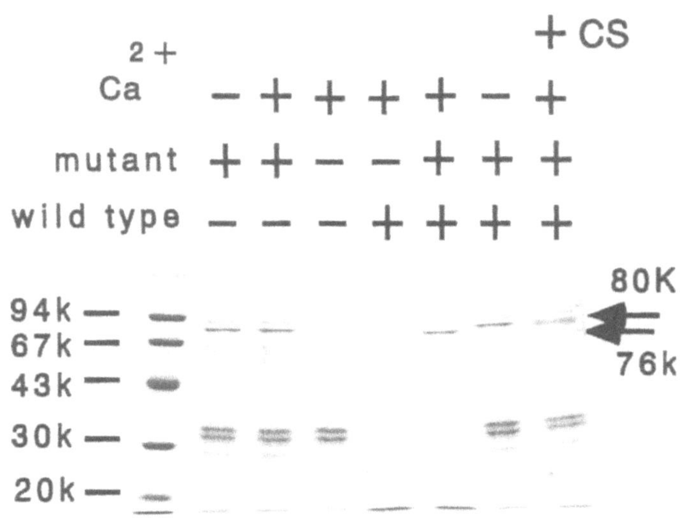

Figure 3. **Intermolecular autolysis of μ-calpain.** The non-active mutant μ-calpain(1 μg) are proteolysed by wild type μ-calpain (50ng) in the presence or absence of 2mM CaCl2. In the reaction mixture, caseins are digested simultaneously as a merkmal of calpain activity. The arrows with 80k and 76k indicates unprocessed (unactivated) and processed (activated) form of large subunit, respectively. Addition of 500ng calpastatin (CS) were added for inhibition of calpain.

3 *Conclusion*

Human calpastatin are highly expressed in baculovirus system and E.coli. His-tagged calpastatin were purified to homogeneity and confirmed to be biologically active.

Human mutant μ-calpain were also expressed in baculovirus system and partially purified. This molecule were non-active but processed by activated calpain, similarly with wild-type calpain. Recombinant μ-calpain interacts with calpastatin inhibitory domain with similar kinetics.

REFERENCES

[1] Suzuki, K., Imajoh S., Emori Y., Kawasaki H., Minami,Y.and Ohno S. (1988) Regulation of activity of calcium activated neutral protease., *Adv.Enzyme Regul.*, **27**, 153-169.

[2] Mellgren,R.L.and Murachi T.(Eds.), (1990) Intracellular calcium-dependent proteolysis., CRCPress, Boca Raton, FL, U.S.A.

[3] Croall D.E.,and DeMartino G.N. (1991) Calcium-activated neutral protease (calpain) system: structure, function, and regulation. , *Physiological Reviews*, **71**, 813-847.

[4] Murachi, T. (1983) Calpain and calpastatin., *Trends in Biochem.Sci.*, **8**, 167-169.

[5] Ariyoshi, E., Shiba, M., Sakon, Kambayashi J., Yoshida K., Kawashima S., and Mori T. (1993) Translocation of human platelet calpain-I., *Biochem.& Mol.Biol.Int.*, **30**, 63-72.

[6] Emori,Y.and Saigo,K. (1994) Calpain localization changes in coordination with actin-related cytoskeletal changes during early embryonic development of Drosophila *J.Biol.Chem.*, **269**, 25137-25142.

[7] Saito K., Elce J.S., Hamos J.E., and Nixon R.A. (1993) Widespread activation of calcium-activated neutral proteinase (calpain) in the brain in Alzheimer disease: a potential molecular basis for neuronal degeneration. *Proc.Natl.Acad.Sci.U.S.A.*, **90**, 2628-2632.

[8] Parkes, (1986) in Preteinase Inhibitors (A.J.,Barrett, and G., Salvesen eds.) Elsevier Science Publishers, B.V., Amsterdam, 571-587.

[9] Maki, M., Takano, E., Osawa, T. , Ooi, T., Murachi, T., and Hatanaka, M. (1988) Analysis of structure and function relationship of pig calpastatin by expression of mutated cDNAs in *Esherichia coli.*. *J.Biol.Chem.*, **263**, 10254-10261.

[10] Baculovirus Expression Protocols,ed.by Pichardson C.D.,(1995) Methods in Molecular Biology, Humana Press, Totowa, New Jersey,.

[11] Hitomi, K.,Yokoyama, A.and Maki, M.(1998) Expression of biologically active human calpastatin in baculovirus-infected insect cells and in *Ecsherichia coli.* *Biochem. Biosci. Biotech.*, **62**,136-141.

[12] Hitomi, K., Utiyama, Y., Ohkubo, I., Kunimatsu, M., Sasaki, M., and Maki, M. (1998) Purification and characterization of the active-site-mutated recombinant human mu-calpain in baculovirus-infected Insect cell. *Biochem. Biophys.Res.Commun.*, **246**, 681-685.

[13] Meyer S.L., Bozyczko-coyne D., Mallya, S.K., Spais C.M.,Bihovsky R.,Kawooya J.K.,Lang D.M., Scott R.W.,and Siman R.,(1996) Biologically active monomeric and heterodimeric recmbinant h uman calpain I produced using the baculovirus expression system. *Biochem.J.*, **314**, 511-519.

EVALUATION OF SPIM INSECT CELLS USING A NOVEL BACULOVIRUS EXPRESSION VECTOR SYSTEM EMPLOYING THE *HYPHANTRIA CUNEA* NPV

YUJI TAKENAKA, JUN KOBAYASHI, YOSHIAKI MATSUDA, CHENGLIANG GONG, MASAHIRO NAGAYA, SHIGETOSHI MIYAJIMA AND TETSURO YOSHIMURA
Laboratory of Molecular Bioengineering, Faculty of Engineering, Mie University, Tsu, Mie 514, Japan

ABSTRACT. Using a lepidopteran insect cell line, FRI-SpIm-1229 (SpIm), and the *Hyphantria cunea* nucleopolyhedrovirus (HycuNPV), we have developed a novel baculovirus expression vector system. When SpIm cell suspension culture was adapted to a serum-free medium, Sf-900 II, and scaled up to 50 ml by the simple shaker culture, no significant reduction in the cell growth as well as HycuNPV replication was observed. We have constructed a HycuNPV polyhedrin promoter-based transfer vector plasmid, pHcMU1, and obtained a recombinant HycuNPV expressing an insect peptide hormone (PTTH). The recombinant PTTH was N-glycosylated and secreted in the culture supernatant, suggesting that SpIm cells could recognize and process correctly both of the N-glycosylation and signal sequences. The expression level of PTTH in HycuNPV/SpIm cell system using serum-free Sf-900 II medium was comparable to that in the BmNPV/BmN4 cell system using TC-100 medium with 10% fetal bovine serum.

1. Introduction

The baculovirus expression vector (BEV) systems, AcNPV/Sf9 (and High Five) cells and BmNPV/BmN4 cells, have become important in the production of recombinant protein [1-4]. However, the qualities as well as quantities of expressed recombinant proteins were often significantly varied among the systems, mainly because of difference in productivity and capability of posttranslational modifications among insect cell lines [2,4]. Thus, we can expect that there would be insect cell lines which possess improved properties in production as well as posttranslational modifications of recombinant proteins.

A lepidopteran insect cell line, FRI-SpIm-1229 (SpIm), which was derived from fat bodies of *Spilosoma imparilis* [5], is susceptible to the *Hyphantria cunea* nucleopolyhedrovirus (HycuNPV). Because SpIm cells can grow well in the low cost, simply formulated Mitsuhashi-Maramorosh (MM) medium [6] supplemented with low concentration (3%) of fetal bovine serum (FBS), HycuNPV/SpIm cell system have been considered as a promising candidate enabling economic scale-up for the large-scale production of recombinant products. Therefore, we have started to construct a novel

Y. Kitagawa et al. (eds.), Animal Cell Technology: Basic & Applied Aspects, Vol. 10, 271–275.

BEV system using SpIm cells and HycuNPV and evaluated its performance in the recombinant protein production.

2. Materials and Methods

2.1 INSECT CELL CULTURE

The SpIm cells maintained in 25- or 75-cm^2 T-flasks (Sumitomo Bakelite) with both Mitsuhashi-Maramorosh (MM) medium supplemented with 3% FBS and Sf-900 II serum-free medium (GIBCO BRL) were used in this study. For comparison, the BmN4 cells [7] derived from *Bombyx mori* maintained in TC-100 medium [8] supplemented with 10% FBS were also used. For shaker culture, SpIm cell suspensions were shaken in 200-mL siliconized glass Erlenmeyer flasks (Iwaki) on an orbital shaker (Bio-Shaker BR-30LF, Taitec). For spinner culture, a 100-mL spinner flask (Bellco) was used. In both shaker and spinner cultures, SpIm cells were seeded at 2.5 x 10^5 viable cells/mL in 50 mL media containing 0.1% Pluronic F-68 (Sigma). Culture of cells was done at 27 °C. Cell density was determined using a hemocytometer and the viability was assessed by 0.2% trypan blue exclusion.

2.2 VIRUS STOCK AND CONSTRUCTION OF RECOMBINANT VIRUS

The A strain of the *Hyphantria cunea* nucleopolyhedrovirus (HycuNPV) was plaque-purified using SpIm cells and a genomic DNA fragment containing the polyhedrin region was cloned and sequenced. According to the determined nucleotide sequence, the cloned DNA fragment was modified to construct a transfer vector, pHcMU1 (Figure 1), for insertion and expression of heterologous genes under control of the HycuNPV polyhedrin promoter. A 500 bp fragment, encoding sarcotoxin 1A signal peptide-*Bombyx mori* prothoracicotropic hormone (PTTH) gene fusion, was cleaved from pPTTHMSig [9] with *Xba*I and *Hin*dIII, blunt-ended with T4 DNA polymerase and cloned into *Sma*I site just downstream of HycuNPV polyhedrin promoter (Fig. 1). The resulting plasmid pHcPTTH and wild-type HycuNPV DNA were cotransfected into cultured SpIm cells using Lipofectin reagent (GIBCO BRL) [10]. One week later, the culture supernatant was collected and used for plaque purification of recombinant viruses, which were identified based on their occlusion-negative phenotype. The replacement of the polyhedrin gene with the PTTH gene in the purified recombinant virus was confirmed by PCR analysis.

2.3 ANALYSIS OF PTTH EXPRESSION

SpIm cells (2.5 x 10^5 viable cells/mL) were infected with the recombinant virus at a multiplicity of infection (MOI) of 1 for one hour and at 9 days post-infection, the culture supernatnats were harvested and analyzed by SDS-polyacrylamide gel electrophoresis (SDS-PAGE) and immunoblotting using anti-PTTH rabbit serum. For comparison, the culture supernatant of BmN4 cells infected with a recombinant BmNPV expressing the same PTTH gene [11] at a MOI of 1 was also analyzed.

3. Results and Discussion

3.1 OPTIMAL CONDITIONS FOR SUSPENSION CULTURE OF SPIM CELLS

When we compared growth kinetics of SpIm cells in suspesion (50 mL) between shaker and spinner cultures, the optimal cell growth, comparable to that in stationary culture (4 mL), was obtained in shaker culture at 100 rpm (Figure 2). In spinner culture, SpIm cells did not grow well at any stirring rates tested. The results demonstrated that spinner culture is not adequete for suspension culture of SpIm cells. Then we compared virus replication rates between shaker culture at 100 rpm (50 mL) and stationary culture (4 mL) and found the similar growth curves of wild-type HycuNPV in the culture supernatants of SpIm cells infected at both MOI of 0.1 and 1 (Figure 3). We further investigated effects of serum-free medium on SpIm cell culture. In both stationary and shaker cultures, growth curves of SpIm cells were not so different between MM medium with 3% FBS and Sf-900 II serum-free medium (Figure 4).

3.2 PRODUCTION OF RECOMBINANT PROTEIN

Immunoblot analysis of the culture supernatants at 9 days post-infection of SpIm cells (4 mL stationary culture) infected with the recombinant HycuNPV expressing PTTH gene revealed that accumulation of secreted PTTHs with ca. 20 kDa (Figure 5). Surprisingly, the expression level of recombinat PTTH was drasticaly increased by cultivation of SpIm cells in Sf-900 II serum-free medium and was comparable to that in BmNPV/BmN4 cell system using TC-100 medium with 10% FBS. The same high level expression was observed in 50 mL shaker culture under the serum-free conditions (data not shown). The immunoblot analysis also showed that there was a single immunoreactive band in BmNPV/BmN4 cell system, while more than two immunoreactive bands in HycuNPV/SpIm cell system. Addition of tunicamycin resulted in detection of a single band with ca. 15 kDa in both BmN4 and SpIm cell culture (data not shown), indicating the N-glycosylation pattern of recombinant PTTH expreesed by SpIm cells was not homogeneous. To reveal the N-glycosylation capability of SpIm cells, analysis of sugar chains formed on the recombinant PTTH are now in progress.

4. Conclusion

We have successufully adapted SpIm cells in suspension culture under the serum-free conditions without any reductions in cell growth and virus replication. Unexpectedly, productivity of the recombinant protein by HycuNPV vector was improved under the serum-free conditions. We will further develop effective methods of scaling-up the HycuNPV/SpIm cell system for large-scale production of recombinant proteins.

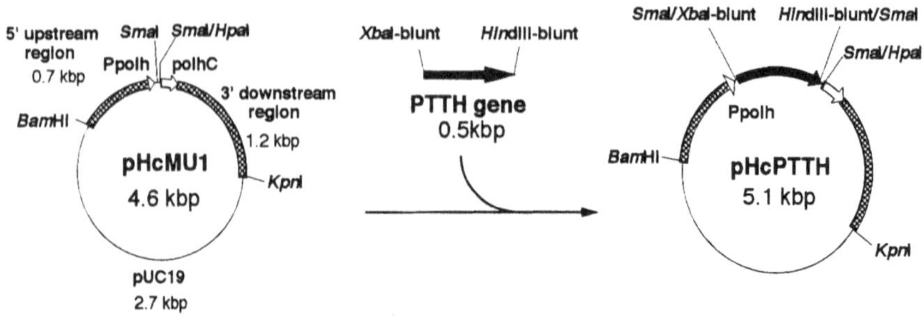

Figure 1. Construction of transfer vector. The shaded boxes of the transfer vector pHcMU1 are both 5' and 3' flanking regions of the HycuNPV polyhedrin gene. The polyhedin promoter (Ppolh) and the C-terminal part of polyhedrin-coding region (polhC) are indicated by white arrows. The recombinant plasmid pHcPTTH was derived by inserting a 500 bp DNA fragment containing PTTH gene (black arrow) into the *Sma*I site of pHcMU1.

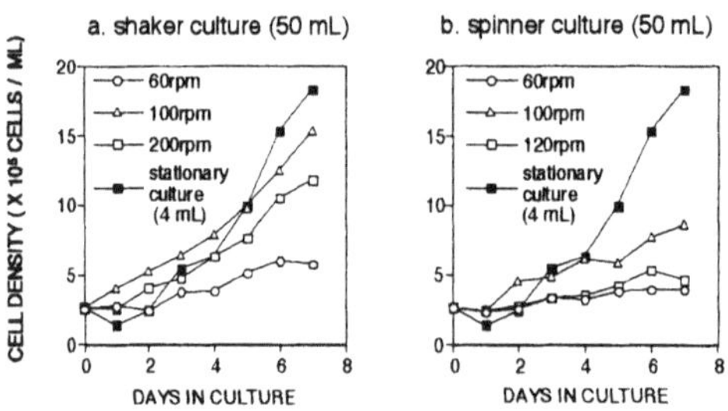

Figure.2. Growth kinetics of SpIm cells in MM medium with 3% FBS using (a) shaker culture methods and (b) spinner culture methods.

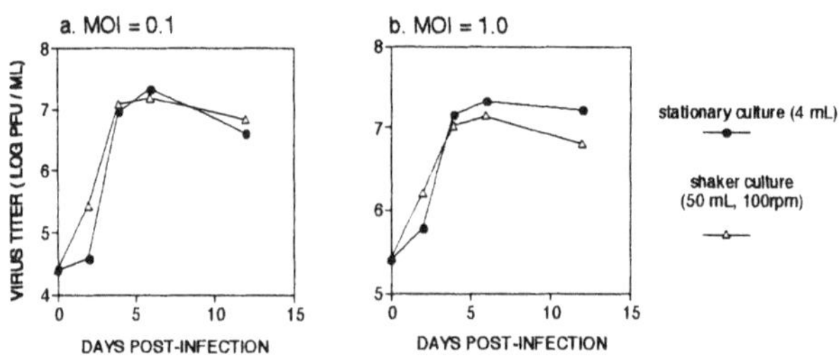

Figure.3. Comparison of growth curves of wild-type HycuNPV on SpIm cells in MM medium with 3% FBS between stationary culture methods and shaker culture methods.

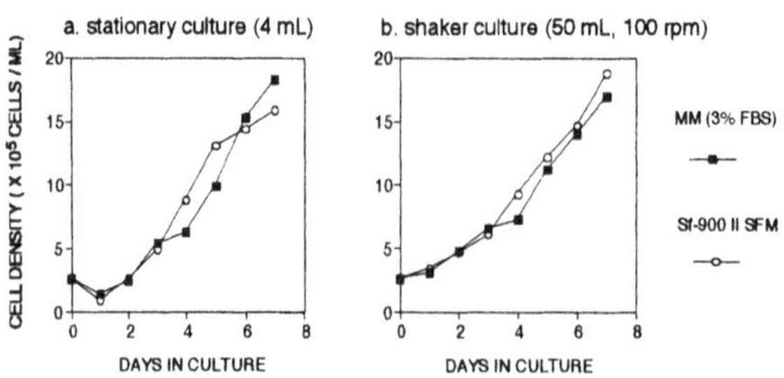

Figure 4. Comparison of growth kinetics of SpIm cells between MM medium with 3% FBS and Sf-900 II serum-free medium using (a) stationary culture methods and (b) shaker culture methods.

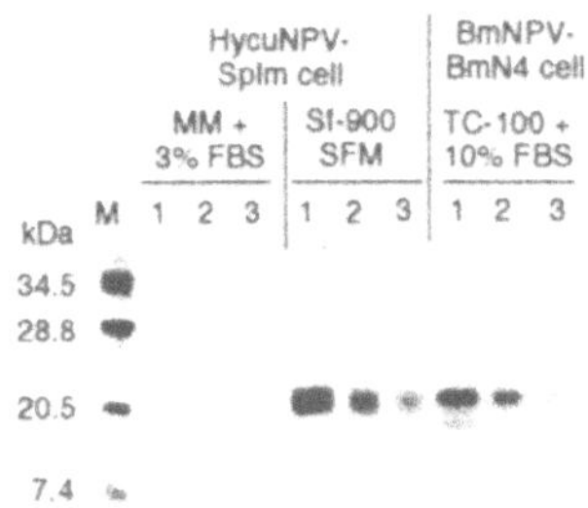

Figure 5. Comparison of PTTH accumulated in culture supernatants at 9 days postinfection between HycuNPV/SpIm cell and BmNPV/BmN4 cell systems. Lane M; prestained MW marker, 1; 5μl of culture supernatant (sup), 2; 2μl of sup, 3; 1μl of sup.

Ackowledgments

This work was supported in part by Grant-in-Aid for Scientific Research on Priority Areas from the Ministry of Education, Science, Culture and Sports of Japan (Grant No. 09265206) and a grant from the Center of Excellence Research Program of the Science and Technology Agency.

References

1. O'Reilly, D.R., Miller, L.K. and Luckow, V.A. (1992) *Baculovirus Expression Vectors : A Laboratory Manual*, W.H. Freeman and Company, New York, NY.
2. Goosen, F.A., Daugulis, A.J. and Faulkner, P. (1993) *Insect Cell Culture Engineering*, Marcel Dekker, New York, NY.
3. Maramorosch, K. and McIntosh, A.H. (1994) *Insect Cell Biotechnology*, CRC Press, Boca Raton, FL.
4. Richardson, C.D. (1995) *Baculovirus Expression Protocols*, Humana Press, Totowa, NJ.
5. Mitsuhashi, J. and Inoue, H. (1988) Obtainment of a continuous cell line from the larval fat bodies of the mulberry tiger moth, *Spilosoma imparilis* (Lepidoptera: Arctiidae), *Appl. Entomol. Zool.*, **23**, 488-490.
6. Mitsuhashi, J. and Maramorosch, K (1964) Leafhopper tissue culture: Embryonic nymphal and imaginal tissues from aseptic insects, *Contrib. Boyce Thompson Inst.*, **22**, 435-460.
7. Maeda, S., Kawai, T., Obinata, M., Chika, T., Horiuchi, T., Maekawa, K., Nakasuji, K., Saeki, Y., Sato, Y., Yamada, K. and Furusawa, M. (1984) Characteristics of human interferon-α produced by a gene transferred by a baculovirus vector in the silkworm, *Bombyx mori*, *Proc. Jpn. Acad.*, **60**, Ser. B, 423-426.
8. Gardiner, G.R. and Stockdale, H. (1975) Two tissue-culture media for production of lepidopteran cells and nuclear polyhedrosis viruses, *J. Invertebr. Path.*, **25**, 363-370.
9. O'Reilly, D.R., Kelly, T.J., Masler, E.P., Thyagaraja, B.S., Robson, R.M., Shaw, T.C. and Miller, L.K. (1995) Overexpression of *Bombyx mori* prothoracicotropic hormone using baculovirus vectors, *Insect Biochem. Mol. Biol.*, **25**, 475-485.
10. Kobayashi, J. and Belloncik, S. (1993) Efficient lipofection method for transfection of the silkworm cell line, NISES-BoMo-15AIIc, with the DNA genome of the *Bombyx mori* nuclear polyhedrosis virus, *J. Seric. Sci. Jpn.*, **62**, 523-526.
11. Kobayashi, J., Ishikawa, H., Fujita, K., O'Reilly, D.R., Maeda, S., Kataoka, H. and Miyajima, S. (1996) Overexpression of *Bombyx mori* prothoracicotropic hormone accerelates the speed of killing of a *Bombyx mori* nuclear polyhedrosis virus mutant lacking the ecdysteroid UDP-glucosyltransferase gene, *Proceedings of XX International Congress of Entomology*, 430.

DEVELOPMENT OF A HYBRID ARTIFICIAL LIVER SUPPORT SYSTEM WITH PORCINE HEPATOCYTE SPHEROIDS FOR PRECLINICAL USE

M. Kaneko*, T. Ito*, K. Nakazawa*, H. Ijima*, T. Matsushita*, T. Gion**, M. Shimada**, K. Shirabe**, H. Hasegawa**, K. Takenaka**, K. Sugimachi** and K. Funatsu*

* Department of Chemical Systems and Engineering, Graduate School of Engineering, Kyushu University

** Second Department of Surgery , Kyushu University Medical School

* 6-10-1 Hakozaki, Higashi-ku, Fukuoka 812-81, Japan

ABSTRACT In this study, we investigated culture conditions of primary porcine hepatocytes for development of a hybrid artificial liver using polyurethane foam (PUF) / spheroid culture. Porcine hepatocytes in the pores of hydrophilic PUF only formed spheroids spontaneously within 1.5 days of culture, and maintained ammonia detoxification activity in Williams' E medium supplemented with 10～20% fetal bovine serum (FBS) for a week. The porcine hepatocyte spheroids in multi-channel (MC) PUF packed-bed module also expressed the highest activities of ammonia detoxification and albumin production at an immobilized cell density of 1.0×10^7 cells/cm^3-module and at an intrachannel medium velocity range from 60 to 80 cm/min.

1. INTRODUCTION

Recently, many researchers have tried to develop an effective hybrid artificial liver module, and some researchers have applied their module to human medical care as a bridge use in liver transplantation[1)-5)]. We also developed a polyurethane foam (PUF) / hepatocyte spheroids packed-bed module and demonstrated its efficiency as a hybrid artificial liver using hepatic failure rats and dogs[6)-8)].

At present, we are trying to develop a hybrid artificial liver support system using porcine hepatocytes for human medical care. In this study, we investigated the periods of spheroid formation and expression of liver specific functions of porcine hepatocytes

Y. Kitagawa et al. (eds.), Animal Cell Technology: Basic & Applied Aspects, Vol. 10, 277–281.

under the various culture conditions *in vitro*, and determined the optimum culture conditions of primary porcine hepatocytes for developing a hybrid artificial liver.

2. MATERIALS AND METHODS

2.1. Preparation of porcine hepatocytes

Porcine hepatocytes were isolated from a domestic pig (body weight : 10 kg) by two-step collagenase treatment technique[9]. Cell viability was more than 85% by trypan blue exclusion method.

2.2. Stationary culture

Polyurethane foam (PUF) We used three kinds of PUF as a cell culture substratum. R-1 and R-4 are hydrophobic PUF and their average pore size is about 500 μm and 250 μm, respectively. W-1 is a hydrophilic PUF and the average pore size is about 250 μm.

Culture mediums Porcine hepatocytes were cultured in five kinds of medium. Williams' E medium (WE) was used as the basal medium. Composition of the serum free medium was WE supplemented with 25 μg/L EGF, 7.23 mg/L insulin, 4 μg/L glucagon, 1 μM dexamethasone, 6.25 mg/L transferrin, 500 mg/L Bovine serum albumin and 20 μg/L LGF. Three kinds of fetal bovine serum (FBS) containing medium consisted of WE supplemented with 5, 10 and 20 % FBS.

Cell culture Isolated hepatocytes were inoculated to a PUF plate ($25 \times 25 \times 1.0$ mm) put into a tissue culture dish of 35 mm diameter at a density of 2×10^6 cells/dish in 2 mL of medium. And hepatocytes were cultured in a CO_2 incubator (5 % CO_2, 95 % air) at 37 ℃. The medium was exchanged at 4 and 24 hours after cell inoculation, thereafter at 48 hour intervals.

2.3. Perfusion culture

We used an MC-PUF packed bed module (module volume:18.8 cm^3) to determine the optimum immobilized cell density and intrachannel medium velocity in a module. We investigated the immobilized cell densities in the MC-PUF module at a range from 0.1×10^7 to 2.1×10^7 cells/cm^3-module, and investigated the intrachannel medium velocities in MC-PUF the module at a range from 30 to 120 cm/min. Isolated hepatocytes were immobilized into a module by centrifugal method[6]. The MC-PUF module was cultured in a perfusion system (total system volume : 250 mL), which was composed of reservoir bottle of medium and gas exchanger in a CO_2 incubator (5 %

CO_2, 95 % air) at 37 ℃. Culture medium was exchanged as above.

2.4. Evaluation of liver specific functions (Albumin production and Ammonia detoxification)

Albumin concentration in the culture medium was measured by the ELISA method. When we evaluated the ammonia detoxification activity, the culture medium was changed to fresh culture medium supplemented with 1 mM-NH_4Cl, and the ammonia concentration was measured using a commercial kit (ammonia test wako ; Wako Pure Chemicals ; Japan) at 6 hours after medium exchange. We measured the ammonia detoxification activity by the difference of ammonia concentration during the culture period. The nucleus numbers in each dish and the MC-PUF module were measured by the nucleus counting method reported by Van Wezel[10]. The activity of albumin secretion and ammonia metabolism were normalized by dividing the nucleus number.

3. RESULTS

3.1. Spheroid formation

Porcine hepatocytes aggregate were already formed at 4 hours after inoculation and formed spheroid with a diameter of 150～200 μm within 1.5 days of culture only in W-1 (Table1). When hepatocytes were cultured in W-1, porcine hepatocytes formed spheroids in all types of culture medium. On the other hand, porcine hepatocytes in R-1 and R-4 maintained a multilayer state during the culture period.

Table 1 Characteristics of PUF and morphology of hepatocytes cultured in each PUF pore.

PUF type	Characteristics	Morphology of hepatocytes	
		1.5 days of culture	7 days of culture
R-1	Hydrophobic	Monolayer	Multilayer
R-4	Hydrophobic	Monolayer	Multilayer
W-1	Hydrophilic	Spheroid	Spheroid

3.2. Expression of liver specific functions under the various culture conditions

Stationary culture Expression of liver specific functions under the various culture conditions are shown in table 2. Albumin production and ammonia detoxification activities in W-1 were expressed and maintained better than that in the others. Especially, the albumin production activity in W-1 was about two times higher than that in the others. On the other hand, although ammonia detoxification activity in serum-free medium rapidly decreased, it increased and was maintained during the culture period by supplementing FBS to the medium. The activity level increased with its concentration.

Perfusion culture Cell activities under the various immobilized cell densities and intrachannel medium velocities in a module are shown in Table 3. The module with an immobilized cell density of 2.1 × 10^7 cells/cm^3-module expressed the highest ammonia detoxification and albumin production activities at 1 day of culture. But its activities decreased rapidly at 3 days of culture. On the other hand, the module with an immobilized cell density of 1.0 × 10^7 cells/cm^3-module expressed ammonia detoxification activity as high as those of 2.1 × 10^7 cells/cm^3-module. They were maintained higher than those of 2.1×10^7 cells/cm^3-module.

The module cultured at an intrachannel medium velocity of 60 cm/min expressed liver specific functions as high as that of 80cm/min, and these levels were higher than that at 30 and 120 cm/min.

Table 2 Expression of ammonia detoxification and albumin production activities of porcine hepatocytes in the various culture mediums in a stationary culture. (n=1)

PUF type	Culture medium	Ammonia metabolizing rate (μmol-NH3/10^6-nuclei/hr)		Albumin secretion rate (μg-alb./10^6-nuclei/day)	
		1 day	5 days	1 day	5days
R-1	WE+hormones	0.0056	-0.0046	11.60	15.41
R-4	WE+hormones	0.011	0.0005	7.78	17.26
W-1	Williams' E (WE)	0.187	0.0277	—	—
	WE+hormones	0.294	0.0793	10.81	31.64
	WE+FBS 5%	0.197	0.3045	—	—
	WE+FBS 10%	0.301	0.388	—	—
	WE+FBS 20%	0.243	0.428	—	—

Table 3 Expression of ammonia detoxification and albumin production activities of MC-PUF module (18.8cm^3) under the various immobilized cell densities and intrachannel velocities. (n=1)

Immobilized cell density (cells/cm^3-module)	Intrachannel velocity (cm/min.)	Ammonia metabolizing rate (μmol-NH3/cm^3-module/hr)		Albumin secretion rate (μg-alb./cm^3-module/day)	
		1 day	3 days	1 day	3 days
2.1×10^7	80	0.475	0.301	82.45	42.16
1.0×10^7	30	0.360	0.306	68.80	31.58
	60	0.361	0.526	70.20	49.57
	80	0.407	0.414	57.39	66.95
	120	0.467	0.337	60.15	42.61
0.5×10^7	80	0.267	0.019	29.45	39.41
0.1×10^7	80	0.008	0.008	3.37	9.27

4. DISCUSSION

When we treat liver failure patients with hybrid artificial liver, we should use human hepatocyte, however since we can not obtain enough, we think porcine hepatocyte is suitable. In this study, we found the optimum culture conditions for porcine hepatocytes to make express and maintain high activities.

In human medical care, we think it is very important to prepare an artificial liver module rapidly. In this study, hepatocytes formed spheroids spontaneously in hydrophilic PUF pores at 1 day of culture. So we think the hydrophilic characteristic is effective for the rapid formation of spheroids. And this result means that we can succeed in the rapid preparation of an artificial liver module.

We think it is very important for an artificial liver to maintain the ammonia detoxification activity, but it was difficult to maintain it in our rat and dog experiment. However we succeeded in its maintenance for more than 1 week by supplementing FBS to the medium. So we think that some component in FBS is effective for the maintenance of ammonia detoxification activity. Either that or that hepatocytes were probably cultured *in vivo* by supplementing FBS, and so, it was maintained. But we don't have enough evidence to be able to define the reason for it. So we need to clarify the reason to develop a more effective artificial liver.

The immobilized cell density in a module determines the module volume. In human medical care, we have to prepare a high performance and compact module. We achieved high cell density culture of a 1.0×10^7 cells/cm^3-module. The module expressed and maintained high ammonia detoxification activity for a week, but albumin production activity decreased with the culture time. So now we have to investigate the maintenance of liver specific functions of the module.

5. REFERENCES

1) Matsumura KN, Guevara GR, Huston H, et al. : *Surgery*: 101, 99-103, 1987

2) Margulis MS, Erukhimov EA, Andreiman LA, et al. : *Res. Surg.* : 2, 99-102, 1990

3) Kelly JH, Sussman NL. : *ASAIO J*: 1, 83-85, 1994

4) SC Chen, J Rozga, AA Demetriou, et al. : *Int. J. Artif. Organs*: 19, 664-669, 1996

5) JC Gerlach. : *Int. J. Artif. Organs*: 19, 645-654, 1996

6) H. Ijima, T. Matsushita and K. Funatsu : *Jpn. J. Artif. Organs*: 23, 463-468, 1994

7) T.Matsushita, H.Ijima, S.Wada and K.Funatsu : *Jpn. J. Artif. Organs*: 24, 815-820, 1995

8) T. Matsushita, H. Ijima, N. Nakazawa, T. Gion, K. Shirabe, M. Shimada, K. Takenaka, K. Sugimachi and K. Funatsu : *Jpn. J. Artif. Organs*: 26, 455-459, 1997

9) K. Nakazawa, H. Ijima, M. Kaneko, T. Ito, T. Matsushita, T. Gion, K. Shirabe, M. Shimada, K. Takenaka, K. Sugimachi and K. Funatsu : *Jpn. J. Artif. Organs*: *in press*

10) van Wezel AL. : *Nature*: 216, 64-65, 1967

IN VITRO SELF-ORGANIZATION OF LIVER CELLS USING ARTIFICIAL MATRIX

KEISUKE YAMADA, MASAMICHI KAMIHIRA, RYUJI HAMAMOTO AND SHINJI IIJIMA
Department of Biotechnology, Graduate School of Engineering, Nagoya University, Chikusa-ku, Nagoya, 464-8603, Japan

ABSTRACT. We developed mass preparation method of hepatocyte spheroids using a water-soluble synthetic polymer as an artificial matrix in cell suspension system. Cell-aggregation was promoted by the addition of a commercially available synthetic polymer, Eudragit (a copolymer of methacrylic acid and methylmethacrylate) into the culture medium. Hepatocytes formed spheroids expressing high liver functions in petridish and spinner flask cultures by adding the polymer.

To achieve further enhanced liver functions, co-culture of liver parenchymal and non-parenchymal cells was examined. The hetero-spheroids consisting of heterogeneous liver cells were also induced by the addition of the polymer. Moreover, the hetero-spheroids expressed enhanced and stable albumin secretion for more than 4 weeks.

1. Introduction

Cell-cell interactions are important in differentiated organs and tissues to express their specific functions [1]. In liver, parenchymal and non-parenchymal hepatic cells interact each other and control the growth and differentiation by autocrine and paracrine mechanisms [2]. It is known that hepatocytes cultured as multicellular aggregate called spheroid, exhibit enhanced liver functions for long-term compared with monolayer [3, 4]. Earlier researches also reported that hepatocytes showed prolonged survival and enhanced liver functions when hepatocytes were co-cultured with liver non-parenchymal cells [5-8].

The hepatocytes expressing high liver functions have been expected to apply for a bioartificial liver support system. But most preparation methods of hepatocyte spheroids require the large surface area for initial cell attachment [9-13]. This is one of the limiting factors for the mass preparation of functional liver cells for the construction of an artificial liver. To solve the problem, some research groups reported spheroid formation in suspension culture [14, 15]. We developed the preparation method of spheroids by using a synthetic polymer Eudragit as an artificial matrix in cell suspension system [16, 17]. In the present study, the preparation of hetero-spheroids by the method was studied.

2. Materials and Methods

Hepatocyte Culture: Hepatocytes were isolated from 6-8 week old Sprague-Dawley rats by conventional *in situ* collagenase perfusion method [18]. Parenchymal and non-parenchymal hepatic cells were separated by centrifugation. Parenchymal hepatocytes were cultured in hormonally defined Williams' medium E (Gibco BRL Co., New York, NY, USA) supplemented with 0.1 μM $CuSO_4{\bullet}5H_2O$, 25 nM Na_2SeO_3, 1.0 μM $ZnSO_4{\bullet}7H_2O$, 0.1 μM insulin, 1.0 μM dexamethasone, 20 μg/l epidermal growth factor

Y. Kitagawa et al. (eds.), Animal Cell Technology: Basic & Applied Aspects, Vol. 10, 283–287.

(EGF), 48 mg/l gentamicin sulfate, 100 mg/l chloramphenicol and 20mg/l yolk lipoprotein. In co-culture with non-parenchymal cells, 100 µM L-ascorbic acid phosphate magnesium salt and 5% fetal bovine serum instead of yolk lipoprotein were added to the medium. Eudragit S100 (a copolymer of methacrylic acid and methylmethacrylate; Röhm Pharma GmbH, Darmstadt, Germany) was added to the medium as an artificial matrix for inducing cell aggregation. Prior to the addition, the polymer was dissolved in water and the pH was adjusted at 7.4 by adding 2M NaOH. For the static petridish cultures, 5×10^5 cells were seeded in a 35 mm-diameter polystyrene non-treated dish (1000-035; Iwaki Glass Works Co., Tokyo, Japan) in 2 ml medium. The medium was changed to the fresh one without the polymer every other day.

Analyses: Albumin concentration in the medium was determined by sandwich solid-phase enzyme-linked immunosorbent assay (ELISA). Ammonia removal and urea synthesis were measured as follows. First, the medium was carefully removed from the cell-cultured dish, and the fresh medium containing 1.0 mM NH_4Cl was added. After the cells were cultured for 4 h, ammonia and urea concentrations in the medium were measured by using diagnostic kits obtained from Wako Pure Chemical Co. (Tokyo, Japan). The rates of ammonia removal and urea synthesis were calculated from the reduction of ammonia concentration and the increase of urea concentration, respectively.

3. Results and Discussion

In our previous study, we found that Eudragit showed low cytotoxicity for animal cells and promoted cell aggregation of hepatocytes [19]. Therefore, we assumed that the polymer could be used as the artificial matrix for inducing the formation of hepatocyte spheroids.

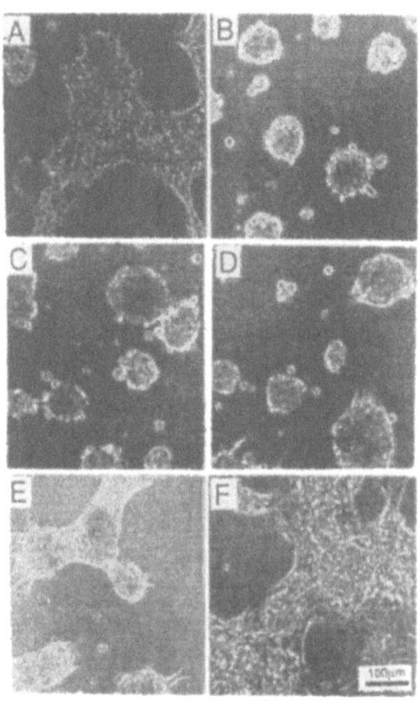

Figure 1. Morphological change of hepatocyte in the presence of various concentration of Eudragit on day 5. Eudragit was added to the medium at the various concentrations, no addition (A), 0.05% Eudragit (B), 0.1% Eudragit (C), 0.18% Eudragit (D), 0.26% Eudragit (E) and 0.5% Eudragit (F).

Eudragit was added to the culture medium at various concentrations. Hepatocytes clearly formed floating spheroids with 0.1% Eudragit within 3-5 days. Above 0.26% Eudragit, cells tended to attach to the dish surface, and formed multilayer flat cell aggregates rather than spheroids (Fig. 1). The albumin secretion increased with 0.1% Eudragit addition. Higher concentration of Eudragit was not effective for enhancement of albumin secretion. The albumin secretion of spheroids induced by 0.1% Eudragit was about 2.5-fold that of monolayer on day 7 (Fig. 2). The other liver functions, ammonia removal and urea synthesis, were also enhanced and the highest rates were observed around Eudragit concentration of 0.1% (data not shown). Therefore, 0.1% Eudragit was selected for large scale production of spheroids using a spinner flask.

In static culture using petridish, it required 3-5 days for formation of spheroids, whereas spheroids was formed within 2 days in spinner flask culture. The spheroids induced in the

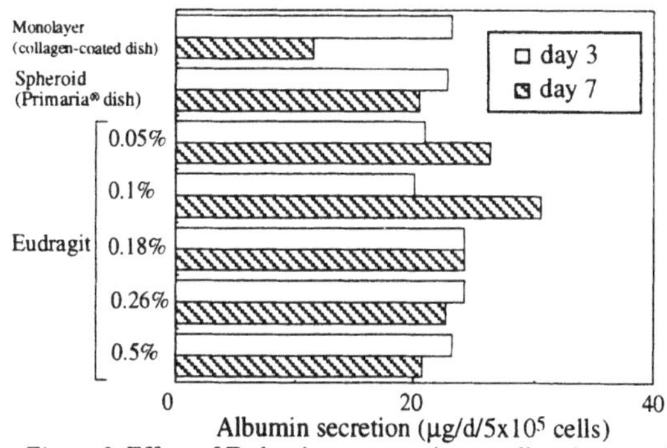

Figure 2. Effect of Eudragit concentration on albumin secretion.

Table 1. The ammonia removal and urea synthesis of cultured hepatocyte.

	Ammonia removal (μmol/h/5x10^5 cells)	Urea synthesis (μmol/h/5x10^5 cells)
Monolayer (static culture)	0.126	0.065
0.1% Eudragit (static culture)	0.399	0.163
No addition (spinner culture)	0.084	Not detectable
0.1% Eudragit (spinner culture)	0.303	0.129

Hepatocytes were cultured in a spinner flask for 48 h in the absence or presence of 0.1% Eudragit and then transferred to a petridish. The ammonia removal and urea synthesis were measured on day 4.

spinner flask also expressed enhanced liver functions (Table 1).

To improve the expression of liver functions and maintain them for long-term, we attempted to form hetero-spheroids by co-culture of parenchymal and non-parenchymal hepatic cells in the medium containing the polymer. First, we examined the effects of non-parenchymal cells addition on the expression of liver functions (Fig. 3). Various numbers of non-parenchymal cells (0.1-2.5x10^6 cells) were added to the culture dishes where parenchymal hepatocytes were seeded at 5x10^5 cells. With the increase in cell number of non-parenchymal cells, albumin secretion increased until non-parenchymal cell number were equal to parenchymal cells. When non-parenchymal cells were added 5-fold excess to parenchymal cells, the further increase in albumin secretion was not observed. Judging from the observation of cell morphology, spheroids were formed with the 1:1 mixture of parenchymal and non-parenchymal cells by adding the polymer. When 5-fold excess number of non-parenchymal cells was mixed, the cells did not form spheroids and spread over on the dish surface. In terms of the cell morphology and the expression of the liver functions, the equal number of parenchymal and non-parenchymal cells was suitable for inducing hetero-spheroids.

We then cultured hepatocytes in various conditions for long-term. Figure 4 shows the time course of albumin secretion during 4 weeks culture. In the presence of non-parenchymal cells, high albumin secretion was observed even without the polymer for 2 weeks but it gradually decreased. Although spheroids consisting of only parenchymal hepatocytes expressed high and relatively stable albumin secretion, the production level

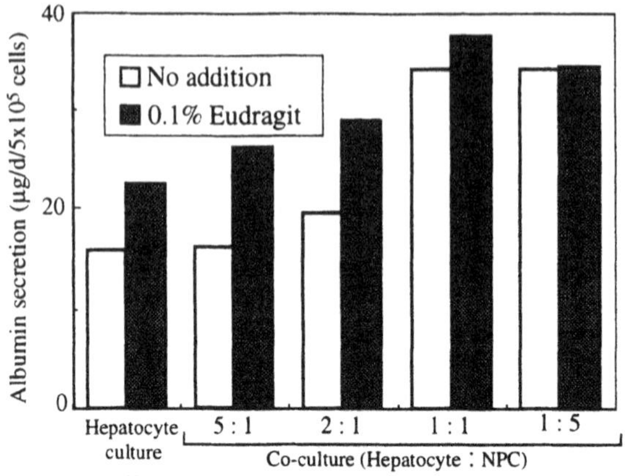

Figure 3. Effect of non-parenchymal cells on albumin secretion on day 5. The hepatocytes were seeded at 5×10^5 cells per petridish and the various ratio of non -parenchymal cells were co-cultured.

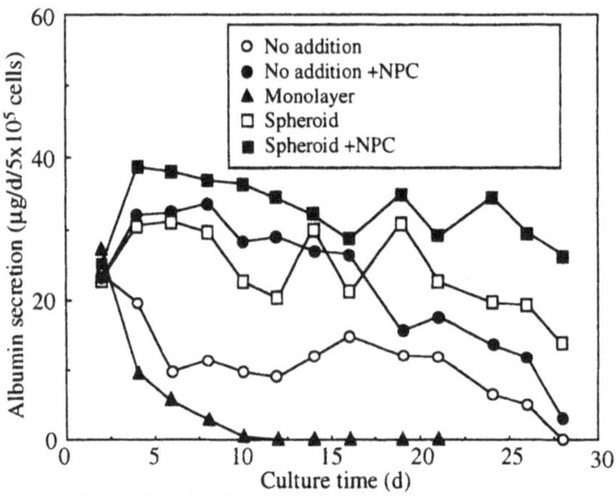

Figure 4. Comparison of albumin secretion in long-term hepatocyte culture The hepatocytes were cultured at 5×10^5 cells per dish in the absence (○, ●) or the presence of 0.1% Eudragit (□, ■). NPC : Non-parenchymal cells

gradually decreased in long term culture around 3 weeks. On the other hand, the hetero-spheroids exhibited higher albumin production which corresponded to 1.5-fold of the parenchymal hepatocyte spheroids, and the level was maintained throughout the culture period of 4 weeks.

In this study, we succeeded in the spheroid formation of heterogeneous liver-consisting cells from their suspension by adding the polymer to the culture medium as an artificial matrix. This method is promising for applying to a hybrid-type bioartificial liver, since the preparation of large amount of high functional hepatocyte spheroids is easy.

4. References

1. Gumbiner, B.M. (1996) Cell adhesion: The Molecular Basis of Tissue Architechture and Morphogenesis, *Cell* **84**, 345-357
2. Martinez-Hernandez, A. and Amenta, P.S. (1995) The extracellular matrix in hepatic regeneration, *FASEB J.*, **9**, 1401-1410
3. Landry, J., Bernier, D., Ouellet, C., Goyette, R. and Marceau, N. (1985) Spheroidal aggregate culture of rat liver cells: histotypic reorganization, biomatrix deposition, and maintenance of functional activities, *J. Cell Biol.* **101**, 914-923
4. Tong, J.Z., Bernard, O. and Alvarez, F. (1990) Long-term culture of rat liver cell spheroids in hormonally defined media, *Exp. Cell Res.*, **189**, 87-92
5. Guguen-Guillouzo, C., Clement, B., Baffet, G., Beaumont, C., Morel-Chany, E., Glaise, D. and Guillouzo, A. (1983) Maintenance and reversibility of active albumin secretion by rat hepatocytes co-cultured with another liver epithelial cell type, *Exp. Cell Res.* **143**, 47-54
6. Fraslin, J.-M., Kneipe, B., Vaulont, S., Glaise, D., Munnich, A. and Guguen-Guillouzo, C. (1985) Dependence of hepatocyte-specific gene expression on cell-cell interactions in primary culture, *EMBO J.* **4**, 2487-2491
7. Yamamoto, N., Imazato, K. and Masumoto, A. (1989) Growth stimulation of adult rat hepatocytes in a primary culture by soluble factor(s) secreted from nonparenchymal liver cells, *Cell Struct. Funct*, **14**, 217-229
8. Yagi, K., Suenobu, N., Serada, M., Tsuda, K., Kondoh, A. and Miura, Y. (1992) Stimulative effect of non-parenchymal liver cells on ability of tyrosine aminotransferase induction in hepatocytes, *Cytotechnol.* **10**, 25-31
9. Koide, N., Shinji, T., Tanabe, T., Asano, K., Kawaguchi, M, Sakaguchi, K., Koide, Y., Mori, M. and Tsuji, T. (1989) Continued high albumin production by multicellular spheroids of adult rat hepatocytes formed in the presence of liver-derived proteoglycans, *Biochem. Biophys. Res. Commun.* **161**, 385-391
10. Koide, N., Sakaguchi, K., Koide, Y., Asano, K., Kawaguchi, M., Matsushima, H., Takenami, T., Shinji, T., Mori, M. and Tsuji, T. (1990) Formation of multicellular spheroids composed of adult rat hepatocytes in dishes with positively charged surfaces and under other non adherent environments, *Exp. Cell Res.* **186**, 227-235
11. Matsushita, T., Ijima, H., Koide N. and Funatsu, K. (1991) High albumin production by multicellular spheroids of adult rat hepatocytes formed in the pores of polyurethane foam, *Appl. Microbiol. Biotechnol.* **36**, 324-326
12. Sakai, Y. and Suzuki, M. (1991) Formation of spheroids of adult rat hepatocytes on polylysine-coated surfaces and their albumin production, *Biotechnol. Tech.* **5**, 299-302
13. Tobe, S., Takei, Y., Kobayashi, K. and Akaike, T. (1992) Receptor-mediated formation of multicellular aggregates of primary cultured rat hepatocytes on lactose-substituted polystyrene, *Biochem. Biophys. Res. Commun.* **184**, 225-230
14. Sakai, Y., Furukawa, K. and Suzuki, M. (1992) Immobilization and long-term albumin secretion of hepatocyte spheroids rapidly formed by rotational tissue culture methods, *Biotechnol. Tech.* **6**, 527-532
15. Wu, F.J., Friend, J.R., Hsiao, C.C., Zilliox, M.J., Ko, W.J., Cerra, F.B. and Hu, W.S. (1996) Efficient assembly of rat hepatocyte spheroids for tissue engineering applications, *Biotechnol. Bioeng.* **50**, 404-415
16. Kamihira, M., Yamada, K., Hamamoto, R. and Iijima, S. (1997) Spheroid formation of hepatocytes using synthetic polymer, *Ann. NY Acad. Sci.* **831**, 398-407
17. Yamada, K., Kamihira, M., Hamamoto, R. and Iijima, S. (1998) Efficient induction of hepatocyte spheroids in a suspension culture using a water-soluble synthetic polymer as an artificial matrix, *J. Biochem.* **123**, 1017-1023
18. Seglen, P.O. (1976) Preparation of isolated rat liver cells, *Methods Cell Biol.* **13**, 29-83
19. Hamamoto, R., Kamihira, M. and Iijima, S. (1996) Specific separation of animal cells using aqueous two-phase systems, *J. Ferment. Bioeng.* **82**, 73-76

THYROID TRANSPLANTATION: POSSIBILITY OF APPLICATION FOR THE TREATMENT OF PERSISTENT HYPOTHYROIDISM AND STUDY OF MECHANISMS OF INTERACTION BETWEEN GRAFT AND HYPOTHALAMIC-PITUITARY AXIS OF RECIPIENT

I.P. PASTEUR, N.D. TRONKO, I.I. DROZDOVICH,
I.S. TURCHIN, I.A. BALLA

Institute of Endocrinology and Metabolism, Acad. Med. Sci. Ukraine, Vyshgorodska Street 69, 254114 Kyiv, Ukraine

1. Introduction

Persistent hypothyroidism resulting from thyroid surgery or radioiodine irradiation requires supplemental treatment with thyroxine for life [1]. Hormonal therapy using thyroid transplantation is an attractive alternative [2,3]. However, transplantation treatment today faces a big and growing problem: shortage of donor tissue and organs [4]. Nevertheless, many researchers have tried to solve this problem by help of such xenotransplantation (e.g. from pig) combined with modern immunosuppressive therapy [4,5]. Hormone replacement therapy by xeno-transplantation is already established in the field of pancreas surgery but not for thyroid disease [6].

The aim of the present study was development of culture technique by which the normal morphological configuration and highest hormonal activity of new-born pig thyroid tissue and some other structural and functional characteristics of thyroid follicles are well preserved.

2. Materials and Methods

Freshly removed newborn pig thyroid glands were freed of the fibrous capsule and the connective tissue. The thyroid glands were cut into small tissue pieces (less than 1 mm^3) and washed several times with cooled medium 199 containing antibiotics (100 IU/ml of penicillin and 100 μg/ml of streptomycin). The thyroid fragments were cultured in medium 199 with a 10% (v/v) fetal bovine serum (Sigma Chemical Co., USA) and antibiotics in culture dishes at 37 C.

Before replacement of medium on days 3, 5, 8, 11, 14, 17 and 20 of culture, aliquots of medium were collected and frozen for determination of thyroxine (T4)

Y. Kitagawa et al. (eds.), Animal Cell Technology: Basic & Applied Aspects, Vol. 10, 289–293.

and triiodothyronine (T3) by radioimmunoassay. Subsequently, 74 kBq of $Na^{131}I$ were introduced in each sample for 90 minutes in order to assess the level of iodine uptake by thyroid tissue using a gamma-counter (Miles Laboratories Inc., USA). The results were expressed as counts per min / thyroid tissue / counts per min / ml medium.

For light microscopy, thyroid fragments were stained with hematoxylin and eosin. Morphometric analyses were performed using Zeiss light microscope. For electron microscopy, thyroid fragments were fixed in formaldehyde-glutaraldehyde fixative, postfixed in 1% unbuffered OsO_4 and embedded in Epson. Ultrathin sections were prepared with knives, stained with 7.5% uranyl acetate and lead citrate and examined in a JEM-100C electron microscope.

Statistical evaluation of the results was performed with Student's *t* test, and $P<0.05$ was taken as significant.

3. Results

A morphometric analysis showed that in 1-day and 3-day organ culture of thyroid gland, as compared to native material, there was a decrease of thyrocyte height, what was probably due to the process of their adaptation (Table 1). In a 5-day organ culture thyrocyte height and volume of their nuclei increases, as compared to the previous period of study. In certain fragments a great surface is occupied by interfollicular tissue.

TABLE 1. Morphometric characteristic of newborn pig thyroid organ culture

Days of culturing	Thyrocyte height (μm)	Thyrocytes nucleus volume (μm)	Follicle area (μm²)	Colloid area (μm²)	CA/FA ratio
0	9.7 ± 0.1	63.6 ± 3.1	1731 ± 166	586 ± 89	0.34
1	6.8 ± 0.2 *	67.5 ± 3.3	810 ± 58 *	390 ± 47 *	0.48
3	7.4 ± 0.3 * **	72.0 ± 4.0 * **	1297 ± 88 * **	566 ± 59 **	0.43
5	10.7 ± 0.2 * **	87.9 ± 2.0 * **	1234 ± 95 *	268 ± 51 * **	0.21
10	9.8 ± 0.2 **	102.1 ± 3.1 * **	1234 ± 90 *	442 ± 54 **	0.35
15	8.9 ± 0.1 * **	96.1 ± 3.1 *	1256 ± 91 *	386 ± 46 *	0.31
20	5.4 ± 0.1 * **	93.1 ± 3.1 *	3322 ± 99 * **	2427 ± 63 * **	0.73
30	5.2 ± 0.2 *	78.3 ± 3.2 * **	3390 ± 90 *	2511 ± 51 *	0.74

Note: Values are mean ± SD. * $P < 0.05$ (compared to 0 day); ** $P < 0.05$ (compared to previous day).

Electron microscopic study shows microvillus on apical cytoplasm thyrocytes. The cavity of microfollicles has a mainly slit shape (Figure 1). Plasmolemmas of adjacent thyrocytes form connective tissue complexes and desmosomic contacts. One notes in endoplasmic reticulum intrasternal inclusions in the form of diffuse osmiophilic content. Golgi complex is located mainly in apical part of cytoplasm, it is represented by rather numerous small lamellas and microvesiculas. There are only isolated secretory granules in perpendicular zone, and in apical zone they are rather numerous. Thus, hypertrophy of endoplasmic reticulum and Golgi complex, numerous secretory granules point out a rather high functional activity of thyrocytes in organ culture.

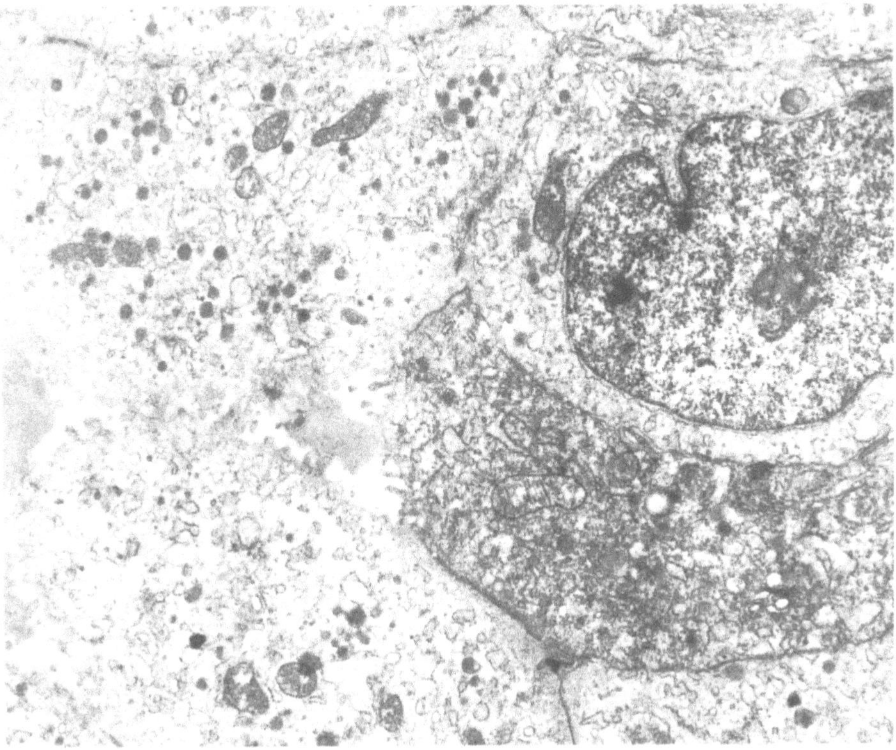

FIGURE 1. Electron micrograph of 5-day newborn pig thyroid organ culture. Secretory granules in the thyrocyte cytoplasm. × 10,000

In 10-day organ culture of thyroid gland one may follow the "vital cycle" of thyrocytes. Follicles with cylindrical epithelium are rather numerous, and, moreover, zones of high functionally most active cells alternate with layers of low ones (Figure 2, left). Mean height of thyrocytes does not differ from height before culturing, and nucleus volume increases as compared to the previous period of study (Table 1). One notes an increase of mitotic activity.

Ultrastructurally, heterogeneity of thyrocyte height is confirmed and flattened cells (because of a prolonged form) have a rather large surface of contact with colloid. Moreover, apical plasma membrane forms numerous microvillus or pseudopodia which capture luminal colloid, and rather numerous micropinocytosis vesiculae appear in cytoplasm. In certain thyrocytes intracellular microfollicles are formed.

In 15 days main structural elements of thyroid gland are preserved in a greater part of fragments, but thyrocyte height and nucleus volume tend to a certain decrease. Secretory granules are the most numerous in apical part of cytoplasm.

In 20-day organ culture thyrocytes are flattened, their height practically does not exceed sometimes the nucleus diameter (Figure 2, right). Along periphery and in the centre of isolated fragments, zones of destruction are noted.

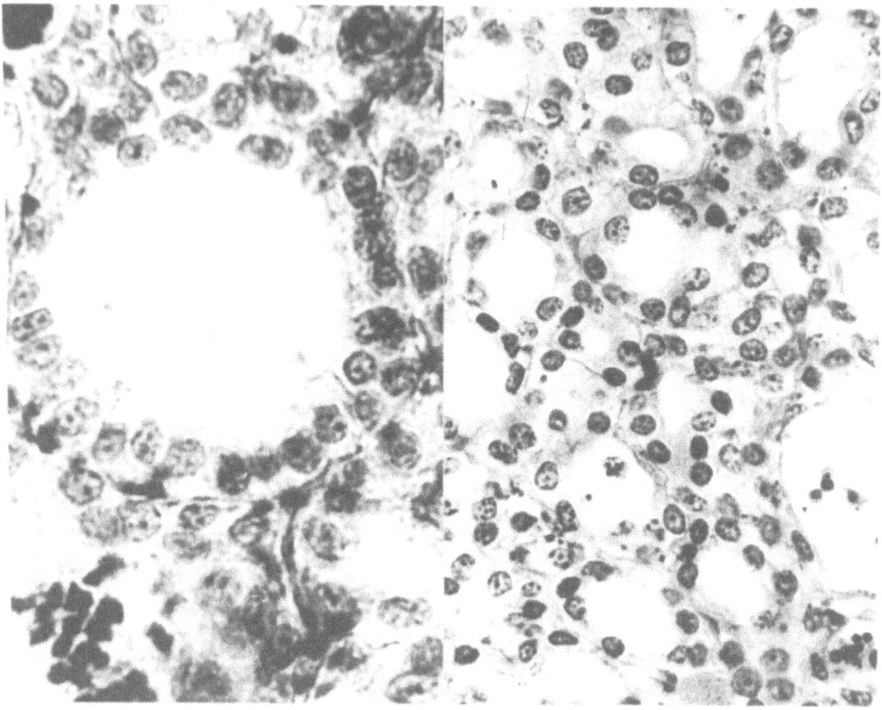

FIGURE 2. 10-day (left) and 20-day (right) newborn pig thyroid organ culture. Thyroid follicle structure. Hematoxylin-eosin stain. × 720

In 30-day organ culture thyrocyte height is practically unchanged, but mean volume of nucleus decreases (Table 1). Moreover, thyrocytes with big clear nuclei are preserved, among this cell population signs of mitotic division are observed.

A functional study showed that newborn pig thyroid organ culture effectively concentrated radioiodine from the surrounding medium and actively secreted thyroid hormones in cultural medium for a prolonged period of time (Table 2).

TABLE 2. Functional activity of newborn pig thyroid organ culture

Days of culturing	Iodine uptake (thyroid/medium)	Medium T4 (pmole/ml)	Medium T3 (pmole/ml)	T4/T3 ratio
3	0.423 ± 0.023	978.15 ± 188.13	167.45 ± 30.56	5.16 ± 2.00
5	0.384 ± 0.053	275.01 ± 19.42 *	50.98 ± 20.29 *	7.85 ± 2.26
8	0.346 ± 0.066	139.66 ± 21.19 * **	33.86 ± 10.16 *	5.96 ± 1.04
11	0.314 ± 0.068	70.66 ± 3.90 * **	11.20 ± 2.20 * **	8.72 ± 1.55
14	0.364 ± 0.044	62.54 ± 3.99 *	5.96 ± 0.80 * **	11.52 ± 1.40 *
17	0.282 ± 0.021 *	56.54 ± 6.74 *	3.70 ± 0.92 *	20.45 ± 3.43 * **
20	0.347 ± 0.072	43.94 ± 1.26 *	1.51 ± 0.38 *	33.81 ± 9.50

Note: Values are mean ± SD. * $P < 0.05$ (compared to 3 day); ** $P < 0.05$ (compared to previous day).

Human thyroid-stimulating hormone (TSH) has a marked and dose-dependent effect on these functional characteristics of newborn pig thyroid organ culture (data not shown). When thyroid in organ culture was exposed to 10 mU/ml of TSH, assay of the culture medium showed that T4 and T3 levels were significantly higher ($P<0.05$) than in medium of non-TSH exposed tissue.

4. Discussion

The study has shown that in the process of culturing of newborn pig thyroid, morphological and functional features of thyrocytes are preserved during a rather long period - one month. Conditions of follicular epithelium, colloid, interfollicular tissue and stromal elements in 5- and 10-day organ culture indicate their highest functional activity.

Our results also confirmed a report on the capacity of pig thyrocytes *in vitro* to actively produce thyroid hormones [7]. As shown before, concentration of intracellular T4 in cultured pig thyrocytes makes 1.890×10^{-8} µg/cell in radioimmunoassay, and 0.708×10^{-8} µg/cell using chromatographic method of assessment, what is equivalent to 4.39×10^{-5} mole/l and 1.78×10^{-5} mole/l, respectively [8]. This level is nearly 1,000 times as high as the corresponding index for circulating blood (3.18 ± 0.59 µg/dl or 3.98×10^{-8} mole/l) [8].

Therefore, the newborn pig thyroid organ culture may be used in experimental transplantology and for treatment of patients with persistent hypothyroidism. We are now trying to study how the transplanted newborn pig thyroid organ culture will affect the immunological background, feedback mechanisms between the hypothalamic-pituitary system, remaining proper thyroid gland of the Wistar rats with postsurgery (thyroidectomy) and postradioiodine hypothyroidism and so on, and, vice versa, how such varying circumstances will affect the transplanted thyroid tissue.

5. References

1. Kinser, J.A., Roesler, H., Furrer, T., et al.: Nonimmunogenic hyperthyroidism - Cumulative hypothyroidism incidence after radioiodine and surgical treatment, *J.Nucl.Med.*, **30** (1989), 1960-1965.
2. Okamoto, T., Fujimoto, Y., Obara, T., et al.: Trial of thyroid autotransplantation in patients with Graves' disease whose remnant thyroid has unintentionally been made too small at subtotal thyroidectomy, *Endocrinol.Japon.*, **37** (1990), 95-101.
3. Shimizu, K., Nagahama, M., Kitamura, Y., et al.: Improvement of thyroid function after autotransplantation of cryopreserved thyroid tissues in rats: clinical application of the procedure to patients with persistent hypothyroid Graves' disease after thyroidectomy, *Thyroidol.Clin.Exp.*, **8** (1996), 55-62.
4. Kemp, E.: Xenotransplantation, *J.Intern.Med.*, **239** (1996), 287-297.
5. Dillner, L.: Pig organs approved for human transplants, *Brit.Med.J.*, **312** (1996), 657.
6. Groth, C.G., Korsgren, O., Wennberg, L., et al.: Xenoislet rejection following pig-to-rat, pig-to-primate, and pig-to-man transplantation, *Transplant.Proc.*, **28** (1996), 538-539.
7. Gruffat, D., Venot, N., Marriq, C., and Chabaud, O.: Thyroid hormone synthesis in thyroglobulin secreted by porcine thyroid cells cultured on porous bottom chambers. Effect of iodide, *Endocrinology*, **131** (1992), 2921-2927.
8. Kawada, J., Nishida, M., Yoshimura, Y., and Yamamoto, T.: The presence of a high concentration of thyroxine in thyroid epithelial cells, *Acta Endocrinol.*, **100** (1982), 393-397.

FIXED-BED REACTORS AS A TOOL FOR ARTIFICIAL ORGANS

R. PÖRTNER[1], S. RÖSSING[1], J. STANGE[2], D. FASSNACHT[1]
[1] *Technische Universität Hamburg-Harburg, Bioprozeß- und Bioverfahrenstechnik, Denicke Str. 15, D-21071 Hamburg, Germany*
[2] *Universität Rostock, Klinik und Poliklinik für Innere Medizin, Ernst-Heydemann-Str. 6, D-18055 Rostock, Germany*

1. Abstract

In this study the immortalised mouse hepatocyte line mHep-R1 was used for cultivation. The cells were first grown in culture flasks and in a small-scale fixed-bed system in order to determine growth characteristics and a suitable carrier type. The cell line was then cultivated in a 40 ml fixed-bed reactor over a period of 75 days at a perfusion rate of 6.25 ml medium per ml fixed-bed and day. A cell density of approx. $8.5 \cdot 10^7$ cells per ml carrier was reached at the end of the experiment proving that a stable cultivation was possible over a long period of time with constant consumption and production rates.

2. Introduction

The immobilisation of animal cells in fixed-bed systems proved to be of advantage in many ways. By immobilising the cells on porous carriers, cells are retained in the system during continuous operation. This is of significance when cultivating cells, which are adherent, have a high demand of substrate and slow growth-rates. This is also accomplished by hollow-fibre modules or systems were cells are immobilised on or into microcarriers, but these systems proved to be difficult in scale-up, a factor which is especially important for the development of artificial organs, where a large amount of cells is needed. A scale-up is easily achieved for the fixed-bed system by pumping the medium radial through the fixed-bed instead of axial and therefore preventing the risk of an oxygen limitation.

Due to the development of new immortalised hepatocyte lines that still show considerable Cytochrome P450 activity, an approach to an artificial liver support system might be accomplished with the aid of a bioreactor containing these cell lines instead of primary hepatocytes which do not proliferate in such a system.

Y. Kitagawa et al. (eds.), Animal Cell Technology: Basic & Applied Aspects, Vol. 10, 295–297.

3. Cell Line and Culture Conditions

The immortalised mouse hepatocyte line mHep-R1 was used for cultivation. It shows in comparison to primary hepatocytes lower but still significant P450 activity of the subenzymes CYP1A2/3A4 and CYP2C (Stange *et al.*, 1995).

For growth low glucose Dulbecco's MEM (Life Technologies, Germany) was used as basal medium. This medium contains 25 mmol l^{-1} HEPES, 4 mmol l^{-1} L-glutamine and 5.5 mmol l^{-1} glucose and was supplemented with 5% (v/w) foetal calf serum. The glucose level was increased in the 40 ml fixed-bed system after the first week by addition of glucose, because growth in preliminary experiments proved to be clearly glucose limited.

For the batch experiments cultivation took place in 25 cm^2 T-flasks. Growth curves were obtained by analysing a doublet of flasks each day.

The 40 ml fixed-bed reactor, a modified reactor from meredos (Germany), contained two DO sensors, automatic pH and temperature control (Pörtner *et al.*, 1997). Cellulose carriers (Cellsnow, Biomaterials, Japan) with a diameter of approx. 5 mm were used to immobilise the cells.

4. Results

4.1. BATCH EXPERIMENTS

The cells grew to a maximum cell density of $1.1 \cdot 10^6$ cells ml^{-1} after 72 hours with a membrane intact index of approx. 95%. The reason for cell death in the following death phase was clearly due to glucose limitation.

4.2. FIXED-BED REACTOR

Continuous operation of the fixed-bed system was initiated 2 days after inoculation. The dilution rate D_{FB} was kept constant at 6.25 ml medium per ml fixed-bed and day throughout the 75 days of operation. Glucose was again the growth limiting factor during the first few days (Fig. 1). To improve growth conditions, the glucose concentration in the feed was raised to 16 mmol l^{-1} on the 10th day, resulting in increased glucose consumption and lactate production rates $q^*_{FB,Glc}$ $q^*_{FB,Lac}$ as well as a higher glucose concentration. Several steady-states with different glucose

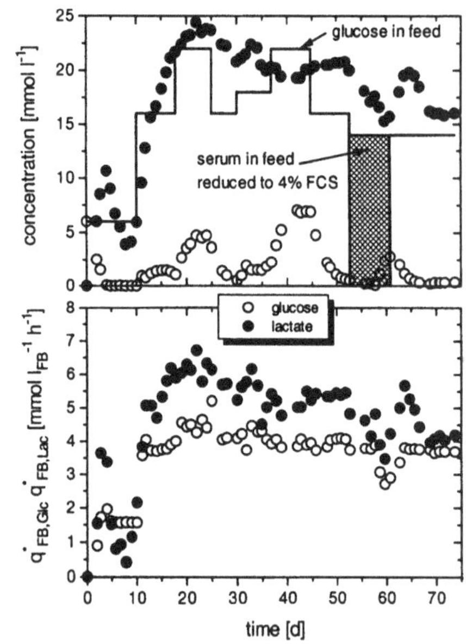

Figure 1. Long-term fermentation in the 40 ml fixed-bed reactor. The top diagram shows the glucose (open circles) and lactate (closed circles) concentration as well as the glucose content of the feed (line). The shaded area indicates the decrease of the serum content in the feed to 4% FCS. The bottom diagram shows the specific glucose consumption (open circles) and lactate production (closed circles) rates related to the fixed-bed volume.

concentrations in the feed were obtained in the following 60 days. A limitation of glucose can be excluded for feed concentrations over 16 mmol l^{-1} because no further increase in the glucose consumption rate $q^*_{FB,Glc}$ was observed. With a high concentration of approx. 1.5 mmol l^{-1}, glutamine was also not limited (Fig. 2). We assume that a substance in the serum is the cause for limited growth because the glucose consumption and lactate production rates decreased rapidly when the serum level was reduced to 4% in the feed on the 53rd day (Fig. 1) and started to recover on the 61st day after returning to the serum level of 5%. The average production rate q^*_{FB} was approx. 0.6 mmol l_{FB}^{-1} h^{-1} for ammonia and 0.08 mmol l_{FB}^{-1} h^{-1} for urea. 4.6 mmol l^{-1} ammonia was added to the feed on the 30th day (Fig. 2) to estimate the stability of the production rates at higher ammonia levels. The ammonia production rate decreased to 0.2 mmol l_{FB}^{-1} h^{-1} and the urea production rate to approx. 0.03 mmol l_{FB}^{-1} h^{-1}.

Several carriers were removed at the end of the experiment, and the total cell density was determined with the crystal violet method to be $5 \pm 2 \cdot 10^7$ cells per ml fixed-bed. This corresponds to $8.5 \pm 3 \cdot 10^7$ cells per ml carrier with an external porosity of 40%.

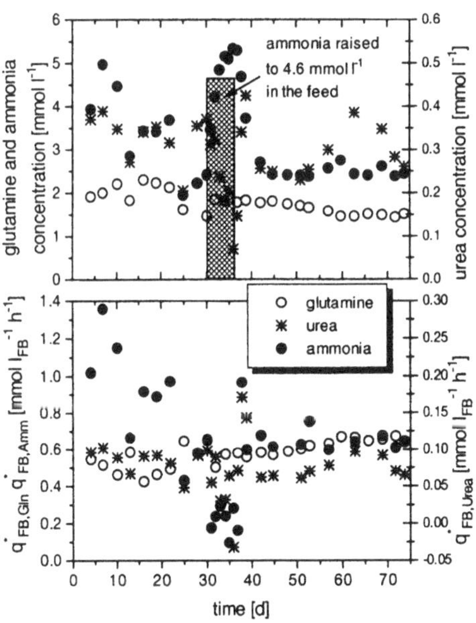

Figure 2. Top diagram: glutamine (open circles), ammonia (closed circles) and urea (stars) concentrations over the cultivation time. The shaded area indicates an increase of the ammonia content of the feed to 4.6 mmol l^{-1}. The bottom diagram shows the corresponding specific glutamine uptake as well as the ammonia and urea production rates related to the fixed-bed volume.

5. Scale-up considerations

For an adult human approx. 300 g of full functional hepatocytes (or 20% of the average liver mass) are able to maintain sufficient function for detoxification. If we assume that a permanent cell line still shows 100% of the function of primary hepatocytes, an extracorporeal liver support system would need a fixed-bed volume of approx. 6.6 litres. Large fixed-bed reactors are operated by pumping the medium in a radial way through the bed rather than axially because of a possible oxygen limitation in fixed-beds over 15 cm of height.

We thank Prof. Strauß of the Max Delbrück Center for Molecular Medicine, Berlin, Germany, for the donation of the mHep-R1 cell line and Biomaterials, Japan for the supply of the Cellsnow carrier.

Pörtner, R., Rössing, S., Koop, M., Lüdemann, I. 1997. Kinetic studies on hybridoma cells immobilised in fixed bed reactors. Biotech. Bioeng. 55: 535-541

Stange, J., Mitzner, S., Strauß, M., Fischer, U., Lindemann, S., Peters, E., Holtz, M., Drewelow, B., Schmidt, R. 1995. Primary or established liver cells for a hybrid liver? Comparison of metabolic features. ASAIO J (United States) 43 (3):M10-M315

TRANSCRIPTIONAL ACTIVITIES OF VIRAL AND CELLULAR PROMOTERS *IN VIVO* IN THE OVIDUCT OF LIVING CHICKENS

H.-M. PARK, J. OKUMURA and T. MURAMATSU
Laboratory of Animal Nutrition, School of Agricultural Sciences, Nagoya University, Chikusa-ku, Nagoya 464-01, Japan

Abstract The experiments described herein were designed to compare the transcriptional activities of promoters *in vitro* and *in vivo* in oviductal cells of chickens with or without steroid hormones. Three viral and cellular promoters fused to the chloramphenicol acetyltransferase reporter gene were transfected *in vitro* into primary cultured oviduct cells from estrogen-stimulated immature chicks, and *in vivo* into oviduct tubular gland cells from laying hens. For transferring DNA to the oviduct of laying hens, *in vivo* electroporation was used. The results indicated that steroid administration induced the *in vivo* transcriptional activity of these promoters with steroid response elements as observed with primary cultures of oviduct cells *in vitro*. The present finding implicates that the localized *in vivo* gene transfer technique by electroporation would serve as a useful tool to elucidate the mechanism of steroid-induced transcription of the oviduct-specific genes in the chicken.

Introduction

In the magnum segment of the chicken oviduct, several tissue-specific egg-white proteins, including ovalbumin (Ov), are being synthesized and secreted at large amounts under the control of steroid hormones. Therefore, chicken oviduct is an attractive model to study the mechanism of transcriptional regulation of tissue-specific and steroid-induced gene expression in higher eukaryotes.

In the past, most of the studies on transcriptional mechanism of Ov have been done with cell lines (Kato *et al.*, 1992; Burbach *et al.*, 1994), or with primary cultures of the oviduct cells from estrogen-stimulated immature chicks, and many putative regulatory elements upstream to the Ov gene have been identified (Muramatsu and Sanders 1995, and refs therein).

However, no studies have been conducted to elucidate the role of such regulatory elements in oviduct tubular gland cells of the laying hen in which those elements originally function. There are at least two major reasons for this: (1) the culture condition of oviduct cells from laying hens was not adequate to obtain the hormonal induction, whereas in the cultured oviduct cells from estrogen-stimulated chicks, gene expression was responsive to hormonal stimuli (Muramatsu et al., 1995), and ; (2) to transfer DNA *in vitro* into cultured oviduct cells from laying hens was not possible by calcium phosphate precipitation or liposomes probably due to large amounts of egg white proteins retained therein. Therefore, it is not known whether the *in vitro* results obtained from cell lines or primary cultures of oviduct cells from estrogen-stimulated immature chicks truly reflect what is happening *in vivo* in oviduct cells of

Y. Kitagawa et al. (eds.), Animal Cell Technology: Basic & Applied Aspects, Vol. 10, 299–302.

laying hens.

Recently, it becomes possible to transfer DNA by using nonviral methods such as microparticle bombardment (Yang *et al.*, 1990, liposomes), liposomes (Muramatsu *et al.*, 1996a) and electroporation (Muramatsu *et al.*, 1996b). Of these, *in vivo* electroporation has been more successful than any other means to transfer DNA into target tissues (Muramatsu *et al.*, 1996b).

In the present study, two series of experiments were conducted. In experiment 1, transcriptional activities of three viral and cellular promoters were compared in the *in vitro* cell culture system with or without steroid hormones to see whether or not steroid induction of transfected genes was attainable. In experiment 2, transcriptional activities of the same viral and cellular promoters were similarly compared *in vivo* in the oviduct of laying hens.

Materials and Methods

The transfected reporter plasmid DNA, pOvCAT900 containing the 900 bp (+9 to -900) length of OV gene promoter, was fused upstream to the chloramphenicol acetyl transferase (CAT) reporter gene. In addition, other CAT reporter plasmid DNAs driven by virus promoters, pSVCAT and pMMTVCAT each containing the SV40 early promoter and MMTV-LTR, were used for transfection. A firefly luciferase (Luc) expression plasmid, pOvLuc100 containing the Ov 100 bp (-87 to +9) promoter fused upstream to the Luc gene, was constructed by substituting this reporter gene for the CAT gene of pOvCAT100.

In experiment 1, primary cultures and gene transfection to oviduct cells were done as reported previously (Park et al., 1997). In experiment 2, White Leghorn laying hens (Aichi Line) at 18 months of age were subcutaneously injected with steroid hormones around the neck for consecutive 3 days, and lightly anesthetized with diethylether. The abdominal cavity was opened surgically, and the magnum portion of the oviduct was exposed. An incision of approximately 3 cm long was made longitudinally in the exposed magnum portion. Into one of the oviduct mucosal folds filled with glandular cells on the mucosa side, the DNA solution containing 20 μg of pMMTVCAT or equal moles of other plasmids was carefully injected together with pOvLuc100 at 5 μg to correct transfection efficiency. After injection, electric square pulses were applied 8 times at 50V for the loading period of 50 msec/pulse with an electro-square porator T820 (BTX, San Diego, USA). Pulses were administered to the tissue using a tweezers type electrode. After 24 h, the hens were euthanatized and the oviduct portions were removed quickly. The tissue samples were subjected to measurements for Luc and CAT activities.

Results

Figure 1 gives the values for CAT activity in experiment 1 where the effect of steroid hormone administration on transcriptional activity of viral andcellular promoters was examined. The results indicated that the MMTV-LTR which contains a glucocorticoid receptor binding site (von der Ahe *et al.*, 1985) showed an increased transcriptional activity in the presence of corticosterone. The Ov900 promoter had roughly the same levels of relative CAT activity with steroid hormones. However, without steroid hormones, transcriptional activity of Ov900 was attenuated as reported

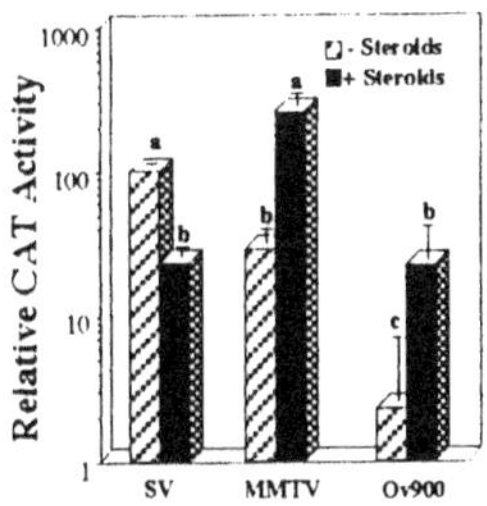

Figure 1. In vitro activity of chloramphenicol acetyltransferase (CAT) in chicken oviduct cells cultured and the extent of steroid induction. *In vitro* transfection was done by calcium phosphate precipitation. Vertical bars stand for means ± SEM of 5 or 6 replicates.
a, b, c Significantly different at P<0.05.

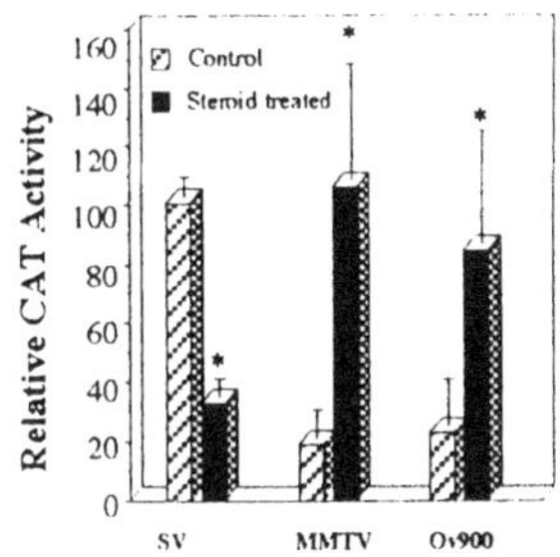

Figure 2. In vivo activity of chloramphenicol acetyltransferase (CAT) driven by viral promoters in the oviduct of laying hens treated with steroid hormones and the extent of steroid induction. Vertical bars stand for means ± SEM of 4 replicates.
* Significantly different from the corresponding no steroid control at P<0.05.

previously (Sanders and McKnight, 1985; Haecker *et al.*, 1995). The extent of steroid induction in the Ov900 promoter was almost the same as that in the MMTV-LTR promoter when steroid hormones were present (Figure 1).

The values for relative CAT activity in experiment 2 are given in Figure 2. The transcriptional activities of MMTV-LTR and Ov900, were significantly increased with exogenous steroid hormones as reported in the *in vitro* experiment (Otten *et al.*, 1988; Haecker *et al.*, 1995). As was expected from the *in vitro* cell culture experiment, no steroid induction was detected *in vivo* in the SV promoter.

Discussion

The experiments described herein suggest that steroid administration induces the *in vivo* transcriptional activities of both viral and cellular promoters with steroid response elements as observed with primary cultures of oviduct cells *in vitro*. However, it remains unknown whether or not the steroid-induced gene transcription observed *in vivo* would truly reflect the gene transcription in the oviduct of laying hens under physiological conditions. Nevertheless, our data implicate that the localized *in vivo* gene transfer technique by electroporation would serve as a useful tool to elucidate further the mechanism of steroid-induced transcription of the oviduct-specific genes in the chicken.

References

Burbach, J. P. H., Silva, S. L. d., Cox, J. J., Adan, R. A. H., Cooney, A. J., Tsai, M.-J. and Tsai, S. Y. (1994) Repression of estrogen-dependent stimulation of the oxytocin gene by chicken ovalbumin upstream promoter transcription factor I, *Journal of Biological Chemistry* **269**, 15046-15053.

Haecker, S. A. E., Muramatsu, T., Sensenbaugh, K. R. and Sanders, M. M. (1995) Repression of the ovalbumin gene involves multiple negative elements including a ubiquitous transcriptional silencer.

ovalbumin gene involves multiple negative elements including a ubiquitous transcriptional silencer, *Molecular Endocrinology* **9,** 1113-1126.
Kato, S., Tora, L., Yamauchi, J., Masushige, S., Bellard, M. and Chambon, P. (1992) A far upstream estrogen response element of the ovalbumin gene contains several half-palindromic 5'-TGACC-3' motifs acting synergistically, *Cell* **68**, 731-742.
Muramatsu, T., Hiramatsu, H. and Okumura, J. (1995) Induction of ovalbumin mRNA by ascorbic acid in primary cultures of tubular gland cells of the chicken oviduct, *Comparative Biochemistry and Physiology* **112B**, 209-216.
Muramatsu, T. and Sanders, M. M. (1995) Regulation of ovalbumin gene expression, *Poultry and Avian Biology Reviews* **6,** 107-123.
Muramatsu, T., Mizutani, Y. and Okumura, J. (1996a) Effects of incubation time and injection sites on gene transfection efficiency in chicken embryos by *in ovo* lipofection, *Animal Science and Technology (Jpn)* **67**, 882-885.
Muramatsu, T., Mizutani, Y. and Okumura, J. (1996b) Live detection of the firefly luciferase gene expression by bioluminescence in incubating chicken embryos, *Animal Science and Technology (Jpn)* **67**, 906-909.
Otten, A. D., Sanders, M. M. and McKnight, G. S. (1988) The MMTV LTR promoter is induced by progesterone and dihydrotestosterone but not by estrogen, *Molecular Endocrinology* **2**, 143-147.
Park, H.-M., Ochiai, H., Okumura, J. and Muramatsu, T. (1997) Transcriptional activity of virus promoters in chicken oviduct cells, in K. Funatsu, Y. Shirai and T. Matsushita (eds.), *Animal Cell Technology : Basic & Applied Aspects,* Vol 8, Kluwer Academic Publishers, Netherlands, pp. 363-365.
Sanders, M. M.and McKnight, G. S. (1985) Chicken egg white genes : multihormonal regulation in a primary cell culture system, *Endocrinology* **116,** 398-405.
von der Ahe, D., Janich, S., Scheidereit, C., Renkawitz, R., Schütz, G. and Beato, M. (1985) Glucocorticoid and progesterone receptors bind to the same sites in two hormonally regulated promoters, *Nature* **313**, 706-709.
Yang, N. S., Burkholder, J., Roberts, B., Martinell, B. and McCabe, D. (1990) *In vivo* and *in vitro* gene tranfer to mammalian somatic cells by particle bombardment, *Proceedings of National Academy of Sciences USA* **87,** 9568-9572.

LIVE DETECTION OF THE FIREFLY LUCIFERASE GENE EXPRESSION IN BOVINE PREIMPLANTATION EMBRYOS

A. NAKAMURA, J. OKUMURA and T. MURAMATSU
Department of Biological Resources and Environmental Sciences, School of Agricultural Sciences, Nagoya University, Chikusa-ku, Nagoya 464-01, Japan

Abstract The present study was done to develop new methodology for measuring bioluminescence activity easily and quickly in live developing bovine embryos by using a photon imaging system. In experiment 1, luciferase expression of fertilized oocytes injected with four gene constructs (pTKEluc, pTK6WEluc, pSVEluc and pMiwluc) was analysed at 2 or 6 days after microinjection. The results indicated that the pMiwluc was expressed more intensively in morulae and blastocysts than three other gene constructs. In experiment 2, the effect of SV40 enhancer was investigated when fused downstream to the luciferase cDNA of the pMiwluc. The results showed that the SV40 enhancer further activated the luciferase activity of the Miw promoter. It was concluded that the bioluminescence activity in live bovine preimplantation embryos could be measured easily and quickly by using a photon imaging system.

Introduction

So far, quantitative analyses of gene expression in preimplantation embryos have been done with bacterial chloramphenicol acetyltransferase (Majumder *et al.*, 1993), bacterial β-galactosidase (Ueno *et al.*, 1987) and firefly luciferase (Thompson *et al.*, 1994, 1995) as reporter proteins. In these quantitative assays, however, fertilized oocytes have to be sacrificed and their cell extracts are to be assayed. During such procedures, some, if not all, of the enzyme activity derived from the reporter gene may be lost. Moreover, once sacrificed, the embryos can no longer be usable for examining the luciferase activity at later developmental stages nor for transplantation to recipient cattle. These problem could be solved by developing a noninvasive monitoring system of gene expression in live developing embryos. The present study was done to develop new methodology for measuring firefly luciferase gene expression easily and quickly in live bovine preimplantation embryos by using a photon imaging system.

Y. Kitagawa et al. (eds.), Animal Cell Technology: Basic & Applied Aspects, Vol. 10, 303–306.

Materials and Methods

Bovine ovaries were collected from cows at a local slaughterhouse. Follicular oocytes with cumulus cells were aspirated from follicles of the ovaries with a syringe, and were cultured as described previously (Nakamura *et al.*, 1995).

In experiment 1, transfer of four different gene constructs (pTKEluc, pTK6WEluc, pSVEluc and pMiwluc) (Figure 1, A, B, C and D) was performed by microinjection into the fertilized eggs as reported previously (Nakamura *et al.*, 1995). After microinjection, the injected embryos were co-cultured for 2 days with the cumulus cells that had been prepared prior to *in vitro* maturation, and were analysed by using a photon imaging system (ARGUS-50/VIM, Hamamatsu Photonics Co., Ltd., Hamamatsu, Japan) at 2 or 6 days following microinjection.

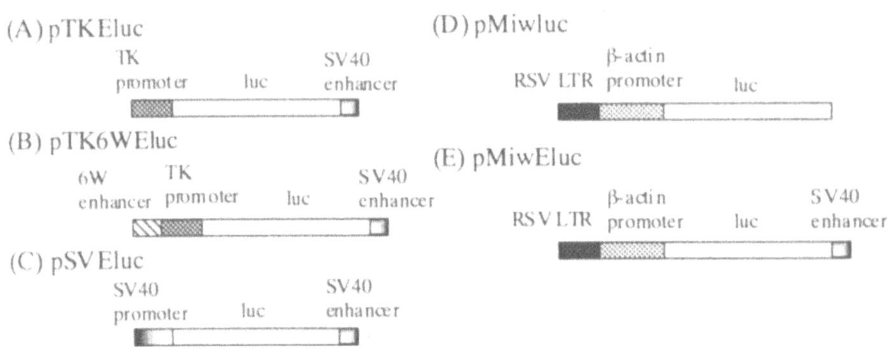

Figure 1. Structures of microinjected transgenes: (A), pTKEluc; (B), pTK6WEluc; (C), pSVEluc; (D), pMiwluc and (E), pMiwEluc. Abbreviations: TK, thymidine kinase; luc, firefly luciferase gene; SV40, simian virus 40; 6W, hexamer of the 1W fragment containing an octamer motif; RSV, Rous sarcoma virus; LTR, long terminal repeat; β-actin, chicken β-actin.

To investigate the effect of SV40 enhancer, experiment 2 was carried out by microinjecting with pMiwluc or pMiwEluc (Figure 1, D and E). The microinjected embryos were cultured for 2 days and assayed for bioluminescence intensity as in experiment 1.

Results

Comparison of luciferase activity obtained by each gene construct is given in Table 1. The results indicated that the pMiwluc was expressed more intensively in bovine embryos at 6 days after microinjection (morulae and blastocysts) than three other gene constructs.

TABLE 1. Comparison of luciferase activity directed by four different gene constructs at 2 or 6 days after microinjection (experiment 1)

	Luciferase activity (photons / embryo / 5 min)	
Transgene	2 Days	6 Days
pTKEluc	632	500
pTK6WEluc	573	397
pSVEluc	1114	471
pMiwluc	812	644

Table 2 presents the effect on gene expression level of the SV40 enhancer at 2 days after microinjection when fused downstream to the luciferase cDNA of pMiwluc. The results showed that the SV40 enhancer further activated the luciferase activity of the Miw promoter.

TABLE 2. The effect of the SV40 enhancer on the firefly luciferase gene expression in bovine embryos at 2 days after microinjection (experiment 2)

Transgene	Luciferase activity (photons / embryo / 5 min)
pMiwluc	649
pMiwEluc	972

Discussion

In the present study, attempts were made to develop a new method for measuring firefly luciferase activity in living bovine embryos by photoncounting with a photon imaging system. So far, it has been reported that by using a luminometer the firefly luciferase gene expression has been analysed quantitatively in the cell extract of mouse preimplantation embryos (Thompson *et al.*, 1994, 1995). However, the embryos can no longer be used for examining the luciferase activity at later developmental stages nor for transplantation to recipients. To overcome this shortcoming, we demonstrated in the present study a noninvasive monitoring system of firefly luciferase gene expression in live bovine embryos.

With the noninvasive photon imaging system, transcriptional activity was compared between gene constructs containing viral and hybrid promoters. The Miw promoter which consists of the chicken β-actin promoter and the RSV LTR is known to be strong in the mouse (Suemori *et al.*, 1990) and bovine embryo (Nakamura *et al.*, 1997). The chicken β-actin promoter, one half of the Miw promoter, is a strong constitutive promoter, while the RSV LTR, the other half, is not only strong but also exhibits minimum variation of transcriptional activity among various cell types. As was expected, the activity of the luciferase gene driven by the Miw promoter tended to be higher than that of three other gene constructs at 6 days after gene transfer (see Table 1). The fact that the SV40 enhancer further increased the transcriptional activity of the Miw promoter suggests that the Miw promoter in combination with the SV40 enhancer may be useful for the detection of reporter gene expression. Taken together, the pMiwEluc plasmid may serve as a good diagnosis tool for the presence of transgenes in preimplantation bovine embryos.

References

Majumder, S., Miranda, M. and DePamphilis, M. L. (1993) Analysis of gene expression in mouse preimplantation embryos demonstrates that the primary role of enhancers is to relieve repression of promoters, *EMBO Journal* **12**, 1131-1140.

Nakamura, A., Okumura, J. and Muramatsu, T. (1995) Early screening of gene expression of SV40 driven lacZ introduced into bovine embryos, *Asian-Australasian Journal of Animal Sciences* **8**, 449-454.

Nakamura, A., Okumura, J. and Muramatsu, T. (1997) LacZ gene expression driven by four promoters in bovine preimplantation embryos, *Animal Science and Technology (Japan)* **68**, 1023-1031.

Suemori, H., Kadodawa, Y., Goto, K., Araki, I., Kondoh, H. and Nakatsuji, N. (1990) A mouse embryonic stem cell line showing pluripotency of differentiation in early embryos and ubiquitous β-galactosidase expression, *Cell Differentiation and Development* **29**, 181-186.

Thompson, E. M., Christians, E., Stinnakre, M.-G. and Renard, J.-P. (1994) Scaffold attachment regions stimulate HSP70.1 expression in mouse preimplantation embryos but not in differentiated tissues, *Molecular and Cellular Biology* **14**, 4694-4703.

Thompson, E. M., Legouy, E., Christians, E. and Renard, J.-P. (1995) Progressive maturation of chromatin structure regulates *HSP70.1* gene expression in the preimplantation mouse embryo, *Development* **121**, 3425-3437.

Ueno, K., Hiramoto, Y., Hayashi, S. and Kondoh, H. (1987) Introduction and expression of recombinant β-galactosidase genes in cleavage stage mouse embryos, *Development Growth and Differentiation* **30**, 61-73.

LOCALIZED *IN VIVO* GENE TRANSFER IN SPERMATOGENIC CELLS OF THE CRYPTORCHID MOUSE TESTIS

T. MURAMATSU, S. RYOKI, Y. OHMORI and J. OKUMURA
Department of Biological Resources and Environmental Sciences, School of Agricultural Sciences, Nagoya University, Chikusa-ku, Nagoya 464-01, Japan

Abstract To examine whether or not expression of foreign genes is maintained in spermatogenic cells for 1 month after *in vivo* DNA transfer, the bacterial lacZ reporter gene was transfected by electroporation to the cryptorchid testis of living mice. The experimental cryptorchidism was induced by surgical retention in the abdominal cavity possibly to increase chances of gene transfer to spermatogonia. Immediately after *in vivo* gene electroporation, the testis was returned to the scrotum, and the mice were maintained for the following 1 month. The lacZ gene expression was detected in some spermatogonia and spermatocytes, though with very low frequency. It was concluded, therefore, that persistent gene expression for 1 month could be attained in spermatogenic cells in the mouse by *in vivo* electroporation.

Introduction

It has been previously demonstrated that the *in vivo* electroporation could serve as an efficient means of gene transfection to the testis of living mice (Muramatsu et al., 1996, 1997a). Foreign gene expression detected in spermatogenic-like cells has implicated that the spermatogenic-cell mediated gene transformation might be possible. However, the gene expression in the above studies is thought to be primarily transient, and therefore brought about by the episomally existing foreign genes within spermatogenic cells. The crucial step to establish spermatogenic-cell mediated gene transformation is to attain stable and long-term gene expression that results from the integration of transfected foreign genes into the chromosome of spermatogenic cells. The long-term in the present study stands for at least 1 month in which mouse spermatogonia can differentiate to mature sperms (Noce, 1994).

For the long-term gene expression, the experimental conditions established for *in vivo* DNA transfection in the testis of living mice may not be optimal as the measurement was done only at 1 or 2 days after transfection (Muramatsu et al., 1996, 1997a). Therefore, the present study was conducted first to examine optimal electric voltages for long-term gene expression by *in vivo* electroporation, and secondly to confirm gene expression in spermatogonia of the mouse testis in which cryptorchidism was surgically induced by retaining in the abdominal cavity for 1 month, followed by descending to the scrotum to get restarted spermatogenesis.

Y. Kitagawa et al. (eds.), Animal Cell Technology: Basic & Applied Aspects, Vol. 10, 307–309.

Materials and Methods

Male ICR strain mice at 4 to 5 weeks of age were used. They were cared for under Guideline of Animal Experimentation, laid down by the Committee of Experimental Animal Care, Nagoya University, Nagoya, Japan. *In vivo* DNA transfer to the testis was done by electroporation as described previously (Muramatsu et al., 1996, 1997a). The plasmid DNAs transfected were pmiwluc (exp. 1) and pmiwZ (exp. 2) at 20 and 50 μg per testis, respectively.

In experiment 1, the mice were sacrificed by decapitation at 1 day or 1 month after transfection, and the testis samples were excised quickly, and weighed. Whole testis samples were then homogenized with three volumes of ice-cold buffer A containing 15 mM Tris, 60 mM KCl, 15 mM NaCl, 12 mM EDTA, 1 mM dithiothreitol, 0.15 mM spermine and 0.4 mM phenylmethylsulfonyl fluoride (pH adjusted to 8.0). The activity of firefly luciferase was determined by bioluminescence intensity with a Pica Gene kit (Wako Pure Chemical Industries, Osaka, Japan) according to the instruction given by the manufacturer.

In experiment 2, at 1 month after surgical cryptorchidism was made on both testes by retaining them into the abdominal cavity, and pmiwZ was transfected by *in vivo* electroporation as in experiment 1. Subsequently, the regressed testes were descended to the scrotum to get restarted spermatogenesis. At 1 month after DNA transfection, the mice were sacrificed, and the testes samples were removed quickly. The testes were fixed with 4% formaldehyde and 0.2% glutaraldehyde solution, rinsed with PBS twice, and stained with X-gal as described elsewhere (Muramatsu et al., 1997a).

The data were treated statistically by analysis of variance, and significance of differences between means was tested by Duncan's multiple range test by using General Linear Model Procedures (SAS, 1985). Where necessary the data were transformed to their common logarithmic values to stabilize error variance.

Results and Discussion

The values for firefly luciferase activities per whole testis are shown in Fig. 1. At

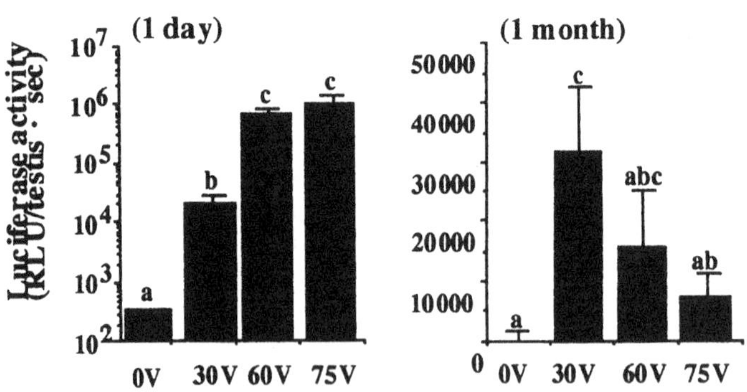

Figure 1. Effect of varying voltages on the luciferase activity (relative light units/whole testis) in the mouse testis to which pmiwluc was transfected at 20 μg/testis by *in vivo* gene electroporation. The values are means ± SEM of 6 replicates. [a,b,c]Significantly different at $P<0.05$.

1 day after transfection, luciferase activity was significantly increased by raising voltages ($P<0.05$), and continuously increased towards 75 volts. When the activity of firefly luciferase per whole testis was measured at 1 month after transfection, there was a significant increase by raising voltages from 0 to 30 volts ($P<0.05$). However, the values decreased gradually towards 75 volts at which no significant increase was observed in comparison with the value at 0 volt.

In our previous study in which long-lasting CAT gene expression was demonstrated in the mouse testis after *in vivo* microparticle bombardment (Muramatsu et al., 1997b), it was not known whether or not spermatogenic cells actually conferred the sustained foreign gene expression. In the present study, *in vivo* gene EP was employed instead of *in vivo* microparticle bombardment, because the latter method did not confer strong gene expression as expected by the former in the mouse testis (Muramatsu et al., 1997a). However, for the detection of the long-term gene expression, the experimental conditions by *in vivo* gene EP had to be optimized. As shown clearly in Fig. 1, high voltages resulted in stronger luciferase activity at 1 day, but not at 1 month after DNA transfection. Thus, for the long-lasting gene expression by *in vivo* EP, the low voltage such as 30 volts would be optimal.

Accordingly, in experiment 2, *in vivo* gene EP was done at 30 V in the surgically induced cryptorchid testis. The experimental cryptorchidism was generated to eliminate meiotic and postmeiotic spermatogenic cells, leading to increased chances of gene transfer to spermatogonia. As reported in the literature (Shikone et al., 1994), most of the testicular cells but spermatogonia disappered by this treatment. Immediately after DNA transfection by *in vivo* EP, the cryptorchid testis was descended to the scrotum to get restarted spermatogenesis. At 1 month after *in vivo* gene EP with pmiwZ, bacterial lacZ gene expression was detected by X-gal staining. A clear signal of lacZ expression was found in some spermatogonia and spermatocytes near the basement membrane of seminiferous tubules, although the frequency was very low (data not shown). Further studies remain to be done to enrich these lacZ-positive spermatogonia and spermatocytes in order to establish spermatogenic-cell mediated gene transformation in the mouse.

References

Muramatsu, T., Shibata, O., Ohmori, Y. and Okumura, J. (1996) *In vivo* electroporation: a convenient method for gene transfer to testicular cells in mice. *Animal Science and Technology (Japan)* **67**, 975-982.

Muramatsu, T., Shibata, O. and Okumura, J. (1997b) Effect of self-replication DNA sequences of Epstein-Barr virus on the expression of a foreign gene transfected *in vivo* to the mosue testis. *Animal Science and Technology (Japan)* (in press).

Muramatsu, T., Shibata, O., Ryoki, S., Ohmori, Y. and Okumura, J. (1997a) Foreign gene expression in the mouse testis by localized *in vivo* gene transfer. *Biochemical and Biophysical Research Communications* **233**, 45-49.

Noce, T. (1994) Gene expression of mouse germ cell. *Protein, Nucleic Acid and Enzyme* **39**, 895-905. (in Japanese).

SAS (1985) *SAS® User's Guide: Statistics.* Version 5 edition. SAS Institute, Cary, NC.

Shikone, T., Billig, H. and Hsueh, A. J. W. (1994) Experimentally induced cryptorchidism increases apoptosis in rat testis. *Biology of Reproduction* **51**, 865-872.

PREPARATION OF A BISPECIFIC ANTIBODY TO OVOMUCOID AND HORSERADISH PEROXIDASE

Hiroshi Shinmoto,[1] Sakiko Miyakoshi,[2] Masuko Kobori,[1] Masami Yonekura,[2] Kazuki Shinohara,[1] and Tojiro Tsushida[1]
[1]National Food Research Institute, Ministry of Agriculture, Forestry and Fisheries, 2-1-2 Kannondai, Tsukuba, Ibaraki 305, Japan; [2] School of Agriculture, Ibaraki University, 3-21-1 Ami, Ibaraki 300-03, Japan

We prepared a bispecific antibody reactive to ovomucoid (OM) and horseradish peroxidase (HRPO) by a chemical recombination method. Hybridomas, HS@03A (anti HRPO) and HS@07A (anti OM), were cultured in a CELLMAX artificial capillary system with MPS module in a serum free medium. Purified antibodies (IgG1) were digested with immobilized ficin and resulting F(ab')$_2$ fragments were reduced with 2-aminoethanethiol hydrochloride followed by treatment with Ellman's reagent to form stable Fab'-TNB-derivatives. Anti-HRPO Fab'-TNB was reduced again, reacted with anti-OM Fab'-TNB and we obtained a bispecific F(ab')$_2$ antibody. The bispecific antibody gave a good standard curve in a competitive ELISA at OM concentrations of 0.1 to 100 μg/ml.

1. INTRODUCTION

Antibodies capable of binding two different antigens are termed bispecific (or bifunctional) antibodies [1-6]. Bispecific antibodies are thought to be useful in detecting biological molecules by ELISA or immuno-stainings, and in treating cancer patients by lymphokine-activated killer cell targeting [7-11].

Well-known technique to prepare bispecific antibodies is chemical recombination. There were several methods for preparing bispecific antibodies by chemical recombination such as simple crosslinking with SPDP (N-succinimidyl-3-(2-pyridyldithiol) propionate) [12], reduction of inter heavy chain disulfide bonds and reoxidation of a mixture of two reduced immunoglobulins [13], and controlled reoxidation of reduced Fab' molecules by chemical blocking of sulfhydryl residues [14].

In this study, we describe the preparation of an antibody bispecific to horseradish peroxidase (HRPO) and ovomucoid (OM), according to Brennan's controlled reoxidation method [14] of reduced immunoglobulin Fab' molecules.

Y. Kitagawa et al. (eds.), Animal Cell Technology: Basic & Applied Aspects, Vol. 10, 311–315.

2. MATERIALS AND METNODS

2.1. CELL LINES

Hybridomas, HS@03A and HS@07A were used (Table 1).

Table 1. Hybridomas used in this study

Hybridoma	antibody	Ig class	light chain
HS@03A	anti-HRPO	IgG1	κ
HS@07A	anti-OM	IgG1	κ

2.2. HIGH-DENSITY CULTURE OF HYBRIDOMAS

Hybridomas, HS@03A and HS@07A, were cultured in a CELLMAX artificial capillary system with MPS module (molecular weight cut off of 30 kDa, GIBCO BRL, Gaithersburg, USA) in a serum-free ITES-ERDF medium (ERDF medium containing 5 μg insulin /ml, 10 μg transferrin /ml, 20 μM ethanolamine, and 25 nM sodium selenite) [15].

2.3. PREPARATION OF F(ab')2 FRAGMENTS OF MONOLCONAL ANTIBODIES

Monoclonal antibodies were purified through a Protein A affinity column with Pierce IgG1 binding buffer system (Pierce, Rockford, USA). Purified IgG1 fractions (6.7 mg for anti-HRPO and 5.9 mg for anti-OM) were digested with a ficin-immobilized column (Pierce) in the presence of 1 mM Cystein at 37°C for 40 h. Resulting $F(ab')_2$ fragments were purified by affinity chromatography through a Protein A column or gel filtration through a Superdex 200 (Pharmacia Biotech, Tokyo, Japan) column (1.2 x 80 cm). We obtained 3.1 mg anti-HRPO $F(ab')_2$ and 3.8 mg anti-OM $F(ab')_2$.

2.4. PREPARATION OF A BISPECIFIC ANTIBODY BY A CHEMICAL RECOMBINATION METHOD

$F(ab')_2$ fragments (3 mg) were reduced in 2 ml of 0.1 M phosphate buffer (pH 6.8) containing 1 mM 2-aminoethanethiol hydrochloride, 1 mM EDTA, and 10 mM sodium arsenite at room temperature for 16 h. Ellman's reagent (4 mg, 5,5'-dithio-bis(2-nitrobenzoic acid)) was added to the reaction mixtures, and the reaction mixtures were then allowed to stand for 3 h to form stable TNB derivatives. Resulting Fab'-TNB fragments were purified through a Superdex 200 column. Yields were 1.5 mg for anti-HRPO Fab'-TNB and 1.2 mg for anti-OM Fab'-TNB. Anti-HRPO Fab'-TNB (0.54 mg) was reduced with 10 mM 2-aminoethanethiol for 30 min to make Fab'-SH form, followed by desalting with a Bio-Rad 10DG desalting column. Anti-HRPO Fab'-SH and anti-OM Fab'-TNB (0.59 mg) were mixed and incubated at room temperature for 16 h. The sulfhydryl residues were blocked with Ellman's reagent again. The resulting Fab'-Fab' (F(ab')$_2$) bispecific antibody was purified with a Superdex 200 column, and 0.64 mg of F(ab')$_2$ fraction (bispecific antibody) was obtained.

2.5. DETECTION OF A BISPECIFIA ANTIBODY

OM solution was pipetted into ELISA plates and held at 4 °C for 16 h. After blocking with Block Ace (Dainippon Seiyaku, Co., Ltd., Osaka, Japan), the bispecific antibody solution (50 μl) was added to the plates. The plates were washed three times, HRPO solution (50 μl, 5 μg/ml) was added, and the enzyme activity immobilized via the bispecific antibody was measured [16].

2.6. COMPETITIVE ELISA OF OM WITH THE BISPECIFIC ANTIBODY

Standard samples of OM (50 μl, 0.1 to 100 μg/ml, n=3), the bispecific antibody (50 μl, 1 μg/ml), and HRPO (50 μl, 5 μg/ml) solutions were mixed and incubated for 2 h at room temperature. Each reaction mixture (50 μl) was added to the OM-immobilized ELISA plates and incubated 2 h at room temperature. The plates were washed six times and HRPO activity immobilized via the bispecific antibody was measured.

3. RESULTS AND DISCUSSION

3.1. PREPARATION OF A BISPECIFIC F(ab')$_2$ ANTIBODY BY A CHEMICAL RECOMBINATION METHOD

Ovomucoid (OM) is a major allergen of hen egg white and causes hen egg allergy [17]. Ovomucoid is also known to be a proteinase inhibitor of hen egg white and shown to have a rigid structure as found in other food allergens [18]. Allergenicities of ovomucoid have been analyzed by human IgE antibodies from patients of egg allergy [19] and mouse monoclonal antibodies [20]. In the course of the study of food allergy, we tried to prepare a mouse hybridoma secreting monoclonal antibody to ovomucoid for detecting egg allergens in food stuffs. In this study, we obtained a mouse hybridoma HS@07A secreting monoclonal antibody to OM. Furthermore, we tried to prepare a bispecific antibody to OM and horseradish peroxidase (HRPO) to improve the detection of OM.

Hybridomas, HS@03A and HS@07A, were cultured in an artificial capillary system with serum-free ITES-ERDF medium. IgG1 was purified from supernatants by Protein A affinity chromatography. As shown in Figure 2, anti-OM antibody was eluted with a mild elution buffer for IgG1 (pH 6.0). Anti-HRPO antibody was not eluted with the mild elution buffer but with an acidic elution buffer (pH 2.8). This result suggested anti-HRPO antibody had stronger affinity to Protein A. Purified antibodies were digested by ficin (a protease of fig latex) digestion. Resulting F(ab')$_2$ fragments were purified with the Protein A column. In this step, anti-HRPO F(ab')$_2$ fragment strongly absorbed on the Protein A column and eluted with acidic buffer.

F(ab')$_2$ fragments were then reduced with 2-aminoethanethiol in the presence of sodium arsenite to prevent reoxidative crosslinking between sulfhydryl residues of heavy chains. Resulting sulfhydryl residues were treated with Ellman's reagent to form

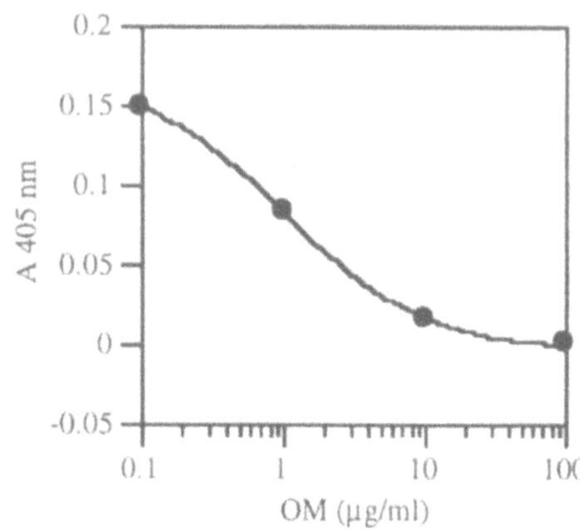

Figure 1 (Left) SDS-PAGE analysis of the bispecific antibody.
Anti-HRPO IgG1 (lane 1), anti-OM IgG1 (lane 2), anti-HRPO F(ab')$_2$ (lane 3), anti-OM F(ab')$_2$ (lane 4), and the bispecific antibody (lane 5) were analyzed by SDS-PAGE under the reducing conditions.

Figure 2 (Right) Standard curve of OM in a competitive ELISA using a bispecific antibody.
Standard samples of OM, diluted bispecific antibody, and HRPO solution were mixed. Each reaction mixture was added to OM-immobilized ELISA plates and the HRPO activity immobilized via the bispecific antibody was measured.

stable TNB derivatives. Anti-HRPO Fab'-TNB was reduced again with 2-aminoethanethiol and mixed with anti-OM Fab'-TNB to form F(ab')$_2$ bispecific antibody.

The result of SDS-PAGE analysis [21] of antibody fragments under the reducing conditions is shown in Figure 1. Anti-HRPO and anti-OM antibody had different molecular weights of light chains (lanes 1 and 2). The molecular weight of ficin-digested heavy chains was similar to that of light chains (lane 3 and 4). The bispecific F(ab')$_2$ antibody (lane 5) had two bands. Because the ficin digested heavy chains had similar molecular weights to the anti-HRPO light chain, it was very difficult to identify the anti-HRPO light chain and ficin-digested heavy chains.

3.2. COMPETITIVE ELISA OF OM WITH THE BISPECIFIC ANTIBODY

We examined a competitive ELISA to measure the concentration of OM. Standard samples of OM, the bispecific antibody, and HRPO solution were mixed and incubated for 2 h at room temperature. Each reaction mixture was added to the ELISA plate coated with OM, and HRPO activity immobilized via the bispecific antibody was measured. Figure 2 shows that the bispecific antibody gave a standard curve at OM concentrations of 0.1 to 100 μg/ml. It was not necessary to prepare an enzyme-labeled antibody. We only mixed enzyme (HRPO) solution and the bispecific antibody to make an enzyme-labeled antibody for the competitive ELISA.

4. ACKNOWLEDGMENT

This study was supported by Grant BMP 98-V-4-1 from the Ministry of Agriculture, Forestry and Fisheries, Japan.

5. REFERENCES

[1] Milstein, C., and Cuello, A. C. (1983) Nature **305**, 537-540.
[2] Reading, C. L. (1983) in Hybridomas and Cellular Immortality (Tom, B. H., and Allison, J. P., Eds), pp. 235-250, Plenum Press, New York.
[3] Ghetie, V., and Moraru, I. (1983) Methods Enzymol. **92**, 523-543.
[4] Milstein, C., and Cuello, A. C. (1984) Immunology Today **5**, 299-304.
[5] Paulus, H. (1985) Behring Inst. Mitt., **78**, 118-132.
[6] Suresh, M. R., Cuello, A. C., and Milstein, C. (1986) Methods Enzymol. **121**, 210-228.
[7] Allard, W. J., Moran, C. A., Nagel, E., Collins G., and Largen, M. T. (1992) Molecular Immunol. **29**, 1219-1227.
[8] Auriol, J., Guesdon, J. -L. Mazie, J. -C., and Nato, F. (1994) J. Immunol. Methods **169**, 123-133.
[9] Hakalahti, L, Vihko, P., Henttu, P., Autio-Harmainen, H., Soini, Y., and Vihko, R. (1993) Int. J. Cancer **55**, 590-597.
[10] Warnaar, S. O., De Paus, V., Lardenoije, R., Machielse, B. N. M., De Graaf, J., Bregonje, M., and Van Haarlem, H. (1994) Hybridoma **13**, 519-526.
[11] Knuth, A., Bernhard, H., Jager, E., Wolfel, T., Karbach, J., Jaggle, C., Strittmatter, W., and Meyer zum Buschenfelde, K. -H. (1994) European J. Cancer **30A**, 1103-1107.
[12] Karpovsky, B., Titus, J. A.,Stephany, D. A., and Segal, D. M. (1984) J. Exp. Med. **160**, 1686-1701.
[13] Nisonoff, A., and Rivers, M. M. (1961) Arch. Biochem. **90**, 460-462.
[14] Brennan, M., Davison, P. F., and Paulus. H. (1985) Science **229**, 81-83.
[15] Murakami, H. (1989) Advances in Biotechnological Processes **11**, 107-141.
[16] Shinmoto, H., urakami, H., Dosako, S., Shinohara, K., and Omura, H. (1986) Agric. Biol. Chem. **50**, 2217-2223.
[17] Bernhisel-Broadbent, J., Dintzis, H. M., Dintzis, R. Z., and Sampson, H. A. (1994) J. Allergy. Clin. Immunol. **93**, 1047-1059.
[18] Kato, I., Schrode, J., Kohr, W. J., and Laskowski, M. Jr. (1987) Biochemistry **26**, 193-201.
[19] Matsuda, T., Nakamura, R., Nakashima, I., Hasegawa, Y., and Shimokata, K. (1985) Biochem. Biophys. Res. Commun. **129**, 505-510.
[20] Kaminogawa, S., Enomoto, A., Kurisaki, J., and Yamauchi, K. (1985) J. Biochem. **98**, 1027-1032.
[21] Laemmli, U. K. (1970) Nature **227**, 680-685.

VITAMIN A ACETATE FOR EFFICIENT PRODUCTION OF MONOCLONAL ANTIBODY BY HUMAN-HUMAN HYBRIDOMAS

YUICHI INOUE[1], MIHOKO FUJISAWA[2], SEIJI KAWAMOTO[2], MASAHIRO SHOJI[3], SHUICHI HASHIZUME[3], YOSHINORI KATAKURA[2], MAKOTO FUJII[1] AND SANETAKA SHIRAHATA[2]

[1]*Department of Biochemical Science and Technology, Faculty of Agriculture, Kagoshima University, 1-21-24 Korimoto, Kagoshima 890, Japan;* [2]*Graduate School of Genetic Resources Technology, Kyushu University, 6-10-1 Hakozaki, Higashi-ku, Fukuoka 812-81, Japan;* [3]*Morinaga Institute of Biological Science, 2-2-1 Shimosueyoshi, Tsurumi-ku, Yokohama 230, Japan*

ABSTRACT Enhancing effect of vitamin A acetate on antibody productivity was examined using human-human hybridoma cell line AE6, BD9 and C5. The effect was maintained for about two weeks by treatment of AE6 hybridomas with 1 μg/ml of vitamin A acetate for a day, and thereafter recovered by its another treatment. The enhancement of antibody production was influenced by cell densities of AE6 hybridomas: two- to four-fold enhancement was observed in cell densities under 10^6 cells/ml, but hardly observed in those over 10^8 cells/ml. Therefore, vitamin A acetate might be suitable for middle scale production of human monoclonal antibody by hybridomas sustained at relatively low cell densities. In addition, enhancing effect of vitamin A acetate was also observed in BD9 hybridomas as well as AE6 hybridomas, but not in C5 hybridomas. These observations indicated that the enhancing effect might be limited by hybridoma cell lines.

1. Introduction

Lung cancer specific human monoclonal antibodies (hMAbs) AE6F4, BD9D12 and HB4C5 produced by human-human hybridoma cell line AE6, BD9 and C5 respectively have been shown potentially useful for the immunocytochemical detection of lung carcinomas (Shoji *et al.*, 1994a; 1994b). We established a hMAb AE6F4 hyper producing recombinant BHK-21 cell line (Inoue *et al*, 1996). On the other hand, we have found the productivities of AE6 hybridomas were enhanced up to about four-fold by the addition of vitamin A (VA) acetate to culture medium. VA acetate seemed to be effective

Y. Kitagawa et al. (eds.), Animal Cell Technology: Basic & Applied Aspects, Vol. 10, 317–321.

for economical hMAb production. Thus, we evaluated the availability of VA acetate to efficiently produce hMAbs using human-human hybridoma cell line AE6, BD9 and C5.

2. Materials and Methods

2.1. CELL LINES AND MEDIA

Human-human hybridoma cell lines AE6 and BD9 were generated by fusing peripheral blood lymphocytes from a healthy adult with the T lymphoblastoid cell line A_4H_{12} derived from Molt4 cells using a *in vitro* immunization method (Kawahara *et al.*, 1992; Shoji *et al.*, 1994a). A human-human hybridoma cell line C5 was generated by fusing lymph node lymphocytes from a lung cancer patient with B lymphoblastoid cell line NAT-30 derived from Namalwa cells (Murakami *et al.*, 1985). All cell lines were maintained in ERDF medium (Kyokuto Pharmaceutical Industrial. Co., Tokyo, Japan) supplemented with 5 μg/ml insulin, 20 μg/ml human transferrin, 20 μM ethanolamine and 25 nM sodium selenite (ITES-ERDF medium), at 37°C in humidified 5% CO_2/95% air. In the experiments, the TES-ERDF medium without insulin was used because insulin did not remarkably affect those cells in a short period of time culture. VA acetate in oil (Wako Pure Chemical Industries Co., Osaka, Japan) was dissolved in ethanol, and added to culture medium at final concentrations of 1 μg/ml.

2.2. ENDURANCE OF ENHANCING EFFECT OF VA ACETATE

AE6 hybridomas were plated in 60 mm plastic dishes at 1 x 10^6 cells/ml and cultured in TES-ERDF medium supplemented with or without VA acetate for a day. After treatment, the cells were washed thoroughly and further cultured in fresh TES-ERDF medium without VA acetate for 14 days. On 14th day, 1 μg/ml of VA acetate was added to culture medium again, the same operations were repeated. During culture, the medium was changed daily. Viable cell density and antibody concentration were examined every day.

2.3. ENHANCING EFFECT OF VA ACETATE IN DIFFERENT CELL DENSITIES

Enhancing effect of VA acetate on antibody production was examined in various cell densities ranging from 10^4 to 10^8 cells/ml. AE6 hybridomas (1 x 10^n cells/ml; n=4-6) were cultured in 35 mm plastic dishes in TES-ERDF medium supplemented with or without 1 μg/ml of VA acetate for 3 days. On the other hand, cells (1 x 10^n cells/ml; n=6-8) were cultured using a hollow fiber bioreactor system Tecnomouse (Integra Biosciences., Germany), as culture in dish. On 3rd day, antibody concentrations were examined, and compared between test and control cultures. Enhancement rate of antibody productivity was calculated by dividing productivity of test by that of control culture without VA acetate.

2.4. ENHANCING EFFECT IN DIFFERENT HYBRIDOMA CELL LINES

Human-human hybridomas (1 x 10^5 cells/ml) were cultured in ITES-ERDF medium supplemented with or without 1 μg/ml of VA acetate for 5 days. Then, cell growth and antibody production were determined, and antibody productivity was evaluated.

3. Results and Discussion

3.1. ENDURANCE OF ENHANCING EFFECT OF VA ACETATE

Enhancing effectof VA acetate was found to keep at least for two weeks by the stimulation with 1 μg/ml of VA acetate for only one day and thereafter decreased productivity could be recovered by another stimulation with VA acetate (Fig. 1). However, the enhancing effect of additional stimulation with VA acetate did not continue for long time but only for three days. This short effect might be due to feedback repression by immediate successive stimulation. These observations suggested that VA acetate might serve as a trigger in consecutive mechanisms for enhancement of antibody productivity.

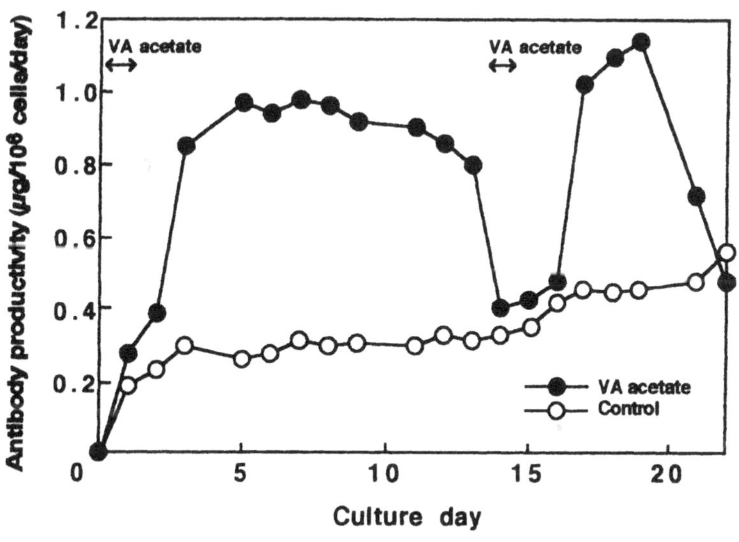

Figure 1. Endurance of enhancing effect by VA acetate. AE6 cells were treated with 1 μg/ml of VA acetate for a day as indicated by the arrows. After treatment, the cells were thoroughly washed and cultured in fresh TES-ERDF medium without VA acetate (solid circle). As control, no treatment culture was also done in the same way (open circle). During culture, the medium was changed daily, and antibody productivity was determined.

3.2. ENHANCING EFFECT OF VA ACETATE AT DIFFERENT CELL DENSITIES

Under 10^6 cells/ml of cell densities, antibody productivities after stimulation with VA acetate were two- to four-fold higher than that of control culture. In 10^6 cells/ml of cell density, the enhancing effect was about three-fold and there was no obvious differences in enhancing effect between dish culture and culture using a hollow fiber bioreacter system. Over 10^8 cells/ml of cell densities, the effects were scarcely observed (Fig. 2). Even under 10^6 cells/ml of cell densities, however, the effect became weaker as cells were growing (datanot shown). These observations suggested that they might be responsible for cell to cell contact due to proliferation of AE6 hybridomas.

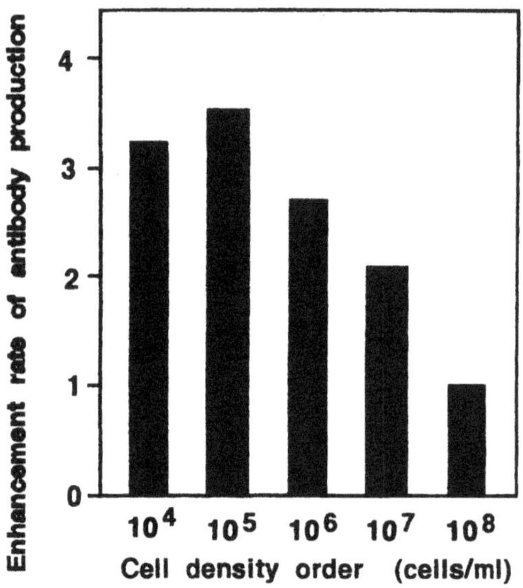

Figure 2. Enhancing effects of VA acetate at different cell densities. AE6 cells (1 x 10^n cells/ml; n=4-8) were cultured in TES-ERDF medium supplemented with or without 1 μg/ml of VA acetate for 3 days. In n<6, cells were cultured in 35 mm plastic dishes. On the other hand, in n>6, cells were cultured using a hollow fiber bioreactor system. In n=6, cells were cultured in both 35 mm plastis dishes and a hollow fober bioreactor system. On 3rd day, Viable cell density and antibody concentration were examined, antibody productivity enhancement rates were calculated.

3.3. ENHANCING EFFECT OF VA ACETATE IN DIFFERENT HUMAN-HUMAN HYBRIDOMAS

Enhancing effects of VA acetate were observed in AE6 and BD9 cell line, but not in HB4C5 cell line (Table 1). These observations indicated that the effect would be limited by hybrodoma cell line. In addition, the effect was found in cell lines derived from the same fusion partner such as A_4H_{12}, regardless of Ig isotypes. Therefore, it was considered that the appearance of enhancing effect might be related to fusion partner rather than B lymphocyte of hybridoma cell line.

TABLE 1. Enhancing effects of VA acetate on antibody productivities of different human-human hybridoma cell lines

Hybridoma	Fusion partner	Immunogloblin class	Enhancement rate of antibody productivity
AE6	A_4H_1 2	IgM	3.5
BD9	A_4H_1 2	IgG	2.3
C5	NAT-30	IgM	1.1

4. Conclusions

We reported here that VA acetate was effective for production of lung cancer specific hMAbs by human-human hybridomas in limited conditions. VA acetate was effective when cell densities were particularly lower than 10^6 cells/ml. Therefore, VA acetate might be available for middle scale production of hMAb using hybridomas sustained at relatively low cell densities.

5. References

1. Aotsuka Y and Naito M (1991) Enhancing effects of retinoic acid on monoclonal antibody production of human-human hybridomas. Cell. Immunol. 133: 498-505.
2. Inoue Y, López LB, Kawamoto S, Arita N, Teruya K, Seki K, Shoji M, Kamei M, Hashizume S, Shiozawa Y, Tachibana H, Ohashi H, Yasumoto K, Nagashima A, Nakahashi H, Suzuki T, Imai T, Nomoto K, Takenoyama M, Katakura Y and Shirahata S (1996) Production of recombinant human monoclonal antibody using *ras*-amplified BHK-21 cells in a protein-free medium. Biosci. Biotech. Biochem., 60: 811-817.
3. Kawahara H, Shirahata S, Tachibana H and Murakami H (1992) *In vitro* immunization of human lymphocytes with human lung cancer cell line A549. Hum. Antibod. Hybridomas 3: 8-13.
4. Murakami H, Hashizume S, Ohashi H, Shinohara K, Yasumoto K, Nomoto K and Omura H (1985) Human-human hybridomas secreting antibodies specific to human lung carcinoma. In Vitro Cell. Dev. Biol. 21: 593-596.
5. Shoji M, Kawamoto S, Sato S, Kamei M, Kato M, Hashizume S, Seki K, Yasumoto K, Nagashima A, Nakahashi H, Suzuki T, Imai T, Nomoto K and Murakami H (1994a) Specific reactivity of human antibody AE6F4 against cancer cells in tissues and sputa from lung cancer patients. Hum. Antibod. Hybridomas 5: 116-122.
6. Shoji M, Kawamoto S, Setoguchi Y, Mochizuki K, Honjoh T, Kato M, Hashizume S, Hanagiri T, Yoshimatsu T, Nakanishi K, Yasumoto K, Nagashima A, Nakahashi H, Suzuki T, Imai T, Nomoto K and Murakami H. (1994b) The 14-3-3 protein as the antigen for lung cancer-associated human monoclonal antibody AE6F4. Hum. Antibod. Hybridomas 5: 123-130.

INDUCTION OF APOPTOSIS BY OSTEOPONTIN IN MDCK CELLS

Hideki Fujimagari, Masamichi Kamihira
and Shinji Iijima
Department of Biotechnology, Graduate School of Engineering, Nagoya University, Chikusa-ku, Nagoya 464-8603, Japan

ABSTRACT. We examined the effects of recombinant osteopontin (OPN) on MDCK (a dog kidney cell line) and RAW (a mouse macrophage cell line) cells to clarify a role of OPN in urinary stone formation. We found that the cell morphology of both MDCK and RAW cells changed during the culture and the cells gradually died. Oxalic acid promoted the cell death by the addition of GST-OPN in the medium, and the cells died at lower OPN concentration than that without oxalic acid. From the electrophoretic analysis of DNA obtained from the dead cells, DNA fragmentation was observed and the cells seemed to give rise to apoptosis. In addition, NO (nitric oxide) production was enhanced by the OPN addition. When OPN was added at the concentration of 0.1-1 μM, both cells produced NO about 2-8 fold that of no addition control. The apoptosis was prevented by the addition of iNOS inhibitor, mercaptoethanol and catalase. These results suggested that the apoptosis induced by OPN was mediated by NO production, and OPN may have an important role not only in calcification but also in cell death in urinary stone formation.

1. Introduction

Osteopontin (OPN) is an acidic calcium-binding phosphoprotein secreted by a variety of cells and tissues (e.g., during bone development, in kidney, mammary gland, macrophage, and smooth muscle) [1]. The protein contains a conserved Arg-Gly-Asp (RGD) sequence [2], and binds to cells via $\alpha_v\beta_3$ integrin [3]. Recently, several OPN receptors have been identified [4]. The protein is now considered to have a key role in some diseases such as urinary stone formation [5], arteriosclerosis [6] and tumor metastasis [7]. However, the functions of OPN in cells are poorly understood.

In this study, in order to clarify a role of OPN in urinary stone formation, a human OPN gene was isolated from HuO9 cells, and the gene was expressed as GST-OPN fusion protein in *Escherichia coli* cells. MDCK and RAW cells were cultured in the medium containing the purified GST-OPN and oxalic acid which may causes urinary stone formation *in vivo*. The effects of the recombinant OPN on viability and NO production of MDCK and RAW cells were examined.

Y. Kitagawa et al. (eds.), Animal Cell Technology: Basic & Applied Aspects, Vol. 10, 323–327.

2. Materials and Methods

Materials
Catalase (SIGMA), superoxide dismutase (SIGMA), mercaptoethanol (Wako Pure Chemical Co.), oxalic acid (Wako Pure Chemical Co.), N^G-methyl-L-arginine (SIGMA) which is an iNOS inhibitor, were used as additives in culture medium.

Cells and cell culture
HuO9 cells (a human osteosarcoma cell line) were cultured in RPMI1640 medium containing 0.2% lactoalbumin hydrolysate and 10% FCS. MDCK cells (a dog kidney cell line) were cultured in MEM medium supplemented with 10% FCS. RAW cells (a mouse macrophage cell line) were cultured in RPMI1640 medium supplemented with 10% FCS. The cells were incubated under 5% CO_2 in a CO_2 incubator.

Measurement of NO production
MDCK and RAW cells were plated into 24-well tissue culture plates ($5X10^4$ cells/well) and cultured until confluent state, and then GST-OPN and various reagents were added to the medium after mitomycin C treatment for 3 h. After 3 days, small amount of the culture medium was removed, and nitrite concentration in the samples was assayed by Griess method [8]. The absorbance at 540 nm for the reaction solution was measured with a spectrophotometer.

Cloning and expression of human OPN cDNA
HuO9 mRNAs were prepared by a procedure using oligo(dT)-cellulose, and synthesis of their cDNA was performed by using a kit obtained from Pharmacia Biotek AB. HuOPN gene was amplified from the cDNA template by PCR, and restriction enzyme sites were added. HuOPN gene was inserted into a pGEX-5X-2 vector (Pharmacia Biotek AB) and expressed in *E. coli* BL21 cells. The HuOPN was produced as a GST-fusion protein and purified by a glutathione-Sepharose column.

DNA isolation and gel electrophoresis for the study of apoptosis
The cells were harvested by centrifugation at 2,500 rpm for 5 min. The cell pellet was lysed with 0.1 ml lysing buffer (10 mM Tris-HCl pH 7.4, containing 10 mM EDTA and 0.5% Triton X-100) and the lysates were centrifuged at 15,000 rpm for 20 min to separate intact chromosomes from DNA fragments. The supernatant was precipitated overnight in 50% isopropanol and 0.5M NaCl. The precipitates were collected by centrifugation at 15,000 rpm for 15 min, air-dried, and resuspended in TE solution. The solution was applied for agarose gel electrophoresis with 1.5% agarose [9].

Measurement of cell viability
The amount of DNA of the cultured cells was determined by DAPI (4, 6-diamidino-2-phenylindole dihydrochloride; Wako Pure Chemical Co.)-DNA fluorometry method after

disruption of the cells by sonication [10].

3. Results and Discussion

Figure 1 shows morphological observation of cultured MDCK and RAW cells. When MDCK and RAW cells were cultured in the medium containing purified GST-OPN and oxalic acid, cell morphology changed during the culture and the cells gradually died (Fig. 1E-H). The addition of oxalic acid, which causes urinary stone formation *in vivo*, promoted the OPN-mediated cell death. In the presence of both OPN and oxalic acid, cells died at lower OPN concentration compared to that without oxalic acid. Without GST-OPN, no change in cell morphology was observed (Fig. 1A-D).

Electrophoretic analysis of DNA obtained from MDCK and RAW cells cultured with or without GST-OPN is shown in Fig. 2. DNA fragmentation was observed when the cells were cultured in the GST-OPN containing medium. Apoptosis is often accompanied by DNA fragmentation caused by the activation of endogenous nucleases. The results indicate that the OPN-mediated cell death of MDCK and RAW cells seems to be apoptosis.

Figure 3 shows viability and NO production of MDCK cells during the culture. NO production increased during first 3 days, and the MDCK cells gradually died in the presence of GST-OPN. When the cells were cultured in the medium containing GST-OPN for 7 days, cell viability decreased about 50% (Fig. 3A). Cell viability further decreased by the addition of oxalic acid. On the other hand, NO production increased about 2-fold, when GST-OPN was added to the medium (Fig. 3B). The NO production was also slightly enhanced by the addition of oxalic acid. These results suggested that the apoptosis induced

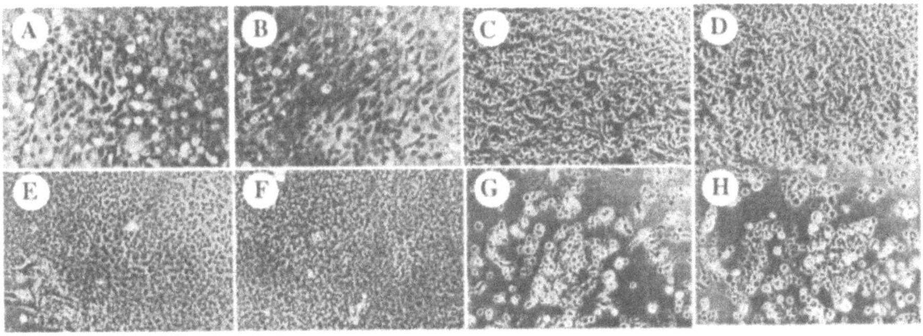

Figure 1. Morphological changes of MDCK and RAW cells by the addition of GST-OPN and oxalic acid on day 3. MDCK (A, B, E, F) and RAW (C, D, G, H) cells were cultured in the normal medium (A, C) or the medium containing 0.125 mM oxalic acid (B, D), 1 μM GST-OPN (E, G) or 0.125 mM oxalic acid and 1 μM GST-OPN (F,H).

by GST-OPN was mediated by NO.

Table 1 summarizes NO production and cell viability of MDCK, when various reagents were added to the medium. With 1 μM GST-OPN in the medium, NO production increased to about 2-fold. In order to inhibit NO production, iNOS (inducible NO synthase) inhibitor was added to the medium. The apoptosis was prevented by the addition. The additions of catalase, mercaptoethanol and superoxide dismutase were also effective for the survival of the cells, although the apoptosis was not completely prevented. Thus, we think that NO is one of the important factors for the OPN-mediated apoptosis. In the case of Raw cells, NO production increased 8-fold in the presence of 1μM GST-OPN. We are now studying about other reactive

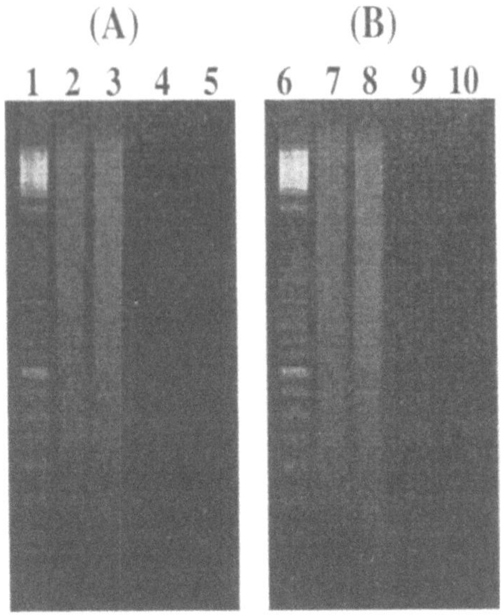

Figure 2. Electrophoretic analysis of DNA from the cells. DNA was extracted from MDCK (A) and RAW (B) cells. Lane 1, 6; 100-base pair DNA ladder Lane 2, 7; 1μM GST-OPN Lane 3, 8; 1 μM GST-OPN and 0.125 mM oxalic acid Lane 4, 9; normal medium (control), Lane 5, 10; 0.125 mM oxalic acid.

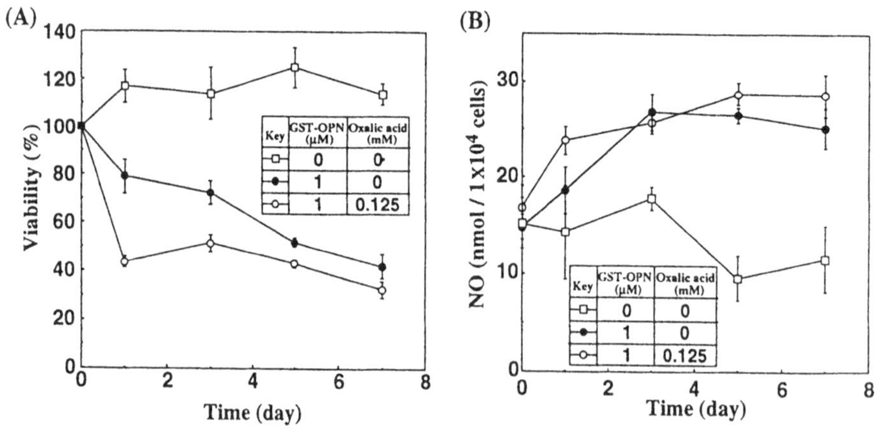

Figure 3. Viability and NO production of MDCK cells, after GST-OPN was added to the medium at the concentration of 1 μM. (A) cell viability, (B) NO production

oxygen species which might associate with OPN-mediated apoptosis.

Osteopontin is considered to have a key role in some diseases such as urinary stone formation and arteriosclerosis, although the functions of OPN in cells are not still clear. The results obtained in this study suggest that OPN may have an important role not only in calcification but also in cell death in urinary stone formation and other diseases.

Table 1. Viability and NO production on MDCK and RAW cells by various reagents. NO production was measured after the cells were cultured in the medium containing various reagents for 3 days. Cell viability was measured at the end of the culture (7 days).

(A) MDCK

	NO production (nmol / 1×10^4 cells)	Viability (%)
No addition	15±2.8	119
GST-OPN (1 μM)	26±3.1	47
Oxalic acid (0.125 mM)	16±0.6	106
GST-OPN (1 μM)+ Oxalic acid (0.125 mM)	26±3.4	34
GST-OPN (1 μM)+ Mercaptoethanol (8.6 μM)	2.2±0.8	73
GST-OPN (1 μM)+ iNOS inhibitor (0.5 mM)	8.6±3.1	98
GST-OPN (1 μM)+ SOD (200 U/ml)	22±3.6	77
GST-OPN (1 μM)+ Catalase (100 U/ ml)	25±4.1	99
GST (1 μM)	14±4.0	121

5. References

1. Butler WT (1989) The nature and significances of osteopontin. *Connect Tissue Res* **23,** 123-136
2. Craig AM, Smith JH, Denhardt (1989) Osteopontin, a transformation-associated cell adhesion phosphoprotein, is induced by 12-O-tetradecanoylphorbol-13-acetate in mouse epidermis. *J. Biol. Chem.* **264**, 9682-9689
3. Yamada KM (1991) Adhesive recognition sequences. *J. Biol. Chem.* **266**, 12809-12812
4. Katagiri YU, Murakami M, Mori K, Iizuka J, Hara T, Tanaka K, Jia WY, Chambers AF, and Uede T (1996) Non-RGD domains of osteopontin promote cell adhesion without involving αv integrins. *J. Cell. Biol.* **62**, 123-131
5. Mckee MD, Nanci A and Khan SR (1995) Ultrastructural immunodetection of osteopontin and osteocalcin as major matrix components of renal calculi. *J. Bone Miner. Res.* **10**, 1913-1929
6. Giachelli C, Bae N, Almeida M, Denhardt D, Alpers CE, Schwartz SM (1993) Osteopontin expression is elevated during neointima formation in rat arteries and in human atherosclerotic plaques. *J. Clin. Invest.* **92,** 1686-1696
7. Craig AM, Bowden GT, Chambers AF, Spearman MA, Greenberg AH, Wright JA, Mcleod M and Denhardt DT (1990) Secreted phosphoprotein mRNA is induced during multistage carcinogenesis in mouse skin and correlate with the metastatic potential of murine fibroblasts. *Int J. Cancer* **46,** 133-137
8. Rollo EE and Denhardt DT (1996) Differential effects of osteopontin on the cytotoxic activity of macrophages from young and old mice. *Immunology* **88**, 642-647
9. Sellins KS and Cohen JJ (1987) Gene induction by γ-irradiation leads to DNA fragmentation in lymphocytes. *J. Immunol.* **139**, 3199-3206
10. Hamada S and Fujita S (1983) DAPI staining improved for quantitative cytofluorometry. *Histochemistry* **79**, 219-226

ANALYSIS OF GLYCOSPHINGOLIPIDS ON DIFFERENTIATION-INDUCTION OF PC12 PHEOCHROMOCYTOMA CELLS BY MANNOSYLERYTHRITOL LIPID

Miki Shibahara[1], Hiroko Isoda[1], Takehide Murata[2], Kazunari K. Yokoyama[2], and Tadaatsu Nakahara[1]

1. Institute of Applied Biochemistry, University of Tsukuba, Ibaraki, Japan
2. Tsukuba Life Science Center, The Institute of Physical and Chemical Research, Ibaraki, Japan

1. Abstract

We found that Mannosylerythritol lipid (MEL), a microbial extracellular glycolipid, induced the neurite outgrowth and the significant enhancement of acetylcholinesterase (AChE) activity in PC12 cells. We also found that the treatment with MEL led to an increase of galactosylceramide (GalCer), whereas treatment with NGF did not increase the content of GalCer.

2. Introduction

Glycosphingolipids (GSLs) are one of the components of the outer leaflet of the plasma membrane of the eukaryotic cells and play a critical role in cellular interactions and control of cell proliferation (Dahms *et al.*, 1983). It has been reported that exogenously administrated GSLs included gangliosides have accelerated the differentiation of the primary neurons and neuroblastomas *in vitro* with neurite sprouting and extension (Dahm *et al.*, 1983). Transfection of ganglioside GD3 (NeuAca2,8NeuAca2,3Galb1,4 Glcb1,1-ceramide) synthase cDNA into neuroblastoma induced cholinergic differentiation with neurite sprouting (Nagai and Tsuji, 1988). These data suggest that ganglioside could modulate signaling pathway of neuronal differentiation, but the molecular mechanisms are still unknown. Mannosylerythritol lipid (MEL), [4-O-(di-

Y. Kitagawa et al. (eds.), Animal Cell Technology: Basic & Applied Aspects, Vol. 10, 329–333.

O-acetyl-di-O-alkanoyl-beta-D-mannopyranosyl)-erythritol], is one of the microbial extracellular glycolipids and was produced by *Candida antarctica* T-34 (Kitamoto *et al.*, 1990). We have recently reported that MEL induced differentiation of HL60 cells (Isoda *et al*, 1997) and melanoma cells B16 (Zhao *et al.*, in this issue). It is noteworthy that the microbial MEL has an ability to induce the differentiation of the vertibrate cells. In the present study, we found that MEL induced the neurite outgrowth in PC12 cells. PC12 cell lines derived from a rat pheochromocytoma, are known to differentiate into neuron-like cells after treatment with nerve growth factor (NGF). Exogenously added MEL led to an increase of a component of GSLs such as galactosylceramide (GalCer). This implies that the functional role of GalCer for the cellular differentiation of PC12 cells.

3. Matelials and Methods

Cell Culture- PC12 cells were obtained from Riken Cell Bank and routinely grown in Dulbecco's modified Eagle's medium (DMEM) supplemented with 5 % fetal bovine serum, 10 % hose serum, 100 mg/ml streptomycin, and 100 units/ml penicillin in the collagen coated culture dish at 37℃ in a humidified 5 % CO_2 incubator.

Induction of PC12 Cellular Differentiation

PC12 cells were incubated with 40 ng/ml NGF or 5 μM MEL for 48 hours to induce cellular differentiation, which was judged by assessing neurite formation and acetylcholinesterase activity as a specific differentiation marker of PC12 cells.

Acetylcholinesterase activity

Cells were washed three times with ice-cold phosphate buffered saline (PBS; pH7.2), suspended into 1 ml of 10 mM phosphate buffer (pH7.0), and then lysed by sonication. AChE activity was measured colorimetrically using acetylthiocholine as a substrate (Ellman *et al.*, 1961). AChE activity is expressed as nmol of thiocholine released/min per mg protein.

Determination of glycosphingolipids

Glycosphingolipids (GSLs) on cell surface membrane during differentiation of PC12 cells were analyzed by the method of High-performance thin-layer chromatography (HPTLC). GSLs were extracted and determined by the modified method of Tsuruoka *et al.* (1993). 10^7 cells were washed twice with PBS and extracted with 0.5 ml of isopropyl alchol/hexane/water (55:25:20, v/v/v) and homogenized by sonication. The cell-pellets were extracted twice and then reextractd three times with 0.5 ml of chloroform/methanol (2:1). All the extracts were evaporated to dryness under nitrogen stream. The sample was dissolved in 0.5 ml of 0.1 M KCl, passed through a SEPPAK PLUS C18 CARTRIDGE (Millipore Corporation, MA, USA) and washed with 2 ml of 0.1 M KCl followed by 5 ml of distilled water and eluted with 2 ml of chloroform/methanol (2:1). The final elutes were evaporated to dryness, and the residue was dissolved in 200 ml of chroroform/methanol /water (40:20:1). The sample was then analyzed on HPTLC plates (HPTLC-Fertigplatten, Kieselgel 60, Merck AG, Darmstadt, Germany) and developed with chloroform/methanol/water (50:40:10) for gangliosides or chloroform/methanol/water (65:25:4) for neutral glycosphingolipids. The GSLs were detected by 0.2 % orcinol in 2 M sulfuric acid.

4. Results

NGF was known to inhibit the growth of PC12 cells and promote neurite extention. In fact, treatment of PC12 with 40 ng/ml NGF elicited extensive fiber-outgrowth as previously reported (Greene *et al.*, 1976). Similarly, addition of 5 μM MEL into the culture medium also caused morphological changes and extention of fibers of PC12 cells. Treatment of PC12 cells with 40 ng/ml NGF induced an increase of AChE (Figure 1). The activity of AChE in PC12 cells induced by MEL showed the similar

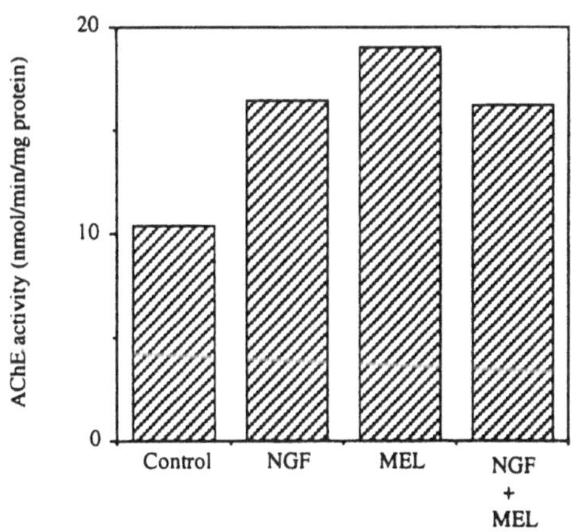

Figure 1. Effects of NGF or MEL on AChE activities of PC12 cells. PC12 cells were cultured in the absence or presence of NGF (40 ng/ml) or MEL (5μM) for 48 hours

extent of the activity after treatment with NGF for 48 hours. The morphology of the longitudinal fiber and the increase of the transmitter-related enzyme AChE activity showed that the incubation of PC12 cells with MEL promoted the outgrowth of neurite. We next examined the content of GSLs of PC12 cells by HPTLC. Thus, we found that treatment with MEL led to an increase of thee specific glycosphingolipid, galactosylceramide (GalCer) and did not change the content of other gangliosides. By contrast, treatment with NGF led to an increase of GM1, GD1a, GD1b, GT1b and no change of GalCer. In addition, the level of MEL remained constant after exposure of MEL until 48 hours.

5. Discussion

Treatment with MEL of PC 12 cells resulted in an increase in the specific activity of AChE and the extention of neurites. PC 12 cells seemed to be differentiated as a chorinergic neuron. Moreover, addition of GalCer in culture medium resulted in the extention of neurite of PC12 cells (Figure 2). Since MEL, a bacterial biosurfactant, has not been known as a substrate of the synthesis of mamalian GalCer, we speculate the increase of GalCer may be resulted by the activation of the biosynthesis of GalCer from ceramide or inhibition of the interconversion of GM4 from GalCer. The molecular mechanism of GalCer mediated cell differentiation of PC 12 cells is the future subject of our research.

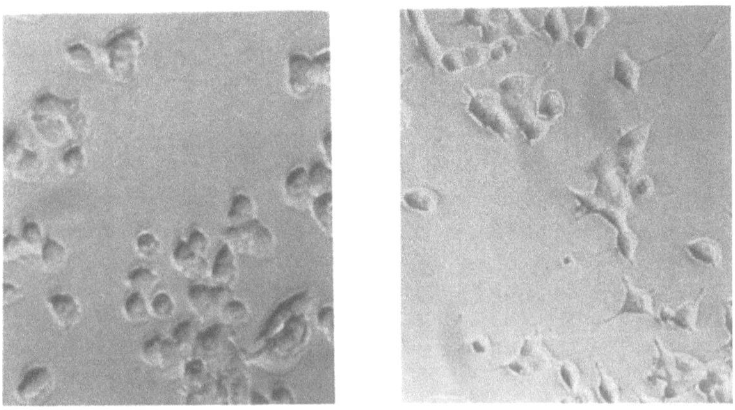

A B

Figure 2. Morphological changes of PC12 cells induced by addition of GalCer. PC12 cells were cultured in a medium without any additions (A), or GalCer (6 μM; B). After 2 days, photographs of each of the cultures were taken (x100).

6. References

Dahms, N. M. and Schnaar, R. L. (1983) Ganglioside composition is regulated during diffentiation in the neuroblastoma X glioma hybrid cell line NG10-15, *J. Neurosci.*, **3** , 806-81

Kitamoto, D., Akiba, S., Hioki, T. and Tabuchi, T. (1990) Extracelluar accumulation of mannosylerythritol lipids by a strain of *Candida antarctica*, *Agri. Biol. Chem.*, **54**, 31-36

Isoda, H., Shinmoto, H., Kitamoto, D., Matsumura, M. and Nakahara, T., (1997) Differentiation of human promyelocytic leukimia cell line HL60 by microbial extracellular glycolipid, *Lipids*, **32**, 263-271

Ellman, G. L., County, K. D., Andres Jr., V. and Featherstone, R. M. (1961) A new and rapid colorimetric determination of acetylcholinesterase activity, *Biochem. Pharm.*, **7**, 88-95

Tsuruoka, T., Tsujii, T., Nojiri, H., Holmes, E. H. and Hakomori, S. (1993)Selection of a mutant cell line based on differential expression of glycosphingolipid, utilizing anti-lactosylceramide antibody and complement, *J. Biol. Chem.*, **268**, 2211-2216

Greene, L. A. and Tischler, A. S. (1976) Establishment of a noradrenergic clonal cell line of rat pheochromocytoma cells which respond to nerve growth factor, *Proc. Natl. Acad. Sci. U.S. A.*, **73**, 2424-2428

Nagai, Y. and Tsuji, S. (1988) Trends in Ganglioside Research, Neurochemical and Neuroregenerative Aspects, Fidia Research Series, **14** (Ledeen, R. W., Hogan, E. L., Tettamanti, G., Yates, A. J., and Yu, R. K., eds.) pp. 329-350, Liviana Press, Padova

MANNOSYLERYTHRITOL LIPID INDUCES ANTI-PROLIFERATION AND DIFFERENTIATION OF MOUSE MELANOMA B16 CELLS

X X ZHAO, H ISODA, T MURATA*, K YOKOYAMA* AND T NAKAHARA
Institute of Applied Biochemistry, University of Tsukuba, Ibaraki 305, Japan
**Tsukuba Life Science Center, The Institute of Physical and Chemical Research, Tsukuba, Ibaraki 305, Japan*

Abstract We report here that the addition of Mannosylerythritol Lipid (MEL), a kind of microbial extracellular glycolipid, demonstrated the inhibition of proliferation on mouse B16 melanoma cells in a dose-dependent manner, the inhibition was significant in the serum-free medium. MEL-treated cells also showed obvious morphological alteration. Incubation of B16 cells with MEL caused 1.7- and 6-fold increases in melanin content and tyrosinase activities, respectively, as compared with those of undifferentiated B16 cells. Western blot study exhibited the increased expression of protein kinase Cα (PKCα) in MEL treated B16 cells and neither PKCβ nor PKCγ was upregulated. Data presented in this report suggest that MEL has an ability to induce the differentiation of B16 melanoma cells.

Introduction

Mannosylerythritol lipid (MEL), a novel microbial extracellular glycolipid, is a biosurfactant composed of both lipophilic and hydrophilic moieties. It was produced by *Candida antarctica* T-34 with a high production of 40 g/l of culture broth (1). We have recently reported that the exposure of MEL could induce the differentiation of human leukemia cells HL60 (2). This observation prompts us the possible use of MEL for the therapeutic reagents for not only leukemia but other cancer cells resistant to chemotherapy. We examine here the effect of MEL on the cell-differentiation of mouse B16 melanoma cells, one of such chemotherapy-resistant cancer cell lines. A number of agents such as retinoic acid, Dimethylsulfoxide and dexamethazone have been reported to induce the differentiation of mouse B16 melanoma cells (3-4). Similary, we found MEL has an ability of differentiation-inducing on melanoma cells via the activation of protein kinase Cα (PKCα).

Materials and methods

Materials Dimethylsulfoxide (DMSO) was obtained from Wako Pure Chemicals (Osaka, Japan). Retinoic acid (RA) and dexamethazone (Dex) were obtained from

Y. Kitagawa et al. (eds.), Animal Cell Technology: Basic & Applied Aspects, Vol. 10, 335–339.

Sigma Chemical CO (St, Louis, Mo, USA). Preparation of MEL and measurement of its concentration were performed by the method of Kitamoto (1).

Cell culture Mouse melanoma B16 4A5 cell line (refereed as B16 cells) was obtained from Riken Cell Bank (Tsukuba, Japan). Cells were maintained in culture with Dulbecco Modified Eagle's Medium (DMEM) supplemented with 10% fetal bovine serum (FBS) or DMEM-ITES medium (serum-free). Number of viable cell was determined by WST-1 cell counting kit (Dojin Laboratories, Kumamoto, Japan).

Melanin content and tyrosinase activity Melanin content of cells was determined by the method of Johnston (5). Tyrosinase activity was analyzed according to the method of Shoji (6) with slight modification. One unit of tyrosinase activity is defined as the activity which could increase 0.001 at A_{280} nm per minute.

Western blot analysis of protein kinase C Cells were incubated in the absence or presence of 20 μm MEL, washed and lysed by lysis buffer. After determination of protein concentration by protein assay (Bio-Rad, Hercules, CA, USA) samples were immunoprecipitated with polyclonal antibodies against PKC (Promega, Madison, WI, USA). Monoclonal antibodies of anti-PKC (Seikagaku, Tokyo, Japan) were used for western blotting. The immunocomplexs were visualized with the ECL detection system (Amersham, Buckinghamshire, England) and autoradiography.

Results

MEL induced anti-proliferation and morphological alteration of B16 cells

The effect of MEL on the cell-proliferation of B16 cells was examined by plating the cells in the absence or presence of MEL. During the 24 h of cultivation, MEL at a

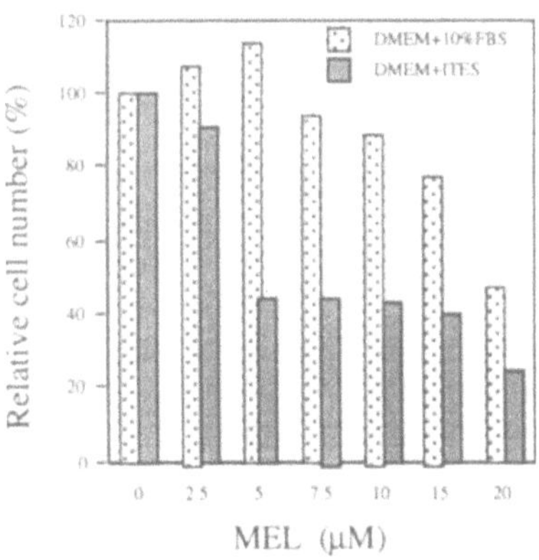

Fig.1 Effect of MEL on proliferation of B16 cells cultured in either serum-free or serum containing medium. Values are the mean of three measurements, standard deviations were below 10%.

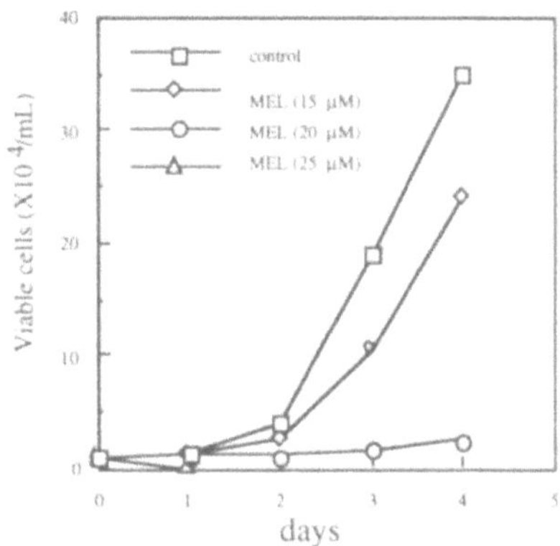

Fig.2 Proliferation of mouse B16 melanoma cells cultured for 4 days with or without MEL in medium of DMEM + 10% FBS, the initial cell number was $1x10^4$/ml. Points are the mean of triplicate cultures. Standard deviations were below 10%

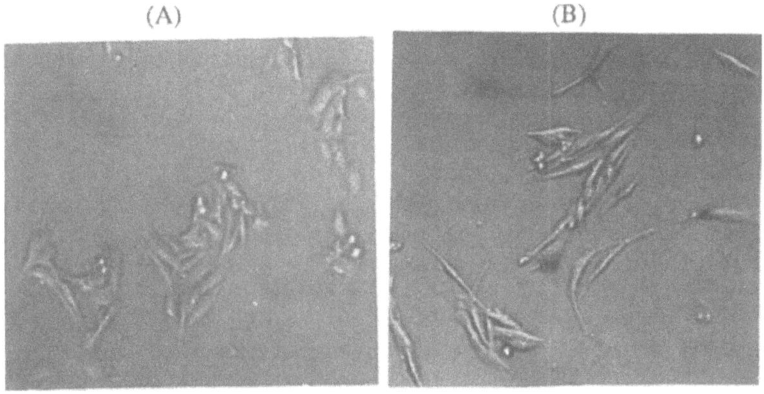

Fig.3 Micrographs of B16 cells cultured in serum containing medium for 2 days with a magnification of 100 x. (A) control B16 cells. (B) B16 cells in the presence of 20 μM of MEL

concentration of 7.5 μM to 20 μM inhibited the growth of B16 cells in a dose-dependent manner. Twenty μm of MEL inhibited cell-proliferation about 50% during 24 h of cultivation in the medium contained 10% FBS (Fig.1). While in serum-free medium, MEL at a concentration of 2.5 μM to 20 μM demonstrated a dose-dependent inhibition on cell-proliferation. This growth inhibition of B16 cells by MEL is significant at the concentration of 5 μM in the serum-free medium (Fig.1). We also investigated the viability of B16 cells for long periods of time (Fig.2). During 4 days of cultivation, 15 μM of MEL reduced the number of viable cells around 70% as compared with that of untreated cells. B16 cells treated with 20 μM of MEL showed almost the complete growth-inhibition with over 95% of cell viability. However, 25 μM of MEL decreased the number of cells significantly within one day-cultivation.

A murine B16 melanoma exhibited the morphological characteristics of fibroblast-like and often appeared in clumps of cells. On the second day of cultivation with 20 μM of MEL, B16 cells undergo morphological changes of differentiated phenotypes, assuming a much more elongated shape and becoming more refractive as shown in Figure 3B.

Measurement of melanin content and the enzymatic activity of tyrosinase

To extend the study of the differentiation of B16 cells, we examined the accumulation of melanin and the tyrosinase activity which are specific for differentiated melanocytes as compared with undifferentiated B16 melanoma cells (7). The level of melanin production of control cells was 9.8 - 12.1 μg/10^6 cells, while B16 cells treated with MEL produced 14.6 - 20.6 μg/ 10^6 cells, of which level was 1.2 - 2.1 fold higher than control cells (Fig.4). Similarly, we measured the tyrosinase activity of B16 cells treated with MEL for 48 h. As illustrated in Figure 4, we found a 6-fold induction of its enzymatic activity. This induction of tyrosinase activity by MEL was higher than that of retinoic acid and dimethylsulfoxide but lower than that of dexamethazone-induction (Fig.4), these three compounds have been reported to induce the cell-differentiation on B16 F-1 melanoma cells (3-4). Results presented here suggest that MEL has the capability of differentiation-induction on B16 cells.

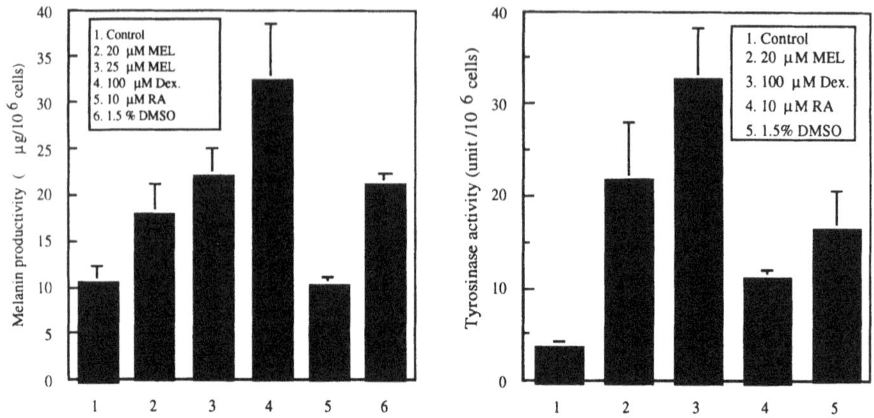

Fig.4 Effect of MEL on melanin content (left) and tyrosinase activity (right) in B16 cells after treatment with 20 µM of MEL for 2 days, all values are the average of results from three experiments and the standard deviation for each value is indicated.

Expression of protein kinase C family

It has been reported that retinoic acid stimulated melanin production in mouse B16 F-1 melanoma cells and induced a 5-8 fold increase in PKCα protein and overexpression of PKCα in these cells resulted in a more differentiated phenotype (8), suggesting that PKCα plays an important role in retinoic acid-induced differentiation, thus we

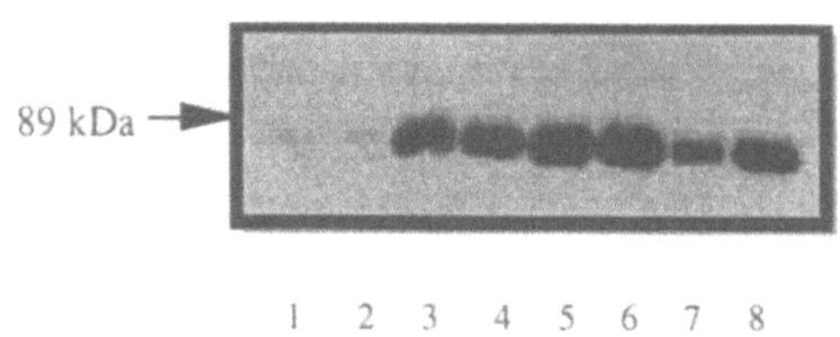

Fig.5 Western blotting analysis of PKCα protein. Lanes 1, 3, 5 and 7 represent control B16 cell lysates collected at 3 h, 14 h, 24 h and 48 h, respectively. Lanes 2, 4, 6 and 8 represent 20 µM of MEL treated B16 cell lysates collected at the same periods of time.

examined the effect of MEL in the expression of PKC. Cells were cultured with MEL for the indicated time and assayed the expression of PKC. As shown in Figure 5, the expression of PKCα were observed at 14 h after treatment with MEL and in control B16 cells. The level of PKCα protein in untreated cells was markedly decreased at 48 h after the cultivation, whereas the MEL-treated B16 cells showed a 2.2-fold increase of PKCα as compared with control cells (Fig.5B). Neither PKCβ nor PKCγ could be induced the cultivation of B16 cells with MEL (data not shown).

Discussion

In this report, MEL was shown to induce anti-proliferative effect in mouse B16 4A5 melanoma cells. Such an inhibition of proliferation was more significant in serum-free medium than in medium supplemented with 10% FBS. The effct of MEL on the growth-inhibition of B16 cells is well corresponded to the critical micelle concentration of MEL in aqueous solution. Thus, we speculate that the formation of micelle of MEL might be critical for this growth inhibition. MEL also induced the morphological changes in B16 cells which reflect a differentiated phenotype. Furthermore, treatment of B16 cells with MEL resulted the significant increase of both expression of melanin and tyrosinase activity which are molecular markers of differentiation of melanoma. MEL, a microbial extracellular glycolipid, is capable to induce the differentiation of cultured B16 cells.

In light of the MEL-induced cell differentiation of B16 cells, we examined whether there was any changes in levels of PKC protein. We found PKCα was expressed in both MEL-treated and control cells. It is the major component of conventional PKC family in B16 4A5 melanoma, because neither PKCβ nor PKCγ was detectable. The treatment of B16 cells with MEL resulted a 2.2-fold increase of level of PKCα protein as compare with control cells. However, the molecular mechanism of this up-regulation of PKCα by MEL remains to be studied.

Our previous studies showed that the differentiation-induction of HL60 cells by MEL was not due to its surfactant character of MEL itself and MEL displayed an influence on the composition of cell surface glycosphingolipids (2). Further studies of membrane associated events induced by MEL in B16 cells are now undertaken. Molecular characterization of the membrane associated events induced by MEL will undoubtedly shed the light on the mechanism controlling the differentiation of B16 melanoma cells.

References

1. Kitamoto, D., Akiba,S., Hioki, T ., and Tabuchi,T. (1990) Extracellular accumulation of mannosylerythritol lipids by a strain of Candida antarctica, *Agric. Biol. Chem* **54**, 31-36.
2. Isoda, H., Shinmoto, H., Kitamoto, D., Matsumura, A., and Nakahara, T. (1997) Differentiation of human promyelocytic leukemia cell line by HL60 by microbial extracellular glycolipids, *Lipids* **32**, 263-271.
3. Niles, R. (1987) *In Vitro* **23**, 803-804.
6. Nordenberg, J., Wasserman, L., Beeery, E., Aloni, D., Malik, H., Stenzel, K, H., and Novogrodsky, A., . (1986) Growth inhibition of murine melanoma by butyric acid and Bimethylsulfoxide, *Exp. cell. Res* **162**, 77-85.
5. Johnston, D., Orlow, S., Levy, E., and Bystryn, J, C. (1992) Induction of B16 melanoma melanogenesis by a serum-free synthetic medium, *Exp. Cell. Res* **201**, 91-98.
6. Shoji, T., Aida, K., Kobri, M., Shimoto, H., and Tsushida, T. (1995) Assays of the inhibitor of melanogenesis, *Tissue Culture* **21**, 293-297.
7. Niles, R, M. and Makarshi, J, S. (1978) *J. Natl. Cancer.Inst* **61**, 523-526.
8. Gruber, J, R., Ohno, S., and Niles, R, M. (1992) Increased expression of protein kinase Cα plays a key role in retinoic acid-induced melanoma differentiation, *J. Biol. Chem* **267**, 13356-13360.

SIGNAL TRANSDUCTION OF DIFFERENTIATION-INDUCTION BY MANNOSYLERYTHRITOL LIPID ON PC12 PHEOCHROMOCYTOMA CELLS

Y. WAKAMATSU, H. ISODA, T. MURATA*,
K. YOKOYAMA* and T. NAKAHARA
Institute of Applied Biochemistry, University of Tsukuba,
1-1-1 Tennodai, Tsukuba, Ibaraki, 305, Japan
**Tsukuba Life Science Center,*
The Instiitute of Physical and Chemical Research
3-1-1 Koyadai, Tsukuba, Ibaraki, 305, Japan

Abstract

We reported here that a microbial extracellular glycolipid, mannosylerythritol lipid (MEL) induced the neuronal differentiation of PC 12 cells and stimulated the action of nerve growth factor (NGF). MEL committed the neurite outgrowth of PC 12 cells even in the presence of antibody against low-affinity NGF receptor ($p75^{NGFR}$) which has been known as the competitive inhibitor of the signal transduction of NGF. This result implies that MEL and NGF could induce the differentiation of neurite outgrowth of PC 12 cells by different mechanisms. The results of Western blotting analysis demonstrated that addition of both NGF and MEL during the cultivation resulted the temporary disappearances of the expressions of $p60^{src}$, focal adhesion protein tyrosine kinase ($p125^{FAK}$) and high-affinity NGF receptor ($p140^{trk}$). When MEL was tested for the regulated expression of these kinases, we found the similar downregulations of expressions of $p60^{src}$ and $p125^{FAK}$, not of $p140^{trk}$. Expression of epidermal growth factor receptor ($p170^{EGFR}$) was unaffected by addition of MEL, irrespective of the presence or absence of NGF. Moreover, MEL also stimulated the phosphorylation of mitogen-activated protein (MAP) kinase. Taken together, these results suggest that MEL and NGF trigger different signal cascades for differentiation of PC 12 cells to commit the neurite outgrowth.

1. Introduction

The PC 12 cell lines, derived from a rat pheochromocytoma, is a useful model to examine the mechanisms underlying neuronal differentiation and signal transduction of various ligands. These cells were committed to differentiate in response to nerve growth factor (NGF) (Greene, L. A. and Tischler, A. S., 1976). The signal transduction of NGF has been known to be mediated by its receptors, $p75^{NGFR}$ and $p140^{trk}$. The addition of NGF allows to form the ligand-receptor complexes with the low-affinity NGF receptor $p75^{NGFR}$ and induced the tyrosine-phosphorylation of $p140^{trk}$. The Shc adapter protein has been implicated in NGF signaling through its binding to activated (tyrosine-phosphorylated) $p140^{trk}$ receptors, which resulted in activation of Ras-

Y. Kitagawa et al. (eds.), Animal Cell Technology: Basic & Applied Aspects, Vol. 10, 341–345.

dependent MAP kinase cascade in neuronal cells such as PC 12 cells (Obermeier *et al.*, 1994 ; Stephens *et al.*, 1994). Involvement of Src in the NGF signaling is implied by the phosphorylation of Shc in the v-src-transformed cells (Pelicci *et al.*, 1992).

Mannosylerythritol lipid (MEL), 4-O-(di-O-acetyl-di-O-alkanoyl-β-D-mannopyranosyl)-erythritol, which is produced by *Candida antarctica* T-34 (Kitamoto *et al.*, 1990), induced neuronal differentiation of PC 12 cells and stimulated NGF-dependent signal cascade. In the present report, we examine the molecular mechanisms of neuronal differentiation of PC 12 cells by MEL and compared with that of NGF.

2. Materials and Methods

2.1. Materials

Nerve growth factor (NGF) 2.5s and monoclonal antibodies against low-affinity NGF receptor ($p75^{NGFR}$) were purchased from Chemicon International, Inc., USA. The monoclonal antibody against $p60^{src}$ and polyclonal antiibody against $p125^{FAK}$ were obtained from Dr. Jun-Ling Guan, Cornell University, USA. The polyclonal antibodies against $p140^{trk}$ and $p170^{EGFR}$ were purchased from Calbiochem., San Diego, CA, USA and Santa Cruz Biotechnology Inc., Santa Cruz, CA, USA, respectively.

2.2. Production of MEL

MEL was prepared as described elsewhere (Kitamoto *et al.* 1990).

2.3. Culture of PC 12 cells

PC 12 cells were obtained from Riken Cell Bank (Tsukuba, Ibaraki, Japan) and maintained in Dulbecco's modified Eagle medium (DMEM) supplemented with 5% fetal bovine serum (FBS), 10% horse serum (HS), 100 μ g/ml streptomycin, and 100 units/ml penicillin at 37 ℃ in a humidified 5% CO_2/95% air. For the assays of neurite outgrowth, PC 12 cells were incubated at 2 x 10^4 cells/cm^2 in a serum-free DMEM-ITES medium, with 5.0 μ M of MEL in the presence or absence of 40 ng/ml of NGF.

2.4. Western blotting analysis

PC 12 cells were lysed in lysis buffer [20 mM Hepes (pH 7.2), 1% Nonidet P-40, 10% (vol/vol) glycerol, 50 mM NaF, 1 mM phenylmethylsulfonyl fuluoride (PMSF), 1 mM Na_3VO_4, 10 μ g/ml leupeptin] (Mutoh *et al.*, 1995). Cell-free lysates were normalized for proteins, resolved by SDS-PAGE and transferred to nitrocellulose membrane. Western blotting analysis was carried out using HRP-conjugated IgG as a secondary antibody and the immune complexes were visualized with the Enhanced Chemiluminescence system (Amersham Japan, Tokyo, Japan).

2.5. MAP kinase Assays

PC 12 cells were lysed in RIPA buffer [phosphate buffered saline (PBS), 1% Nonidet P-40, 0.5% sodium deoxycholate, 0.1% SDS, 1mM PMSF, 1mM Na3VO4, 30 μ l/ml aprotinin]. Cell-free lysates were normalized for proteins, resolved by SDS-PAGE and transferred to nitrocellulose membrane. Western blotting was carried out using PhosphoPlus MAPK Antibody Kit (New England Biolabs Inc., Beverly, MA, USA).

3. Results

3.1. Effect of MEL on neurite outgrowth

MEL has induced the neurite initiation of PC 12 cells which is similar to the case of NGF. An optimal concentration of MEL on the induction of neurite outgrowth was 5.0 μM. A maximal response was observed at 48 hr after exposure with MEL and approximately 30% of the PC 12 cells was extended the neurites. We found that extensive neurite outgrowth promoted by NGF was enhanced further by addition of MEL (H. Isoda *et al.*, submitted).

Moreover, we examined the effect of antibody against low-affinity $p75^{NGFR}$ on the neurite outgrowth. The appearance of cells with neurites was decreased 4- to 5-fold when the $p75^{NGFR}$-specific antibody was added in the culture of PC 12 cells in the presence of NGF (8% vs 38%; number of staining cells with antibody). By contrast, the addition of this antibody did not affect the neurite extension induced by MEL (25% vs 28%; number of staining cell with antibody). Treatment with the $p75^{NGFR}$-specific antibody alone did not inhibit the neurite outgrowth (H. Isoda *et al.*, submitted).

3.2. Effect of MEL on expressions of $p60^{src}$, $p125^{FAK}$, $p140^{trk}$ and $p170^{EGFR}$

The expression of cell-surface membrane proteins such as $p60^{src}$, focal adhesion protein tyrosine kinase ($p125^{FAK}$), high-affinity NGF receptor ($p140^{trk}$) and epidermal growth factor receptor ($p170^{EGFR}$) were examined by Western blotting analysis during the differentiation of PC 12 cells induced by either MEL, NGF alone or both NGF and MEL. The treatment of PC 12 cells with both NGF and MEL, caused a temporary disappearance of the expression of $p125^{FAK}$. When the effect of MEL alone was tested, the similar decreased expressions of $p60^{src}$ and $p125^{FAK}$ were observed (H. Isoda *et al.*, submitted). Expression of $p140^{trk}$ was affected by simultaneous addition of NGF and MEL together, but not affected by addition of MEL alone (Fig.1-a). The level of expression of $p170^{EGFR}$ was not changed significantly by addition of MEL, irrespective of the presence or absence of NGF (Fig. 1-b).

3.3. Phosphorylation of MAP kinase

We observed that the neurite outgrowth and the induced phosphorylation of MAP kinase were paralleled each other after addition of NGF in the cultivation of PC 12 cells. We also examined the effect of MEL on phosphorylation of MAP kinase and found the similar activation of phosphorylation of MAP kinese. p44/p42 MAP kinase appeared to start to be autophosphorylated at 2 hr after treatment with MEL and its maximal phosphorylation was observed at 4 hr (Fig.2). However, the onset of phosphorylation of MAP kinase induced by MEL was delayed in PC 12 cells as comapared with treatment with NGF, which was detected at 10 min.

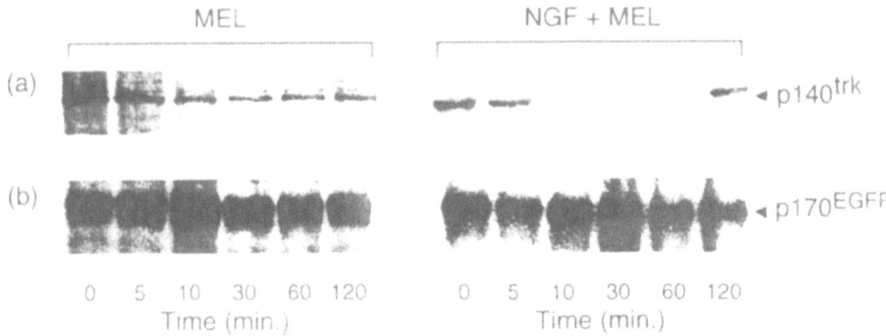

Fig. 1. Effect of MEL on the expressions of (a) $p140^{trk}$ and (b) $p170^{EGFR}$. PC12 cells were incubated at the indicated period of cultivation with NGF (40 ng/ml) and MEL (5 μ M), or MEL alone.

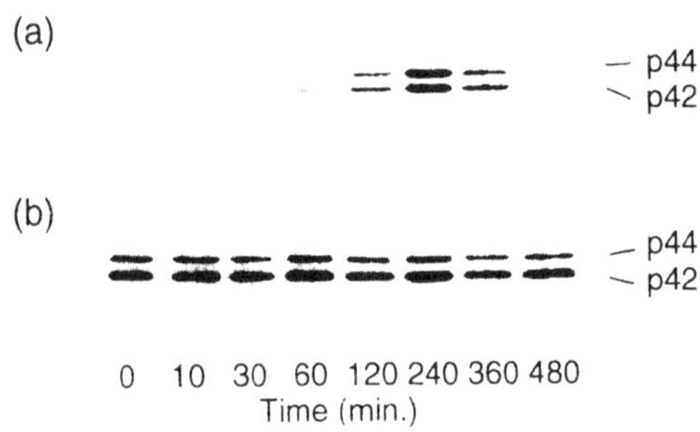

Fig. 2 Effect of MEL on the phosphorylation of MAP kinase. PC 12 cells were treated with MEL (5 μ M) for the indicated period of culture. Western blots were carried out as described in Materials and Methods. (a) phosphotyrosine-specific antibody recognized tyrosine residue at 204 of MAP kinase. (b) control antibody specific for MAP kinase.

4. Discussion

We reported here that MEL has induced the neurite outgrowth of PC 12 cells even in the presence of antibody against low-affinity NGF receptor ($p75^{NGFR}$) which was known to inhibit the signal transmittance by NGF. This indicates that the neurite outgrowth promoted by MEL is not depended on $p75^{NGFR}$-mediated signaling pathway. Alternatively, it might be possible that MEL has an ability to stimulate the $p140^{trk}$-mediated signal cascade or to affect the factors downstream of $p75^{NGFR}$. Werstern blotting analysis demonstrated the expressions of both $p60^{src}$ and $p125^{FAK}$ disappeared temporarily in PC 12 cells by treatment of MEL and $p140^{trk}$ also disappeared temporarily immediately after addition of NGF and MEL. However, we do not know the exact mechanisms of such temporarily disappearences of these molecules at present. We have now undertaken to study whether or not ther downregulation of such molecules are due to the transcriptional (post-transcriptional) regulation of the genes, unstability of the mRNAs, translational regulation, or unstability of these proteins. Similarly, in the case of the signal transduction by NGF alone, we asssume that $p140^{trk}$, $p60^{src}$ and $p125^{FAK}$ might possibly contribute to the neuronal differentiation of PC 12 cells by stimulation of phosphorylation of these proteins. However, the expressions of $p140^{trk}$, $p60^{src}$ and $p125^{FAK}$ were not affected by NGF. The phosphorylation of MAP kinase was also promoted by the treatment of MEL, although the onset of the phosphorylation was delayed as compared with that of NGF. We also observed that the onset of the phosphorylation of MAP kinase and the restoration of the expression of $p60^{src}$ and $p125^{FAK}$ were comparable. We predict that the restoration of $p60^{src}$ and $p125^{FAK}$ after the temporal disappearance by the treatment with MEL might be involved in the phosphorylation of MAP kinase during the MEL-induced differentiation of PC 12 cells. We now examine the possibility that the restorations of the expressions of $p60^{src}$ and $p125^{FAK}$ are really correlated with the comittment of neuronal differentiation of PC 12 cells induced by MEL using the specific kinase-inhibitors, antibodies against these proteins or antisense oligonucleotides to these genes.

References

Kitamoto, D., Akiba, S., Hioki, T. and Tabuchi, T. (1990) Extracellular accumulation of mannosylerythritol lipids by a strain of Candida antarctica, *Agric. Biol. Chem.* **54**, 31-36.

Greene, L. A. and Tischler, A. S. (1976) Establishment of a noradrenergic clonal cell line of rat pheochromocytoma cells which respond to nerve growth factor, *Proc. Natl. Acad. Sci. USA* **73**, 2424-2428.

Mutoh, T., Tokuda, A., Miyadai, T., Hamaguchi, M. and Fujiki, N. (1995) Ganglioside GM1 binds to the Trk protein and regulates receptor functtion, *Proc. Natl. Acad. Sci. USA* **92**, 5087-5091.

Obermeier, A., Bradshaw, RA., Seedorf, K., Choidas, A., Schlessinger, J. and Ullrich, A. (1994) Neuronal differentiation signals are controlled by nerve growth factor receptor/Trk binding sites for SHC and PLC-gamma, *EMBO J.* **13**, 1585-1590.

Stephens, RM., Loeb, DM., Copeland, TD., Pawson, T., Greene, LA. and Kaplan, DR. (1994) Trk receptors use redundant signal transduction pathways involving SHC and PLC-gamma 1 to mediate NGF responses, *Neuron* **12**, 691-705.

Pelicci, G., Lanfrancone, L., Grinnani, F., McGalde, J., Cavallo, F., Forni, G., Nicoletti, F., Pawson, T. and Pelicci, PG. (1992) A novel transforming protein (SHC) with an SH2 domein is implicated in mitogenic signal transduction, *Cell* **70**, 93-104.

ALTERED EXPRESSION OF THE CIP/KIP FAMILY CDK INHIBITORS IN TUMOR PHENOTYPE-SUPPRESSED HUMAN LUNG CANCER CELLS

Seiji Kawamoto[1,2], Yoshinori Katakura[2], Kazuhisa Ono[1] & Sanetaka Shirahata[2]
[1]*Department of Molecular Biotechnology, Graduate School of Engineering, Hiroshima University, 1-4-1 Kagamiyama, Higashi-Hiroshima 739, Japan.*
[2]*Graduate School of Genetic Resources Technology, Kyushu University, 6-10-1 Hakozaki, Higashi-ku, Fukuoka 812-81, Japan.*

Abstract: The cyclin-dependent kinase inhibitor (CDKI) plays pivotal roles in the control of mammalian cell cycle progression as well as in tumor suppression. However, there is relatively few evidences describing an involvement of these inhibitors in the suppression of tumor development in humans. Here we focused on the Cip/Kip family CDK inhibitors, and investigated their roles in the tumor suppressive phenotypes observed in human lung cancer-derived cells. We unexpectedly found that $p21^{Waf1/Cip1}$ protein level was decreased concomitant with tumor suppression, although mRNA level of $p21^{Waf1/Cip1}$ was conversely up-regulated. Observed transcriptional activation of $p21^{Waf1/Cip1}$ was partly attributable to increased p53 activity. These data imply that post-transcriptional $p21^{Waf1/Cip1}$ down-regulatory mechanism is activated in the tumor phenotype-suppressed lung cancer cells. We also found that serum starvation-triggered $p27^{Kip1}$ induction pathway is exclusively up-regulated concomitant with tumor suppression, whereas cell contact- or TGF-β-responsible $p27^{Kip1}$ induction pathway was unaffected. These data suggest that restriction point control-associated $p27^{Kip1}$ induction is an important pathway for tumor suppression, and that frequently observed $p27^{Kip1}$ down-regulation in human tumors might be attributable to inactivation of this mitogen starvation-responsible pathway.

1. Introduction

The CDKI is a critical negative regulator of the G1/S phase cell cycle checkpoint. It specifically interacts with a series of cyclin-CDK complexes, and inhibits phosphorylation of the retinoblastoma (RB) tumor suppressor gene product (1). Hypophosphorylated RB captures the E2F molecule, a family of transcription factors which positively regulates the G1/S cell cycle transition. Gene disruption studies have suggested that malfunction of these CDKIs leads to impaired cell differentiation or acceleration of tumor development. These

Y. Kitagawa et al. (eds.), Animal Cell Technology: Basic & Applied Aspects, Vol. 10, 347–351.

evidences have suggested that the CDKI is an candidate for designing effective antitumor pharmaceuticals. The mammalian CDKI is now classified into two families; the Cip/Kip family ($p21^{Waf1/Cip1}$, $p27^{Kip1}$, and $p57^{Kip2}$) and the Ink4 family ($p15^{Ink4A}$, $p16^{Ink4B}$, $p18^{Ink4C}$, and $p19^{Ink4D}$). While mutations in the *Ink4* family genes are frequently found in a variety of tumors, there are few mutations in those of the *Cip/Kip* family. However, several lines of evidence have implicated an involvement of the Cip/Kip family members in tumor suppression. $p21^{Waf1/Cip1}$ expression is under the control of tumor suppressors or cellular senescence program (2, 3). In addition, down-regulation of $p27^{Kip1}$ protein is observed in several malignant tumors (4, 5). To address a possible role of the Cip/Kip CDKIs in the control of tumor development, we have here compared the mode of the Cip/Kip family expression between A549 human lung cancer cells and their subline A5DC7 cells which exhibit acquired contact inhibition, reduced tumorigenicity (6), and telomere shortening with impaired telomerase activity (7).

2. Materials and Methods

2.1. CELL CULTURE AND REAGENTS

Human lung cancer cell line A549 was cultured in ERDF medium (Kyokuto Seiyaku, Japan) containing 5% fetal bovine serum (FBS, Whittaker, USA) at 37°C in 5% CO_2/95% atmosphere. A5DC7, a A549 cell-derived subline, was established as described (6). For TGF-β treatment, cells were plated at 2 x 10^5 cells per 35 mm dish (Falcon 3001, Becton Deckinson, USA), and preincubated for 24 hr. Then medium was changed to fresh 5% FBS/ERDF containing various concentrations of human recombinant TGF-β (nacalai tesque, Japan). After 72 hr treatment, cells were harvested and subjected to Western blot analysis as described below. For serum deprivation, cells were precultured in 5% FBS/ERDF for 24 hr, and then culture medium was changed to 0.1% FBS/ERDF. Western analysis was performed 24 hr after serum depletion.

2.2. WESTERN BLOT ANALYSIS AND RT-PCR ASSAY

3 x 10^6 cells were harvested, and washed twice with phosphate-buffered saline (PBS). Then the cells were lysed in an extraction buffer (1% Nonidet-P40 in 1 mM PMSF, 150 mM NaCl, 50 mM Tris-HCl [pH 8.0]) on ice for 15 min. After centrifugation, supernatant was mixed with equal volume of Laemmli's SDS-PAGE buffer, and boiled for 5 min. Samples were then subjected to SDS-PAGE in a mini gel apparatus (BIO-RAD, USA). Fractionated proteins were electronically transferred onto a nitrocellulose filter (BIO-RAD). The filter was blocked using skim milk (BlockAce, Dainippon Seiyaku, Japan) and immunoblotted with primary monoclonal antibody against human $p21^{Waf1/Cip1}$, $p27^{Kip1}$, RB (PHARMINGEN, USA), or p53 (Oncogene Science, USA), followed by incubation with

secondary peroxidase-conjugated anti-mouse IgG polyclonal antibody (Tago, USA). Protein bands were visualized by the ECL system (Amersham, UK). For RT-PCR analysis, total RNA was extracted from 3 x 10^6 cells using the Trizol reagent (BRL, USA), and was then reverse-transcribed to single-strand cDNA. Then the cDNA was subjected to the PCR reaction (94°C for 40 sec., 56°C for 1 min., and 72°C for 1 min., 35 cycles). PCR primers for detecting *p21*$^{Waf1/Cip1}$ or control glyceraldehyde dehydrogenase (*GAPDH*) gene expression was constructed as follows; p21 forward: ACTGTGATGCGCTAATGGC, reverse: AATCTGTCATGCTGGTCTGC, GAPDH forward: AACAGCCTCAAGATCATCAGC, reverse: CATGAGTCCTTCCACGATACC. Data from RT-PCR or western analysis were quantified by densitometry.

2.3. REPORTER GENE ASSAY

The p53 reporter construct pGVPFA5, a luciferase expression vector under the control of p53 responsible element Fragment A, is a gift from Dr. M. V. Kato (RIKEN, Japan). Transfection and luciferase assay were performed as described (8). Experiments in each cell line were performed in triplicate.

3. Results and Discussion

3.1. DOWN-REGULATION OF P21 PROTEIN IN THE TUMOR PHENOTYPE-SUPPRESSED A5DC7 CELLS

A A549 human lung cancer-derived cell line, A5DC7, shows some tumor suppressive phenotypes, such as restored contact inhibition, loss of anchorage-independent growth, impaired *in vivo* tumorigenicity, and down-regulation of telomerase activity (6, 7). To address an involvement of the Cip/Kip family in the tumor suppressive phenotypes, we first compared p21$^{Waf1/Cip1}$ expression within A549 cells and A5DC7 cells. RT-PCR assay indicated that *p21*$^{Waf1/Cip1}$ mRNA is up-regulated in A5DC7 cells (Fig. 1A). However, to our surprise, western analysis indicated that p21$^{Waf1/Cip1}$ protein level was conversely down-regulated in these cells (Fig. 1B). These results suggest that post-transcriptional p21$^{Waf1/Cip1}$ down-regulatory mechanism is activated in A5DC7 cells, although *p21*$^{Waf1/Cip1}$ mRNA level is increased concomitant with tumor suppression.

3.2. INCREASED P53 ACTIVITY IN A5DC7 CELLS

Transcriptional activation of p21$^{Waf1/Cip1}$ is partly mediated via tumor suppressor p53 (2). To test whether increased *p21*$^{Waf1/Cip1}$ mRNA in A5DC7 subline is attributed to p53 activation, we next compared p53 activity within A5DC7 cells and parental A549 cells by luciferase reporter gene assay. We found that p53-responsible transcriptional activity in

A5DC7 cells is about three times higher than that of A549 cancer cells, indicating that p53 is activated in these cells during acquisition of tumor-suppressive phenotypes (data not shown). Western analysis indicated that intracellular p53 protein was equivalently detectable within A5DC7 cells and A549 cells (data not shown), suggesting that increased p53 activity was not due to p53 induction.

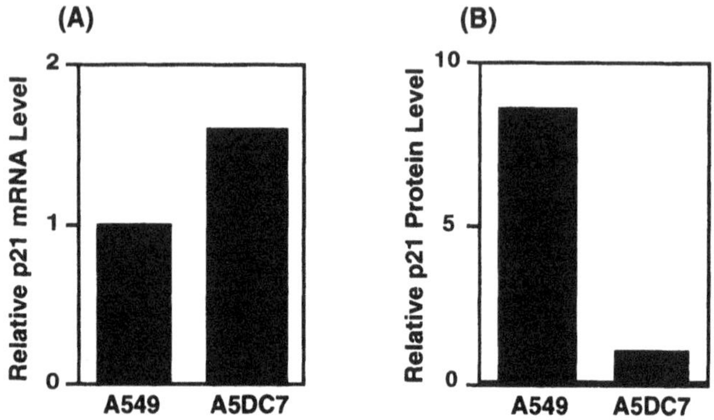

Fig. 1. Down-regulation of p21 protein in tumor phenotype-suppressed A5DC7 cells. (A) Densitometric analysis of p21 mRNA levels detected by RT-PCR assay, (B) Western blot analysis of p21 protein levels.

3.3. SERUM STARVATION-TRIGGERED P27 INDUCTION PATHWAY IS EXCLUSIVELY RESTORED IN A5DC7 CELLS

We next examined $p27^{Kip1}$ induction property whether this pathway altered in tumor phenotype-suppressed A5DC7 cells. Western analysis indicated that basal $p27^{Kip1}$ protein level was indistinguishable between A5DC7 cells and original A549 cells (Fig. 2A). Although A5DC7 cells have been shown to exhibit increased susceptibility for contact inhibition, we found that cell contact-triggered $p27^{Kip1}$ induction was equivalently occurred within these cells (Fig. 2A), suggesting that increased contact inhibition susceptibility in A5DC7 cells was not due to up-regulation of cell contact-associated $p27^{Kip1}$ pathway. TGF-β, a well-known $p27^{Kip1}$-inducing cytokine, had less effect for $p27^{Kip1}$ induction in both of these cell lines (Fig. 2B). In contrast, we found that $p27^{Kip1}$ was more inducible in A5DC7 cells when cultured in serum-depleted medium (Fig. 2C). These data suggest that, in multiple known $p27^{Kip1}$ induction pathways we have tested, restriction point control-associated $p27^{Kip1}$ induction pathway is exclusively restored concomitant with loss of transformed phenotypes in A5DC7 cells. This *in vitro* evidence also raises a possibility that frequently observed $p27^{Kip1}$ down-regulation in human malignant tumors might be attributable to inactivation of this mitogen starvation-responsible pathway.

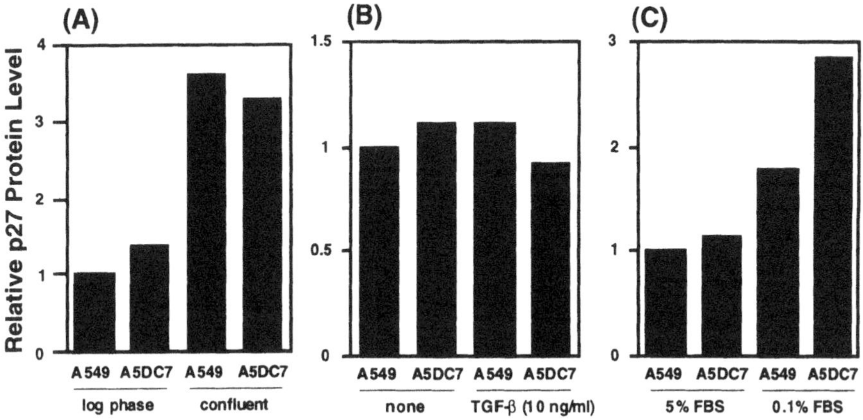

Fig. 2. Western blot analysis of p27 induction levels within A5DC7 cells and parental A549 lung cancer cells. (A) Cell contact-triggered pathway, (B) TGF-β-responsible pathway, (C) Serum starvation-triggered pathway.

3.4. CONCLUSION

In the present study, we have shown that the Cip/Kip family CDK inhibitors are positively or negatively regulated during suppression of tumor phenotypes in human lung cancer-derived cells. Restoration of serum depletion-associated $p27^{Kip1}$ induction pathway found here implicates a molecular basis for explaining frequently observed $p27^{Kip1}$ reduction *in vivo* tumors. In addition, observed post-transcriptional down-regulation of $p21^{Waf1/Cip1}$ is intriguing, because this type of p21 regulation might play an important role *in vivo* tumor progression.

4. References

1. Weinberg R. A. (1995) *Cell* **81**: 323-330.
2. El-Deiry W. S. *et al.* (1993) *Cell* **75**: 817-825.
3. Tahara H. *et al.* (1995) *Oncogene* **10**: 835-840.
4. Porter P. L. *et al.* (1997) *Nature Med.* **3**: 222-225.
5. Loda M. *et al.* (1997) *Nature Med.* **3**: 231-234.
6. Kawamoto S. *et al.* (1995) *Biochem. Biophis. Res. Commun.* **215**: 280-285.
7. Katakura Y. *et al.* (1997) *Biochem. Biophis. Res. Commun.* **237**: 313-317.
8. Sato H. *et al.* (1995) *Biochem. Biophys. Res. Commun.* **214**: 468-474.

SCREENING FOR BIOACTIVE COMPOUNDS THAT INDUCE THE DIFFERENTIATION OF BONE MARROW-DERIVED PREADIPOCYTIC CELL LINE, ST2, INTO THE CELLS OF OSTEOBLASTIC PHENOTYPE

TOMOHIRO SHIMIZU, JE-TAE WOO AND KAZUO NAGAI
Department of Bioengineering, Tokyo Institute of Technology, 4259 Nagatsuta-cho, Midori-ku, Yokohama 226-8501, Japan

We established an assay system for bioactive compounds which induce the differentiation of bone marrow derived preadipocytic cells, ST2, into osteoblastic phenotype by measuring alkaline phosphatase (ALP) activity which is known as an early marker of osteoblasts. In the course of screening we found that protein kinase inhibitors, K-252a, KT5926 and staurosporine, stimulated ALP activity in ST2 cells.

1. Introduction

Osteoporosis is a disease resulted from imbalance between bone formation by osteoblasts and bone resorption by osteoclasts. It is well appreciated that a fundamental problem in the disease is an oversupply of osteoclasts and undersupply of osteoblasts [1]. Osteoblasts originate from bone marrow stromal stem cells which can differentiate into adipocytes, fibroblasts and chondrocytes [2]. Clinical and in vitro observations document an inverse relationship between adipogenesis and osteoblastogenesis in bone marrow. In osteoporotic patients, increased bone marrow adipose tissue correlates with decreased trabecular bone volume [3]. In rodent stromal cells, enhanced expression of adipocytic markers such as production of lipid vacuoles are paralleled by decreased expression of osteoblastic markers such as osteopontin and osteocalcin [4, 5]. These findings suggested the possibility that the excess of the differentiation of multipotent stromal stem cells into adipocyte results in the decrease of osteoblasts and bone formation. Therefore low molecular weight compounds that induce the differentiation of bone marrow stromal stem cells into osteoblasts would be developed as the stimulators of bone formation and also would be used to gain insights about the mechanism concerning the differentiation of adipocytes and osteoblasts.

We established an assay system for bioactive compounds which induce alkaline phosphatase (ALP) activity, an early marker of osteoblasts, in a mouse bone marrow-derived preadipocytic cell line, ST2. In the course of screening we found that some

Y. Kitagawa et al. (eds.), Animal Cell Technology: Basic & Applied Aspects, Vol. 10, 353–356.

bioactive compounds stimulated ALP activity in ST2 cells. In this paper we report that protein kinase inhibitors, K-252a, KT5926 and staurosporine, stimulated ALP activity in ST2 cells.

2. Materials and methods

2.1. CHEMICALS

K-252a, KT5926 and staurosporine were purchased from CALBIOCHEM (La Jolla, CA). Retinoic acid was purchased from Wako Pure Chemicals Co. (Osaka, Japan). Forskolin was purchased from Sigma Chemical Co. (St. Louis, MO).

2.2. CELL CULTURE

A mouse bone marrow-derived cell line, ST2, and a mouse myoblastic cell line, C2C12, were obtained from RIKEN Cell Bank (Tsukuba, Japan), a mouse fibroblastic cell line, C3H10T1/2, and a mouse osteoblastic cell line, MC3T3-E1 were gifted from Dr. Tsuji (Sagami Chemical Research Center).

ST2 and C2C12 were maintained in RPMI1640 and DMEM medium containing 10% FBS and antibiotics (100U/ml of penicillin-G and 100μg/ml of streptomycin), respectively. MC3T3-E1 and C3H10T1/2 were maintained in α-MEM containing 10% FBS and antibiotics (100U/ml of penicillin-G and 100μg/ml of streptomycin). All cells were inoculated in 96-well tissue culture plates at $4x10^3$/well. One day after the inoculation, the culture media were changed to fresh media containing various bioactive compounds, and the cells were cultured for 3 or 6 days.

2.3. ASSAY OF ALKALINE PHOSPHATASE (ALP) ACTIVITY

After removing the culture medium, the cell layers were washed with PBS and then were fixed with EtOH for one minute. ALP activity in the cells was assayed at 37℃ in the buffer containing 0.1M glycine, 1mM $MgCl_2$ and 0.1mM $ZnCl_2$, pH 10.5 for 30 minute using p-nitrophenylphosphate as a substrate. The reaction was terminated by addition of 100μl of 0.05N NaOH. The ALP activity (U) was expressed as nanomoles of p-nitrophenol produced per minute per mg of protein. The protein content was determined with protein assay reagent (Bio-Rad Lab.,Tokyo, Japan).

3. Results and Discussions

We examined the effects of retinoic acid, which is known to induce osteogenic differentiation in immature osteoblast-like cells, on induction of ALP activity in ST2, C3H10T1/2, C2C12 and MC3T3-E1 cells. Figure 1 shows the dose- and time-dependent effects of retinoic acid on cellular ALP activity in ST2 cells. In control cells ALP activity was very low. By contrast, in cells treated with retinoic acid ALP activity increased remarkably in dose- and time-dependent manner. The stimulatory effect of retinoic acid was observed at concentrations higher than 100nM. Although whether retinoic acid can induce the differentiation of undifferentiated mesenchymal cells into mature osteoblasts has not been studied, retinoic acid has been reported to increase

ALP activity in undifferentiated mesenchymal cell lines C3H10T1/2 and ST2 cells [8,9]. In our assay system we confirmed this effects of retinoic acid. As shown in Figure 2, retinoic acid also increased ALP activity in C2C12, but not in MC3T3-E1 cells. This result suggests that the stimulatory effect of retinoic acid is specific on undifferentiated mesenchymal cell lines, not osteoblast-like cells.

It remains to be determined whether the bioactive compounds that stimulate ALP activity in ST2 cells can induce the differentiation of ST2 cells into mature osteoblastic phenotypes.

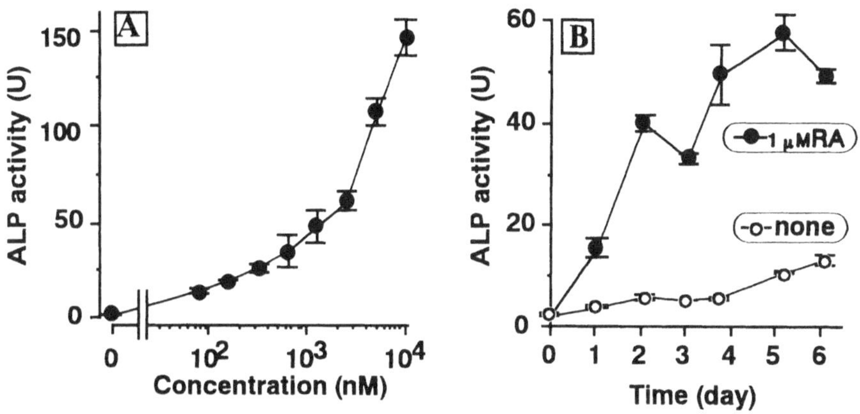

Figure 1. A: Retinoic acid induced ALP activity at day 3 dose dependently.
B: ST2 cells were cultured with 1μM retinoic acid for indicated days.

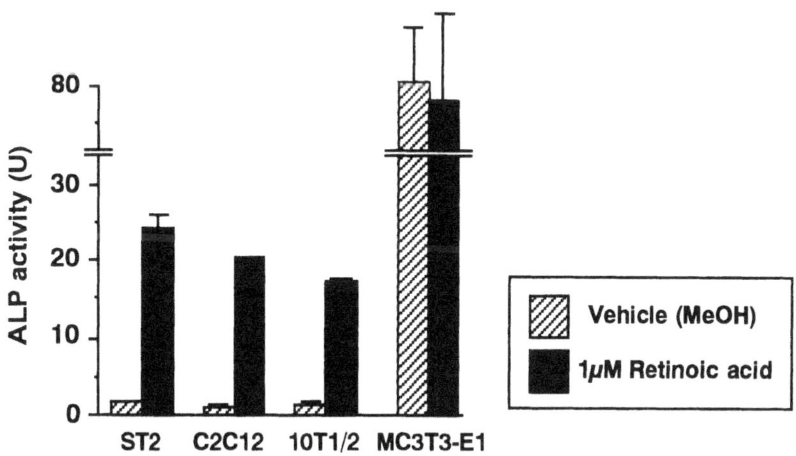

Figure 2. Several cell lines were cultured with 1μM retinoic acid for 3 days.

In the course of screening for bioactive compounds that stimulate ALP activity in ST2 cells, we found that K-252a, KT5926 and staurosporine, which are known to inhibit protein kinases [10,11,12], stimulated the enzyme activity in ST2 cells. The stimulation of ALP activity by these compounds was accompanied by inhibition of cell proliferation. Further study of the stimulatory effects of these compounds on the other osteoblastic phenotypes and on bone formation will be needed.

TABLE 1. The effects of low molecular weight compounds on ALP activity in ST2 cells.

Treatment	Concentration	ALP activity (U) 3 days	6 days
K-252a	300 (nM)	10.34±1.65	34.06±3.08
KT5926	100 (nM)	5.285±0.205	6.989±0.129
Staurosporine	0.10 (nM)	3.607±0.634	10.43±0.718
Retinoic acid	1.0 (μM)	60.89±4.82	104.6±16.8
Vehicle (MeOH)	------------	1.385±0.180	2.676±0.303

ST2 cells were cultured with compounds in 96 well culture plates for 3 or 6 days.

4. References

1. Manolagas SC and Jilka RL. Mechanism of disease: Bone marrow, cytokines, and bone remodeling-emerging insights into the pathophysiology of osteoporosis. *N Engl J Med* 1995; 332: 305-311.
2. Triffitt JT. The stem cell of the osteoblast. In: Bilezikian JP, Raisz LG, Rodan GA (eds). *Principles of Bone Biology*. Academic Press, San Diego, CA, USA, 1996; 39-50.
3. Meunier P, Aaron J, Edouard C and Vignon G. Osteoporosis and the replacement of cell populations of the marrow by adipose tissue. A quantitative study of 84 iliac bone biopsies. *Clin Orthop* 1971; 80: 147-154.
4. Beresford JN, Bennett JH, Devlin C, Leboy PS and Owen ME. Evidence for an inverse relationship between the differentiation of adipocytic and osteogenic cells in rat marrow stromal cell cultures. *J Cell Sci* 1992; 102: 341-351.
5. Dorheim MA, Sullivan M, Dandapani V, Wu X, Hudson J, Segarini PR, Rosen DM, Aulthouse AL and Gimble JM. Osteoblastic gene expression during adipogenesis in hematopoietic supporting murine bone marrow stromal cells. *J Cell Physiol* 1993; 154: 317-328.
6. Ng KW, Gummer PR, Michelageli VP, Bateman JF, Mascara T, Cole WG and Martin TJ. Regulation of alkaline phosphatase expression in a neonatal rat clonal calvarial cell strain by retinoic acid. *J Bone Miner Res* 1988; 3: 53-61.
7. Heath JK, Rodan SB, Yoon K and Rodan GA. Rat calvarial cell lines immortalized with SV-40 large T antigen: constitutive and retinoic acid-inducible expression of osteoblastic features. *Endocrinology* 1989; 124: 3060-3068.
8. Katagiri T, Yamaguchi A, Ikeda T, Yoshiki S, Wozney JM, Rosen V, Wang EA, Tanaka H, Omura S and Suda T. The non-osteogenic mouse pluripotent cell line, C3H10T1/2, is induced to differentiate into osteoblastic cells by recombinant human bone morphogenetic protein-2. *Biochem Biophys Res Commun* 1990; 172: 295-299.
9. Yasuda Y, Inazu M, Kodama H, Takeshita S and Kudo A. BMP-2-producing L cells induce osteogenesis in vivo and in vitro. *J Bone Miner Res* 1996; 14: 15-20.
10. Kase H, Iwahashi K, Nakanishi S, Matsuda Y, Yamada K, Takahashi M, Murakata C, Sato A and Kaneko M. K-252 compounds, novel and potent inhibitors of protein kinase C and cyclic nucleotide-dependent protein kinases. *Biochem Biophys Res Commun* 1987; 142: 436-440.
11. Nakanishi S, Yamada K, Iwahashi K, Kuroda K and Kase H. KT5926, a potent and selective inhibitor of myosin light chain kinase. *Mol Pharmacol* 1990; 37: 482-488.
12. Schachtele, C., Seifert, R. and Osswald, H. Stimulus-dependent inhibition of platelet aggregation by the protein kinase C inhibitors polymyxin B, H-7 and staurosporine. *Biochem Biophys Res Commun* 1988; 151: 542-547.

STRUCTURE AND FUCTION OF SUCCINOYL TREHALOSE LIPID ON DIFFERENTIATION-INDUCTION OF HUMAN PROMYELOCYTIC LEUKEMIA CELL LINE HL-60

T. SUDO, H. ISODA, M. KOBORI*, T. MURATA**, K. YOKOYAMA**
AND T. NAKAHARA
Institute of Applied Biochemistry,
University of Tsukuba, 1-1-1 Tennodai, Tsukuba, Ibaraki, 305, Japan,
**National Food Research Institute,*
Ministry of Agriculture, Forestry and Fisheries, 2-1-2 Kannondai, Tsukuba, Ibaraki 305, Japan,
***Tsukuba Life Science Center, The Institute of Physical and Chemical Research, 3-1-1 Koyadai, Tsukunba, Ibaraki 305, Japan*

Abstract

Four analogues of Succinoyl Trehalose Lipid-3 (STL-3) that have even- or odd-chain, unsaturated or halogenated fatty acids were compared to examine their abilities of inhibiting the growth and of inducing the differentiation of a human promyelocytic leukemia cell line HL-60. The optimal concentration of STL-3 for such activities was around the critical micelle concentration (CMC) of STL-3. The STL-3 analogues with even- or odd-chain or unsaturated fatty acids showed the potent growth inhibition and the inductions of differentiation of HL 60 cells, as evaluated by NBT reducing enzymatic activity and appearence of CD36 antigen. By contrast, the STL-3 analogue with halogenated fatty acids exhibited the weakest, but significant activities of inhibiting the cell growth and inducing of differentiation of HL-60 cells. Collectively, these results indicate that the induced activities of STL-3 for growth-inhibition and cell-differentiation depend on the structure of hydrophobic moiety.

1. Introduction

Many microorganisms are known to produce surface active compounds generally termed biosurfactants. STL-3 (Fig. 1) is glycolipid type biosurfactant produced by

Y. Kitagawa et al. (eds.), Animal Cell Technology: Basic & Applied Aspects, Vol. 10, 357–361.

Rhodococcus sp. TB-42 from *n*-alkane as a carbon source. We have recently reported that STL-3 inhibited the growth of HL-60 cells and induced the monocytic differentiation[1]. Hydrophobic moiety of STL-3 is derived from the components of carbon source in the culture medium of *Rhodococcus* sp. TB-42. In this study, we modified the hydrophobic fatty acid moiety of STL-3 and prepared the STL-3 analogues which contained the even-chain, odd-chain, unsaturated and halogeated fatty acids from *n*-hexadecane, *n*-pentadecane, 1-hexadecene and 1-chlorohexadecane as a carbon source, respectively. We reported here the comparative studies of structure and function of the differentiation of HL-60 cells.

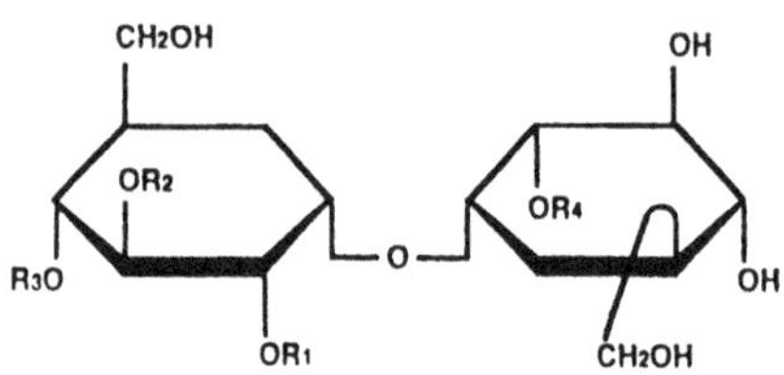

R1-4=1 x Succinic acid, 3 x Fatty acid

Fig. 1 Structures of succinoyl trehalose lipid-3 analogues (STL-3).

Table 1. Composition of fatty acids of STL-3 analogues and their CMC

	Fatty acid composition (%)	CMC (μM)
Even-chain STL-3	Octanoic acid (27) Decanoic acid (67)	3.6
Odd-chain STL-3	Nonanoic acid (56) Undecanoic acid (39)	3.4
Unsaturated STL-3	Octenoic acid (15) Decenoic acid (51)	3.9
Halogenated STL-3	Chlorooctanoic acid (38) Chlorodecanoic acid (38)	4.6

2. Materials and Methods

2.1 PREPARATION OF STL-3 ANALOGUES

STL-3 analogues were extracted with ethyl acetate from culture broth of *Rhodococcus* sp. TB-42 and purified by the preparative thin layer chromatography. The composition of fatty acids of these analogues was shown in Table 1.

2.2 CELL CULTURE

HL-60 cells were incubated in RPMI 1640 medium (Nissui Pharmaceutical Co., Tokyo, Japan) containing 10% fetal bovine serum in a 95% air and 5% CO2 atomosphere at 37 ℃. In some case, HL-60 cells were cultured in a serum-free ERDF (Kyokuto Pharmaceutical Kogyo Co., Tokyo, Japan)-ITES medium, in which ITES contains 5 μg/ml insulin, 10 μg/ml iron-free human transferrin, 25 μM ehanolamine and 25 nM selenite (RD-1; Kyokuto Pharmaceutical Kogyo Co., Tokyo, Japan) as described elsewhere[1]. Cells were counted using a hemacytometer and the viability of cells was estimated by trypan blue dye exclusion method. STL-3 analogues were sonicated, sterilized by passage through a filter unit (0.45 μm) and were added to the culture medium at the indicated concentration.

2.3 DETERMINATION OF CRITICAL MICELLE CONCENTRATION (CMC)

CMC was caluculated from surface tension-concentration plot as described elsewhere[2]. Surface tension was determined using a Whilhelmy type automatic tensiometer CBVP-A3 (Kyowa kaimenkagaku, Tokyo, Japan) at 25 ℃.

2.4 INDEX OF CELL DIFFERENTIATION

The activity of reduction of nitroblue tetrazolium (NBT) was determined by the modified method of Takeda *et al*[3]. The percentage of cells containing blue-black formazan deposits was determined by counting at least 200 cells under microscope. The surface membrane antigen, CD36 was assessed by flow cytometry using antibody against CD36 (Serotec Co., Oxford, England).

3. Results

3.1 CELL GROWTH OF HL-60

Cells were seeded at an initial concentration of 2×10^5 cells per ml with or without STL-3 analogues. Viable cells were counted at day 2. Each sample showed growth inhibition in a dose-dependent manner (Fig. 2). The optimal concentrations of STL-3 analogues for growth inhibition were around CMC (Table 1).

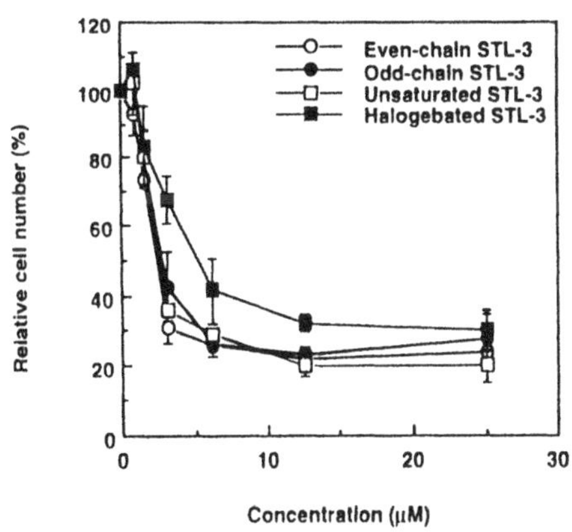

Fig. 2 Effect of STL-3 analogues on cell-growth of HL-60.

3.2 NBT REDUCING ACTIVITY

Effects of STL-3 analogues on the NBT reducing activity was examined (Fig. 3). In the control experiment, the cultivation of HL-60 cells without STL-3 analogues showed 8% of cell was positive with NBT. In contrast, cells incubated with 3 μM STL-3 analogues showed the significant increase of NBT reducing activity (32 to 38%). Especially, treatment of HL-60 cells with STL-3 contained saturated fatty acids (even-chain STL-3 and odd-chain STL-3) resulted in 4.0- to 4.8-fold increase of NBT reducing activity as compared with that of control HL-60 cells (Fig. 3).

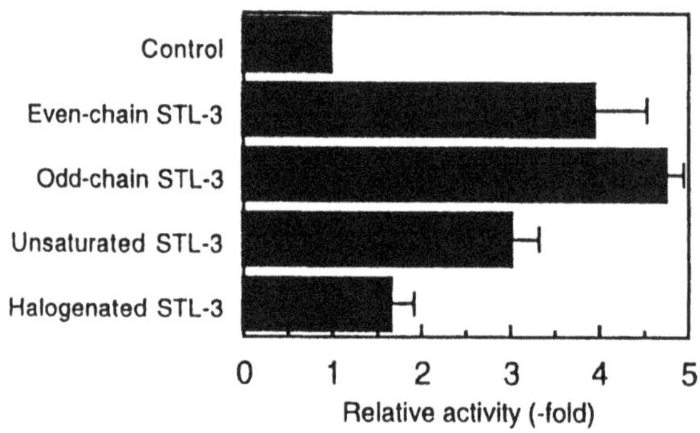

Fig. 3 NBT reducing activity of HL-60 cells treated with STL-3 analogues.

3.3 EXPRESSION OF CD36

The expression of CD36 (membrane glycoprotein expressed by monocytes and macrophages) was assesed on day 2 after incubation with STL-3 analogues by fluorescent activated cell sorter (FACS). Treatment with 3 μM of STL-3 analogues increased expression of CD36. The significant enhancement of expression of CD36 was observed in HL-60 cells cultured with STL-3 analogues with even- or odd-chain saturated fatty acids (Fig. 4). By contrast, the treatment with STL-3 analogue with halogenated fatty acids did not significantly increase the numbers of CD36 positive cells.

4. Discussion

All STL-3 analogues have inhibited the growth of HL-60 cells in a dose-dependent manner and showed the characteristic feature of differentiation which was evaluated by NBT reducing activity and expression of CD36. Thus, we found that STL-3 analogues tested had potential activities to induce the monocytic differentiation of HL-

60 cells. The optimal concentration of STL-3 analogues for induction of cell-differentiation and anti-proliferation were around CMC, suggesting these analogues form micelles which might interact with the membrane of HL-60 cells. Further studies on the molecular relationships between the micelle formation with their activities are now being investigated.

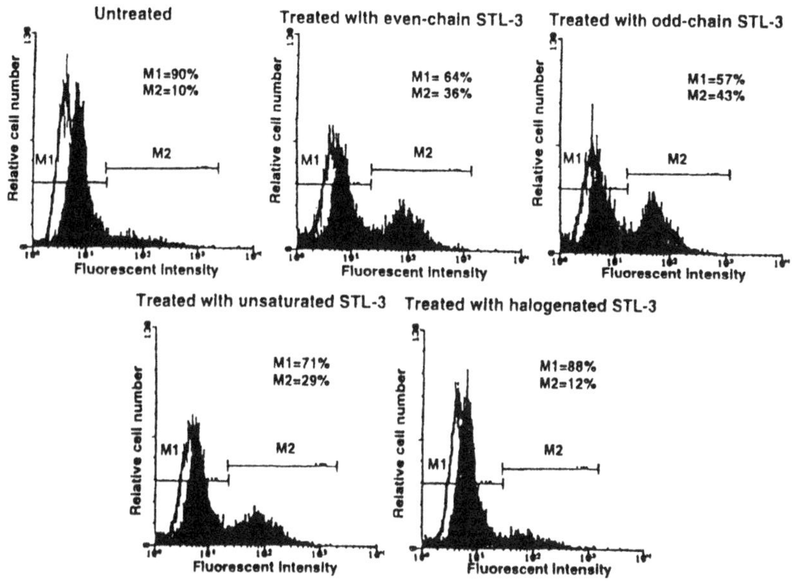

Fig. 4 Expression of CD36 on HL-60 cells treated with STL-3 analogues.
Open histgrams; negative control with irrelevant mAbs, Solid histgrams; staining by the FITC-labeled anti-CD36.

5. References

1. Isoda, H., Kitamoto, D., Shinmoto, H., Matsumura, M. and Nakahara, T.: Microbial extracellular glycolipid induction of differentiation and inhibition of the protein kinase C activity of human promyelocytic leukemia cell line HL-60, *Biosci. Biotech. Biochem.*, **61**, 609-614 (1997).
2. Kwan, C. C. and Rosen, M. J.: Relationship of structure to properties in surfactants. *J. Phys. Chem.*, **84**, 547-551 (1980).
3. Takeda, K., Hosoi, T., Noda, M., Arimura, H. and Konno, K.: Effect of fibroblast-derived differentiation inducing factor on the differentiation of human monocytoid and myeloid leukemia cell lines, *Biochem. Biophys. Res. Commun.*, **155**, 24-31 (1988).

THE ESTABLISHMENT OF AN ASSAY SYSTEM FOR THE COMPOUNDS AFFECTING ACTIN RING FORMATION OF MURINE OSTEOCLAST-LIKE CELLS

The screening assay system for actin ring formation of osteoclast-like cells

SHINYA KASAI, MASAMICHI TAKAMI, JE-TAE WOO AND KAZUO NAGAI
Department of Bioengineering, Tokyo Institute of Technology, 4259 Nagatsuta-cho, Midori-ku, Yokohama 226, Japan

This study was carried out to establish an assay system and to screen low molecular weight compounds that affect actin ring formation of murine osteoclast-like multinucleated cells (OCLs). OCLs prepared from a coculture system of calvarial cells and bone marrow cells were cultured in the presence of low molecular weight compounds for 24h. Actin ring of OCLs was detected by staining actin filaments with rhodamine conjugated phalloidin. During our screening of active compounds, we found that compactin (ML236B), gramicidin and lavanducyanin disrupted actin ring of OCLs.

1. Introduction

OCLs are multinucleated cells that play a key role in the degradation of the mineralized bone matrix. During the process of bone resorption, OCLs attach to the bone surface via integrins and actin filaments and resorb mineralized bone by secreting acid and enzymes to the extracellular compartment between the ruffled border and the bone surface [1]. Actin ring formed by polymerization of actin filaments which are localized at the intracellular region is necessary for OCLs to attach to the bone surface and to resorb the bone. [2] Therefore, the low molecular weight probes that affect actin ring formation would be useful to elucidate the mechanism of actin ring formation of OCLs for bone resorption. This study was carried out to establish an assay system and to screen low molecular weight compounds that affect actin ring formation of OCLs.

2. Materials and methods

2.1. ANIMALS AND MATERIALS

Y. Kitagawa et al. (eds.), Animal Cell Technology: Basic & Applied Aspects, Vol. 10, 363–367.

Newborn Std.ddY mice and 6- to 9-weeks old male Std.ddY mice were purchased from Japan SLC, Co. (Hamamatsu, Japan). Fast violet LB salt and naphthol AS-MX phosphate were purchased from Sigma Chemical Co. (St. Louis, MO). HerbimycinA, wortmannin, forskolin and gramicidin were purchased from Wako Pure Chemicals Co. (Osaka, Japan). Collagen gel solutions (cell matrix, type I-A) were purchased from Nitta Gelatin Co. (Osaka, Japan). Rhodamine conjugated phalloidin was purchased from Eugene Co. (Oregon, U.S.A.). Compactin (ML236B) was a generous gift from Dr. A. Endo (Tokyo University of Agriculture and Technology). Lavanducyanin was generously gifted from Dr. H. Seto (Institute of Molecular and Cellular Biosciences, The University of Tokyo).

2.2. CELL CULTURE

OCLs were prepared from a coculture system as previously described [3]. Briefly, bone marrow cells were obtained from the tibiae and femora of 6- to 9-weeks old male mice and primary osteoblastic cells were obtained from the newborn mouse calvaria. Bone marrow cells were cocultured with the osteoblastic cells on 100×20 mm style culture dishes (CORNING Co., NY, U.S.A.) precoated with collagen gel in α-minimal essential medium (α-MEM) containing 10% FBS in the presence of 10nM 1 $\alpha,25(OH)_2D_3$ (Wako Pure Chemicals, Osaka, Japan) for 7 days. At the end of the culturing, dishes were treated with 0.2% collagenase (Wako Pure Chemicals, Osaka, Japan) and 0.1% dispase (Godo Shusei, Tokyo, Japan), and cells were collected.

2.3. TRAP STAINING

Morphological change of OCLs was observed by staining tartrate resistant acid phosphatase (TRAP) activity. TRAP activity-staining was carried out as described previously [4]. In brief, the cells fixed with 10% formalin and ethanol were reacted for 30min in 0.1% sodium acetate buffer (PH 5) containing 50mM sodium tartrate, 0.1mg/ml naphthol AS-MX phosphate and 1mg/ml fast violet LB salt. After washing with distilled water and drying, TRAP-positive cells were counted as OCLs.

2.4. ACTIN FILAMENTS STAINING

Actin ring of OCLs were detected by staining actin filaments with rhodamine conjugated phalloidin. Collected OCLs fixed with 10% formalin and ethanol were reacted for 30min with 1 drop of 200U/ml rhodamine conjugated phalloidin in the dark. And then, cells were washed with cold phosphate buffered saline (PBS) and visualized under UV. Actin ring of OCLs were identified by fluorescence microscopy with a ×10 objective lens (Olympus, DPlanApo 10UV PL, Tokyo, Japan).

3. Results and Discussion

We cocultured bone marrow cells (2.0×10^7 cells/culture dish) and osteoblastic cells

(1.0×10^6 cells/culture dish) on collagen gels for 7 days in the presence of 10nM 1 α,25$(OH)_2D_3$. At the end of culturing, we obtained multinuclear OCLs with the activity of TRAP which is known to be a typical marker enzyme of osteoclastic cells. OCLs collected from the culture dish were replaced into 96-well culture plates and allowed to settle down in α-MEM for about 24h. After the OCLs were cultured in the

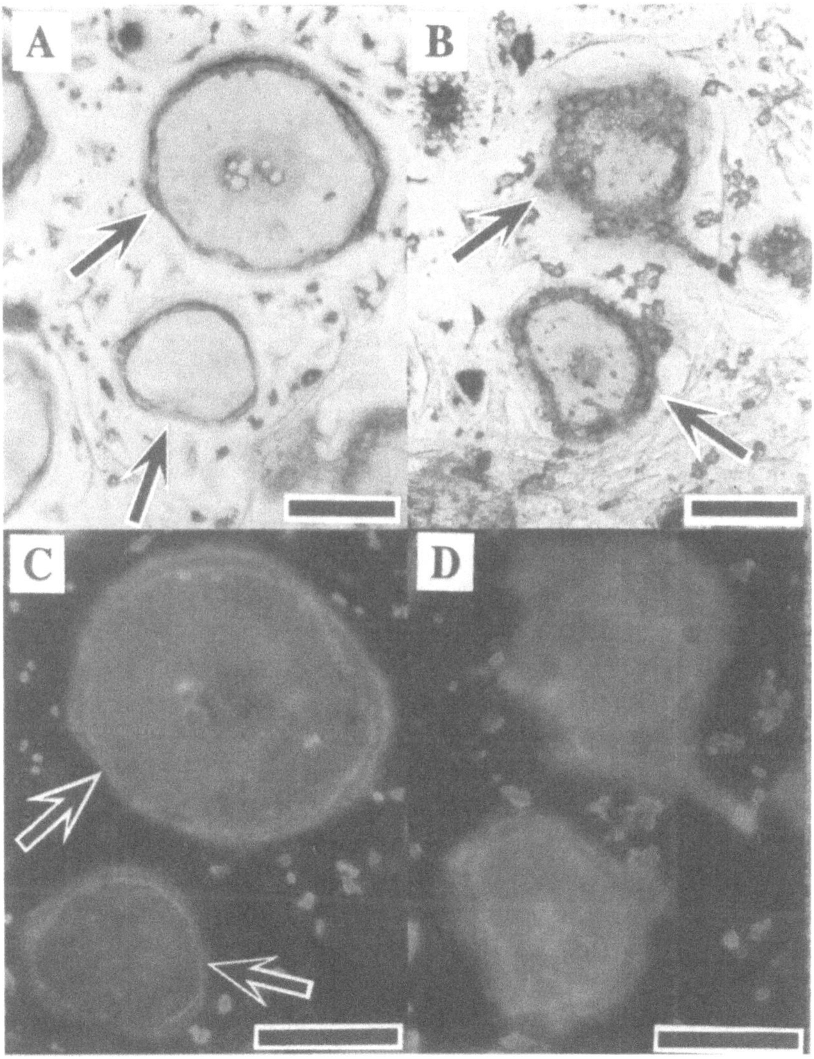

Fig. 1 OCLs were cultured in the presence of the vehicle (A, C) or 7.0μM herbimycinA (B, D) in 96 well culture plates for 24h, and were fixed and stained for TRAP activity (A, B) and actin filaments (C, D). Arrows indicate TRAP(+) OCLs on upper panel, and actin ring of OCLs on lower panel. All bars indicate 100μm.

TABLE 1. The effects of the compounds, which are known to disrupt actin ring, on morphology of OCLs.

Treatment	Concentration (μM)	% of shrunken OCLs
HerbimycinA	7.0	82.86 ± 1.57
Wortmannin	0.07	42.19 ± 1.43
Forskolin	10.0	61.74 ± 1.86
Vehicle	-----	24.94 ± 2.57

OCLs were cultured in the presence of compounds in 96 well culture plates for 24h. At the end of culturing, cells were fixed and stained for TRAP activity. Data are expressed as means ± SD of 4 cultures.

TABLE 2. The effects of low molecular weight compounds on actin ring formation of OCLs.

Treatment	Concentration	% of actin ring (+) OCLs
Compactin	7.0 (μM)	4.08 ± 1.36
Gramicidin	0.7 (μg/ml)	44.12 ± 4.86
Lavanducyanin	1.1 (μg/ml)	34.97 ± 4.11
Vehicle	-------	85.48 ± 2.38

OCLs were cultured in the presence of compounds in 96 well culture plates for 24h. At the end of culturing, cells were fixed and stained for actin filaments. Data are expressed as means ± SD of 4 cultures.

presence of low molecular weight compounds for 24h, morphological change and actin ring disruption of OCLs were examined.

HerbimycinA, wortmannin and forskolin are known to disrupt actin ring of OCLs [2, 5, 6, 7]. OCLs were shrunken after the treatment with these compounds at the concentration which disrupt actin ring (Fig. 1, TABLE. 1). However, these compounds did not affect the viability of osteoblasts and OCLs at the same concentration. In the case of the vehicle (MeOH), about 80% of total OCLs showed normal morphology, OCLs of remaining 20% were shrunken. The number of OCLs with shrunken morphology was increased by about 80% after the treatment with herbimycinA. The shrunken OCLs did not have actin ring. Therefore, we regarded the shrinking of OCLs without the loss of viability of OCLs and osteoblasts as an indicator of disruption of actin ring. OCLs shrunken by active compounds were stained with rhodamine conjugated phalloidin to confirm the disruption of actin ring.

In the course of screening for low molecular weight compounds that affect actin ring formation of OCLs, we found that compactin (ML236B), gramicidin and lavanducyanin disrupted actin ring of OCLs (TABLE. 2). We expect that these three compounds would be useful to gain insights about the regulatory mechanisms of actin ring formation of OCLs.

4. References

1. Rifkin, B. R., and Gay, C. V.: Biology and Physiology of osteoclast, *CRC Press, Boca Raton* (1993), 259-314.
2. Suda, T., Nakamura, I., Jimi, E., and Takahashi, N.: Regulation of osteoclast function, *J. Bone. Miner. Res. Rev.* **12** (1997), 869-879.
3. Takahashi, N., Akatsu, T., Udagawa, N., Sasaki, T., Yamaguchi, A., Moseley, J. M., Martin, T. J., and Suda, T.: Osteoblastic cells are involved in osteoclast formation, *Endocrinology*. **123** (1988), 2600-2602.
4. Takahashi, N., Yamana, H., Yoshiki, S., Roodman, G. D., Mundy, G. R., Jones, S. J., Boyde, A., and Suda, T.: Osteoclast-like cell formation and its regulation by osteotropic hormones in mouse bone marrow culture, *Endocrinology*. **122** (1988), 1373-1382.
5. Tanaka, S., Takahashi, N., Udagawa, N., Murakami, H., Nakamura, I., Kurokawa, T., and Suda, T.: Possible involvement of focal adhesion kinase, $p125^{FAK}$, in osteoclastic bone resorption, *J. Cell. Biochem.* **58** (1995), 424-435.
6. Hall, T. J., Schaeublin, M., and Missbach, M.: Evidence that c-src is involved in the process of osteoclastic bone resorption, *Biochem. Biophys. Res. Commun.* **199** (1994), 1237-1244.
7. Nakamura, I., Takahashi, N., Sasaki, T., Tanaka, S., Udagawa, N., Murakami, H., Kimura, K., Kabuyama, Y., Kurokawa, T., Suda, T., and Fukui, Y.: Wortmannin, a specific inhibitor of phosphatidylinositol-3 kinase, blocks osteoclastic bone resorption, *FEBS Lett.* **361** (1995), 79-84.

PROTECTION OF HUMAN MELANOMA CELLS FROM UV DAMAGE BY A FERMENTED MILK, KEFIR

TSUTOMU NAGIRA[1], JUNKO NARISAWA[1], KENICHI KUSUMOTO, KIICHIRO TERUYA[1], YOSHINORI KATAKURA[1], DAVID W. BARNES[2], SENNOSUKE TOKUMARU[3], and SANETAKA SHIRAHATA[1]

Graduate School of Genetic Resources Technology, Kyushu University, Fukuoka 812-8581, Japan; [2]Department of Biochemistry and Biophysics, Oregon State University, Corvallis, OR 97331 U.S.A.; [3]Nihon Kefir Co. Ltd., 13-6 Asahi-machi, Fujisawa 251, Japan.

ABSTRACT UV irradiation produces reactive oxygen species from H_2O and O_2 in skin cells and damage melanocytes and other skin cells, causing liver spots, freckles, wrinkles and skin cancer. We studied the scavenging effect of the Kefran–Kefir against superoxide radicals and the protective effect of the Kefran–Kefir extract on UV damage of human melanoma HMV-1 cells. The Kefran–Kefir extracts exhibited strong scavenging effect against superoxide radicals produced by the hypoxanthine-xanthine oxidase system. When HMV-1 cells were treated with the Kefran-Kefir extracts before, during or after UV irradiation, the alive cell number was remarkably increased compared to the control values. These results suggest that Kefran–Kefir can protect human melanomas from UV damage.

1. Introduction

Recently it has been revealed that reactive oxygen species (ROS) such as $^1O_2$, $O_2^{\cdot-}$, H_2O_2 and $\cdot OH$ cause aging and a variety of diseases including diabetes, arteriosclerosis and cancer. Kefir is a fermented milk drink originating in Caucasus mountains and is known to have numerous benefits including anti-cancer effect and stimulation of the immune system. So far it has been demonstrated that sphingomyeline derived from the Kefran-Kefir stimulates the interferon β production of human osteosarcoma MG-63 cells [1], [2]. Here we report the scavenging effect of Kefran–Kefir against superoxide radicals and the protective effect of Kefran–Kefir extract on the UV damage of human melanoma cells.

Y. Kitagawa et al. (eds.), Animal Cell Technology: Basic & Applied Aspects, Vol. 10, 369–373.

2. Materials and Methods

2.1. Cells

The human melanoma cell line HMV-1 cells were provided by JCRB (Japanese Cancer Research Resources Bank). This cell line can not produce melanin pigment and is relatively resistant to UV irradiation.

2.2. Preparetion of Kefran–Kefir extract samples

Kefran–Kefir powder was supplied by Nihon Kefir Co. Ltd.. The Kefran–Kefir powder was suspended into distilled water or F12 medium and centrifuged to remove indissoluble fraction. The supernatant was sterilized by filtration. The dry weight of the Kefran-Kefir extract was determined by lyophilization and a stock solution of 100 mg/ml was prepared.

2.3. Measurement of the scavenging effect of Kefran–Kefir extract for superoxide radicals

The scavenging effect of Kefran–Kefir extract for superoxide radicals was examined by using the hypoxanthine-xanthine oxidase system [3]. Briefly, the reaction mixture (1 ml) for measuring chemiluminiescence (CL) intensity specific to $O_2^{\cdot -}$ contained 500 μM hypoxanthine (HX), 200 μM EDTA, 40 mM sodium phosphate buffer (pH 7.0), 0.6 ml of the Kefiran-Kefir extract, 2.5 mM 2-methyl-6-phenyl-3,7-dihyroimidazo[1,2-a] pyrazine-3-one (*Cypridina* luciferin analog (CLA), Tokyo Kasei Industrial Co., Tokyo) and 0.5 U/l of xanthine oxidase (SOD) (Wako Purechemical Industries, Tokyo). The CL intensity of the reaction mixture (0.85 ml) except for XOD solution was measured in a glass-tube in a CL reader (Aloka, type BLR-301) at 26 ℃. AT 18 seconds time point, 0.15 ml of XOD solution was injected into the tube and the CL intensity was continuously measured for 120 sec.

2.4. WST-1 assay

WST-1 reagent is prepared by dissolving 65mg 2-(4-Iodophenyl)-3-(4-nitro-phenyl)-5-(2, 4-disulfophenyl)-2H-tetrazolium, sodium salt (WST-1)(Dojin kagaku laboratory) and 7mg 1-methoxy-5-methylphenazinium methylsulfate (PMS) per 100ml phosphate buffered saline (PBS)(pH7.4). By the action of respiratory enzyme in mitochondrion WST-1 reagent is changed to WST-1 phorumazann color of which is yellow in presence of 1–methoxy PMS and enhance the O.D value. So the cell number is determined by measuring the O.D value at 450 nm using a plate reader.

2.5. Determination of the protective effect of Kefran–Kefir extract on UV damage of human melanoma cells

HMV–1 cells (4.0×10^3 cells) were inoculated in each well of 96-well plates and cultured in 10% fetal bovine serum (FBS)–F12 medium for 15 hours. Then cells were cultured in 10% FBS–F12 medium containing the Kefran-Kefir extract sample for 5 hours. After washing cells and changing medium to fresh one containing no extract sample, UV was irradiated for various minutes. Then medium was changed to fresh one and the number of alive cells was determined by WST–1 assay after cultivation for 10 hours. To examine the protective effect of Kefran–Kefir extract during and after UV irradiation, the cells were treated with Kefran–Kefir extract was also done only during or after the UV irradiation.

3. Results

3.1. Scavenging effect of Kefran–Kefir extract for superoxide radicals

XOD changes HX to xanthine, producing superoxide radicals from dissolved oxygen molecules. When the Kefiran-Kefir extract was added in the HX-XOD system, superoxide radicals were scavenged in a dose-dependent manner (Fig. 1). The extract of 18 mg/ml completely suppressed the production of superoxide radicals. This result suggested that the components of the Kefran-Kefir may be utilized as anti-oxidants for protection of the skin and the body.

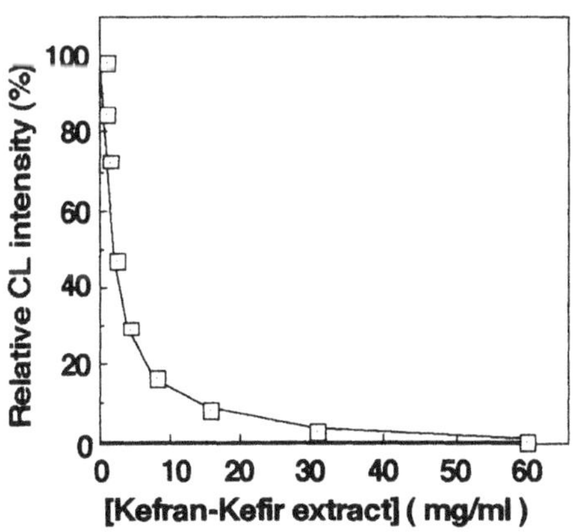

Fig.1. Scavenging effect of Kefran-Kefir extract for superoxide radicals by the hypoxanthine-xanthine oxidase system.

3.2. Protective effect of the Kefran-Kefir extract on human melanoma cells from UV damage: effect of pretreatment.

Human melanoma HMV-1 cells were precultured in 10% FBS–F12 medium containing the Kefran-Kefir extract sample for 5 hours before UV irradiation. Alive cell number was determined after UV irradiation for 1 min. As shown in Fig. 2, alive cell number was increased dependent upon the amount of the extract until 1.2 mg/ml. When UV was irradiated for 10 min, cell viability was more decreased and the protective effect of the Kefran-Kefir extract became smaller. Higher concentration of the extract decreased the viability. From these results, the Kefran-Kefir extract was expected to exhibit the protective effect from weak UV damage by pretreatment of cells with the extract.

3.3. Protective effect of the Kefran-Kefir extract on human melanoma cells from UV damage: effect of treatment during UV irradiation.

HMV-1 cells were cultured in 10% FBS–F12 medium containing the Kefran-Kefir extract sample only during UV irradiation. Cell number after UV irradiation remarkably increased dependent upon the amount of the extract.(Fig. 3) This result suggests that the Kefran-Kefir extract existing out of cells can protect human melanoma cells from the UV damage, perhaps by scavenging ROS or by cutting UV.

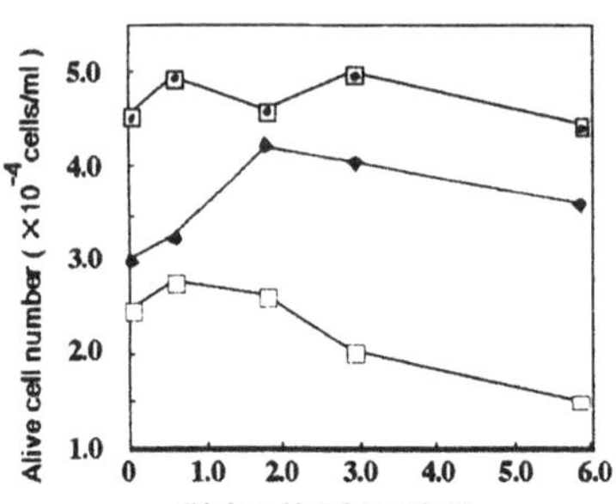

Fig.2. The protective effects of the Kefran-Kefir extract from UV damage by treatment of the human melanoma cells before UV irradiaition.

HMV-1 cells were cultured in 10%FBS-F12 medium containing the Kefran-Kefir extract for 5 hours before UV irradiaiton. —▣—, control; —◆—, UV 1min; —□—, UV 10min.

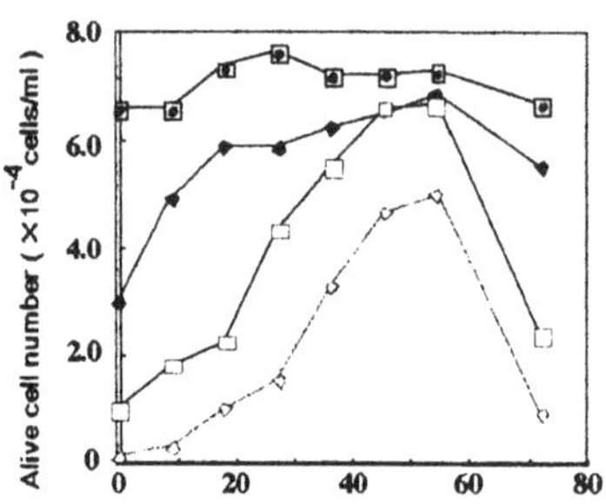

Fig. 3. The protective effects of the Kefran-Kefir extract from UV damage by treatment of the human melanoma cells during UV irradiaition.

HMV-1 cells were treated with the Kefiran-Kefir extract only during UV irradiation. —▣—; control; —◆— ,UV 1min; —□—, UV 5min; -◇- , UV 20min.

3.4. Protective effect of the Kefran-Kefir extract on human melanoma cells from UV damage: effect of treatment after UV irradiation.

HMV-1 cells were cultured in 10% FBS–F12 medium containing the Kefran-Kefir extract sample for 10 hours only after UV irradiation. Cell number after UV irradiation increased depending upon the amount of the extract (data not shown). This result suggest that the Kefran-Kefir extract may activate the repair system of DNA and membrane or prevent the apoptosis triggered by UV damage.

4. Discussion

It was revealed that the Kefran-Kefir extract could protect human melanoma cells from UV damage in all cases that the cells were treated with the extract before, during and after UV irradiation. By pretreatment of the cells with the Kefran-Kefir extract, absorbed components of the extract may scavenge ROS or activate the repair system. The extract may cut UV or scavenge ROS during UV irradiation. The extract exhibited protective effect on the cells even after UV irradiation. This suggests that the Kefran-Kefir extract activates the repair system in the cells and prevent apoptosis. Further investigation on the UV damage protective effects of the Kefran-Kefir extract is desirable.

References

1. Osada, K., Nagira, K., Teruya, K., Tachibana, H., Shirahata, S. and Murakami, H. (1994) Enhancement of interferon- β production with spingomyelin from fermented milk, *Biotherapy*, **7**, 115-123.
2. Kabayama, S., Osada, K., Tachibana, H., Katakura, Y., and Shirahata, S. (1997) Enhancing effects of food components on the production of interferon β from animal cells suppressed by stress hormones, *Cytotechnology*, **23**, 119-125
3. Shirahata, S., Kabayama, S., Nakano, M., Miura, T., Kusumoto, K., Goto, M., Hayashi, H., Otsubo, K., Morisawa, S., and Katakura, Y. (1997) Electrolyzed-reduced water scavenges active oxygen species and protects DNA from oxidative damage. *Biochem. Biophys. Res. Commun.*, **234**, 269-274.

IDENTIFICATION OF SPECIFIC GENES INVOLVED IN CELLULAR SENESCENCE BY USING HUMAN LUNG ADENOCARCINOMA DERIVED CELLS

N. UEHARA, Y. KATAKURA, S. NISHIMOTO, K.TERUYA, AND S. SHIRAHATA

Graduate School of Genetic Resources Technology, Kyushu University, Hakozaki 6-10-1, Higashi-ku, Fukuoka 812-81, Japan

ABSTRACT Telomerase is a ribonucleoprotein enzyme that adds hexameric TTAGGG repeats onto the telomeres to compensate for progressive loss of telomere which occurs with each round of DNA replication (end replication problem). To identify the genes involved in cellular senescence, telomere length maintenance and telomerase regulation, we performed a differential screening by using cDNA libraries prepared from human lung adenocarcinoma derived cell lines. Consequently, we found several differentially expressed genes and analyzed the expression pattern of these genes by dot blot hybridization.

1. INTRODUCTION

Telomerase activity has been detected in germ line cells, immortal cells, but not in somatic cells (1). Repression of telomerase caused telomere shortening, which signals cellular senescence (2). Then repression mechanism of telomerase must be clarified to understand cellular senescence. Although, several kinds of genes involved in cellular senescence have been identified, function of these genes have not been clarified (3-8). In order to develop potential tools for the anti-cancer therapy and/or diagnosis, we must understand the regulation mechanism of telomerase. We have established several kinds of sublines from the human lung adenocarcinoma cell line A549. A5DC7 cells were isolated from IFN-

Y. Kitagawa et al. (eds.), Animal Cell Technology: Basic & Applied Aspects, Vol. 10, 375–379.

γ-treated A549 cells and showed normal senescent cell phenotypes with restored ability of contact inhibition and anchorage-dependent growth as well as lost tumorigenicity (9). We also established CK cells from senescent A5DC7 cells by raising serum concentration from 5% to 10%. CK cells regained telomerase activity and telomere maintenance mechanism as well as resumed proliferation (10). Furthermore, we established AST-9 cells from A549 cells treated with hydrogen peroxide. Although, AST-9 cells has telomerase activity equal to A549 cells, AST-9 cells showed the passage number-dependent telomere shortening. These cells are derived from A549 cells, but have quite different properties. Thus, these cell lines are thought to be useful for screening the genes involved in induction and/or maintenance of cellular senescence, maintenance of telomere length and telomerase regulation.

In this study, we attempted to screen and identify the genes specifically expressed in these cells by using PCR-select subtraction method. Consequently, we found several genes that specifically expressed in A549, A5DC7, CK and AST-9 cells, respectively.

2. MATERIALS AND METHODS

2.1 Cell culture. The A5DC7, CK and AST-9 cell lines were established from the human lung adenocarcinoma cell line A549. AST-9 cell was obtained by treating with 200 μM of hydrogen peroxide (H_2O_2) for 4 days, and detached from the culture dishes and cloned by standard limiting dilution. All cell lines were grown at 37°C in 5% CO_2 atmosphere. A549, A5DC7 and AST-9 cells were cultured in ERDF medium (Kyokuto Pharmaceutical, Tokyo, Japan) supplemented with 5% fetal bovine serum(FBS; Irvine Scientific, Santa Ana, CA, USA). CK cell were cultured in ERDF medium supplemented with 10% FBS (10).

2.2 Assay for telomerase activity. Telomerase activity was measured by TRAP(Telomeric Repeat Amplification Protocol) assay with some modifications (1,11). Briefly, 10^6 cells were pelleted and lysed with lysis buffer. Cell lysates were centrifuged at 15,000 rpm for 20 min., and supernatants were stored at -80°C. TRAP assay products were analyzed by electrophoresis in 1×Tris-borate EDTA buffer on 10% polyacrylamide nondenaturing gels.The gel was stained with SYBR Green I (TAKARA, Shiga, Japan).

2.3 DNA extraction and analysis. Genomic DNA was prepared using the DNA Extractor WB Kit (Wako, Kyoto, Japan) according to the manufacturer's protocol. Length of the terminal restriction Fragments (TRF) was determined by southern blot analyses with a telomeric sequence probe and measured the length of telomere according to the procedure of KATAKURA *et al.* (10).

2.4 RNA preparation. Total cellular RNA was isolated from cell pellets of A549 derived cell lines and prepared by TRIzol (GIBCO BRL, Gaithersburg, MD, USA). Poly

(A) + RNA was then selected from total RNA by using oligo d(T) cellulose (Collaborative Biomedical Products, Bedford, MA, USA).

2.5 Construction of the cDNA library and subtractive hybridization. cDNA libraries were constructed from poly (A) + RNA of each A549 derived cell lines, and synthesized tester and driver cDNA. We then subtracted driver cDNA from tester cDNA in all combinations of cell lines and amplified the specifically expressed cDNA fragments in tester cDNA pool by using the PCR-Select cDNA Subtraction Kit (CLONTECH Laboratories, Inc., Palo Alto, CA, USA) according to the manufacturer's protocol.

2.6 Dot blot hybridization. We then analyzed expression level of selected cDNA fragments in each A549 derived cell line, and evaluated their specificities.

3. RESULTS AND DISCUSSION

We have established several kinds of cell lines, which was derived human lung adenocarcinoma cell line A549. Then we assessed the phenotypes of A549 derived cell lines, A5DC7, CK and AST-9 (telomerase activity, telomere length and cellular senescence)(Table 1). Higher telomerase activity was detected in A549, CK and AST-9 cells, but that of A5DC7 cells was greatly reduced. TRF (telomere restriction fragment) length of A5DC7, CK and AST-9 cells were analyzed by southern blot hybridization. TRF length of A5DC7 and AST-9 cells shortened as compared to A549, and that of CK cells lengthened during the culture. Senescence marker of β-galactosidase activity was observed only in A5DC7 cells. Furthermore, we analyzed the expression of hTRT and TP1 mRNAs. hTRT is the catalytic subunit of human telomerase, and correlated with the telomerase activity. TP1 is the homolog of the *tetreahymena* telomerase-associated protein p80. Correlated with telomerase activity, expression of hTRT mRNA in A549, CK and AST-9 cells was downregulated. TP1 mRNA was equally expressed in all kinds of A549 derived cells.

Table 1 Phenotypes of A549-derived cell lines

	A549	A5DC7	CK	AST-9
Telomerase Activity	+	-	+	+
Telomere Shortening	-	+	-	+
Senescence Marker (β-galactosidase)	-	+	-	-
hTRT	+	-	+	+
TP1	+	+	+	+

In this study, we attempted to screen and identify the genes specifically expressed in A549 derived cells which down-regulate telomerase activity or have a defect in telomere maintenance mechanism by using PCR-select subtraction method. Consequently, we found 110 genes specifically expressed in A549, A5DC7, CK and AST-9 cells (data not shown). Secondary differential screening was done by dot blot analyses (Fig. 1). We are now sequencing the genes selected by dot blot analyses.

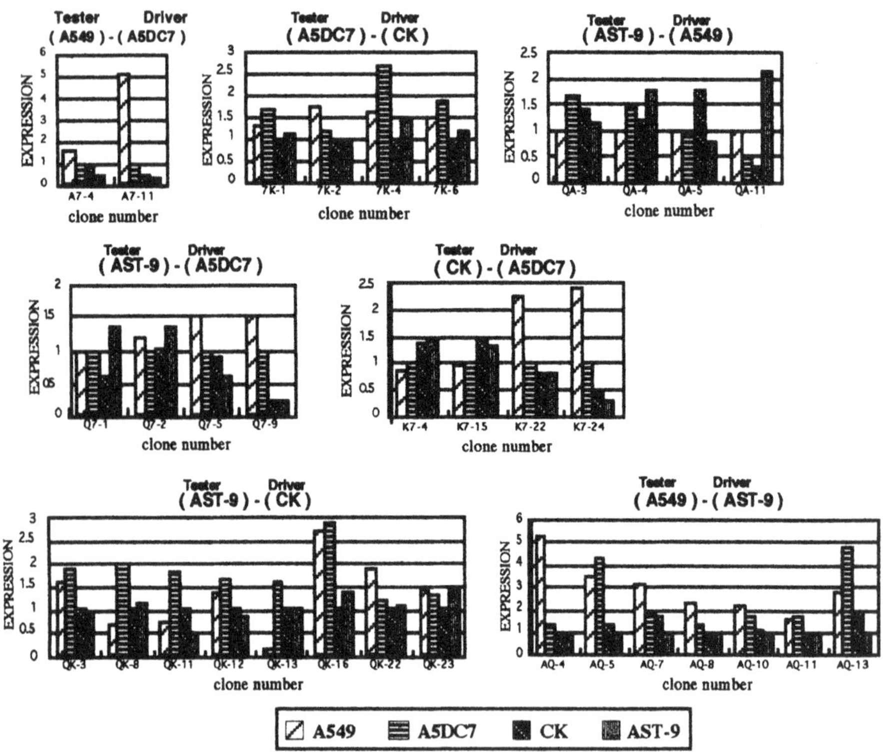

Fig. 1 Dot blot hybridization of subtracted DNA clones against A549-derived cell lines.

4. REFERENCES

1. Kim, N. W., Piatyszek, M. A., Prowse, K.R., Harley, C.B., West, M. D., Ho,P. L. C., Coviello, G. M., Wright, W. E., Weinrich, S. L., and Shay, J. W. (1994) *Science* **266,** 2011-2015.
2. Levy, M. Z., Allsopp, R. C., Futcher, A. B., Greider, C. W., Harley, C. B. (1992) *J. Mol. Biol.* **225,** 951-960.
3. Shibanuma, M., Mashimo, J., Kuroki, T., Nose, K. (1994) *J. Biol. Chem.* **269,** 26767-26774.
4. Neull, M. J., Stewart, D. A., Walker, L., Friedman, V., Wood, C. A., Owens, G. A., Smith. J. R., Schneider, E. L., Dell'Orco, R., Limpkin, C. K., Danner, D. B., McClung, J. K. (1991) *Mol. Cell. Biol.* **11,** 1372-1381.
5. Wang, E. (1985) *J. Cell. Biol.* **100,** 545-551.

6. Wistrom, C., Villeponteau, B. (1992) *Exp. Cell. Res.* **199,** 355-362.
7. Schneider, C., King, R. M., Philipson, L. (1988) *Cell* **54,** 787-793.
8. Fornace, A. J. Jr. (1992) *Ann. Rev. Geenet.* **26,** 507-526.
9. Kawamoto, S., Inoue, Y., Shinozaki, Y., Katakura, Y., Tachibana, H., Shirahata, S., Murakami, H. (1995) *Biochem. Biophys. Res. Commun.* **215,** 280-285.
10. Katakura, Y., Yamamoto, K., Miyake, O., Yasuda, T., Uehara, N., Nakata, E., Kawamoto, S. and Shirahata, S. (1997) *Biochem. Biophys. Res. Commun.* **237,** 313-317.
11. Piatyszek, M. A., Kim, N. W.,Weinrich, S. L., Hiyama, K., Hiyama, E., Wright, W. E. and Shey, J. W. (1995) *Meth. Cell Sci.* **17,** 1-15.
12. Hastie, N. D., Dempster, M., Dunlop, M. G., Thompson, A. M., Green, D. K. and Allshire, R. C. (1990) *Nature* **345,** 866-868.

LAMININ CHAINS EXPRESSED IN HUMAN KERATINOCYTES, MOUSE 3T3-L1 ADIPOCYTES AND MOUSE EMBRYONAL CARCINOMA F9

Tomoaki Niimi[1], Chino Kumagai[2], Mojgan Azimi[2] and Yasuo Kitagawa[1,2]

[1]*Nagoya University BioScience Center,* [2]*Graduate Program of Biochemical Regulation, Graduate School of Agricultural Sciences, Nagoya University, Nagoya, 464-01, JAPAN*

Abstract: Laminin chain genes expressed in three cultured cell lines were explored by reverse transcription-polymerase chain reaction (RT-PCR) using paired primers designed based on reported cDNA sequences. RT-PCR on mRNA extracted from human keratinocytes gave amplified fragments for γ2, α3, γ1, β1 and β3. mRNA from mouse 3T3-L1 adipocytes gave the fragments for α4, β1 and γ1 suggesting the assembly of only laminin-8 (α4β1γ1) and the amount of these mRNAs increased depending on the adipocyte differentiation. mRNA from mouse embryonal carcinoma F9 cells gave the fragments for α1, α4, α5, β1 and γ1, and the result remained unchanged by the differentiation of F9 cells into primitive, parietal or visceral endoderm-like cells.

1. INTRODUCTION

Laminins are a family of glycoproteins composed of α, β and γ chains. As the major components of basement membranes, laminins have a strong effect on proliferation, differentiation and migration of various tissue cells and have crucial role in organogenesis. Five α (α1–α5), three β (β1–β3) and two γ (γ1 and γ2) chains have been cloned from human and mouse. The family is still growing and the presence of fourth β and third γ is recently suggested. Replaceable assembly of α, β and γ chains into αβγ heterotrimers produces the variety of laminins and every laminin variant may have distinct function at various stage of animal development and at various site of animal body. Identification of laminin variants expressed in various cell lines is important to understand their physiological role. We selected human keratinocytes, mouse 3T3-L1 adipocytes and mouse embryonal carcinoma F9 for this analysis and found the expression of specific laminin variants by RT-PCR on RNA extracted from the cells.

2. MATERIALS AND METHODS

2.1 Cell culture- Human keratinocyte Hk-f cells (Kyokuto Seiyaku Co.) were cultured in 60 mm dishes by feeding K-110 (type II) medium every other day at 37°C under humidified 95% air and 5% CO_2. 3T3-L1 cells were cultured and induced to adipocytes as described (1). embryonal carcinoma F9 were cultured and induced to differentiate as previously described.

2.2 RT-PCR Analysis - Total RNA was extracted from the cells by an acid guanidinium thiocyanate/phenol/chloroform extraction method(1). For reverse transcription of mRNAs, 2 μg of total RNA extract was incubated for 10 min at 70 °C and chilled on ice, added with 4 μl of 5x first strand synthesis buffer (GIBCO BRL), 1 μl of a 25 mM mixture of four

Y. Kitagawa et al. (eds.), Animal Cell Technology: Basic & Applied Aspects, Vol. 10, 381–384.

deoxynucleotide triphosphates, 2 μl of 0.1 M dithiothreitol and 100 ng of random primer to make the final volume of 20 μl. The mixture was incubated for 2 min at 42 °C, added with 200 units of Super Script II RNase H⁻ reverse transcriptase (GIBCO BRL) and incubated for 60 min at 42 °C. Reverse-transcript corresponding to 0.4 μl of the reaction mixture was amplified by 20-30 cycles of PCR using recombinant Taq DNA polymerase (TAKARA) under a condition of 94 °C for 1 min, 52 °C for 1 min and 72 °C for 1 min. PCR products were separated on 1.0 % agarose gels using a λ-DNA fragments digested with *Hind*III and *Eco*RI as size markers. To amplify laminin chain cDNA, paired primers designed based on reported sequences were used.

To amplify human α1 cDNA fragment of 675 bp corresponding to nucleotides 7175-7847, a paired primers of 5'GGCAGAGTGAAGTTATGAC3' and 5'AATGCGCTTGTCCAT CACTG3' were used. To amplify human α2 cDNA fragment of 646 bp corresponding to nucleotides 8759-9403, a paired primers of 5'ATAGCATTGATGGCTGCGTC3' and 5'ATGATACAGGTTGAACGCCC3' were used. To amplify human α3 cDNA fragment of 619 bp corresponding to nucleotides 4524-5140, a paired primers of 5'GAAGAACTTTCAGCTGGA TTC3' and 5'ACTGGTCAGGACAACCATTC3' were used. To amplify human α4 cDNA fragment of 648 bp corresponding to nucleotides 4124-4769, a paired primers of 5'GTCTCAGC AATCTCCAGCTC3' and 5'TTGGGCAGCTGTGCTCTGTC3' were used. To amplify human β1 cDNA fragment of 690 bp corresponding to nucleotides 4653-5340, a paired primers of 5'CCAGGATAGTGCTGATTTGG3' and 5'TTTGAGCAGTTGCAGCTTGC3' were used. To amplify human β2 cDNA fragment of 680 bp corresponding to nucleotides 4918-5595, a paired primers of 5'CAGCTACTGCAGGATGCACG3' and5'GTGCTAGGAACTGGGGTAGG3' were used. To amplify human β3 cDNA fragment of 687 bp corresponding to 2970-3654, a paired primers of 5'CAAGCAGGACATTGCGCGTG3' and 5'AACGGGCTGGAACGTGTAGC3' were used. To amplify human γ1 cDNA fragment of 675 bp corresponding to nucleotides 4436-5108, a paired primers of 5'GCCATCAACCAGACCATCAC3' and 5'CCTTCCAGCCC TAAAGACAC3' were used. To amplify human γ2 cDNA fragment of 672 bp corresponding to nucleotides 2709-3378, a paired primers of 5'GTCCTTTCAGGTGGAAGAAG3' and 5'TCTGGTATCAACCTTCTGGG3' were used.

To amplify mouse α1 cDNA, a paired primers of 5'GCGCGTAAAGATTTCCAG CC-3' and 5'-GTCTCTGTCCAAAGCTCCTG-3' (nucleotides 5113-5736,) were used expecting a 626-bp product. For mouse α2 cDNA, a paired primers of 5'GGGTCGGATCAATCATGC TG-3' and 5'CGTGGAACATAAGCTTCTCG-3' (nucleotides 8457-9043) were used expecting a 589-bp product. For mouse α3 cDNA, a paired primers of 5'-AAAGGTGCACCTGGTGGT GG-3' and 5'-TTCGGTGGGAAGGAAAGCTG-3' (nucleotides 6944-7528,) were used expecting a 587-bp product. For mouse α4 cDNA, a paired primers of 5'-CATGGGATCCTATTGGCC TG-3' and 5'-CACATAGCCGCCTTCTGTGG-3' (nucleotides 197-858 Ref.1) were used expecting a 664-bp product. For mouse α5 cDNA, a paired primers of 5'-ACGGCTCAGAAG GTTTCCCG-3' and 5'-CTTCAGCGGGTATGACTTCC-3' (nucleotides 9673-10221) were used expecting a 551-bp product. For mouse β1 cDNA, paired primers of 5'GCTGGATCCGCTTG CAGCAGAGTGCAGCTGA-3' and 5'-CGCGAATTCGCTAAGCAGGTGCTGTAAACCG-3' (nucleotides 4858-5535) were used expecting a 702-bp product. For mouse β2 cDNA, a paired primers of 5'-CTGCAGCGGGTATGACTTCC-3' and 5'-GGCTTGTGCAGCCAGAACAC-3' (nucleotides 4475-5022) were used expecting a 550-bp product. For mouse β3 cDNA, a paired primers of 5'-CAGGTAGATGATGTGGTCGG-3' and 5'-ACTGCGGATCTGCTCCACAC-3' (nucleotides 3079-3621) were used expecting a 545-bp product. For γ1 cDNA, a paired primers of 5'GCGGGATCCCCAATGACATTCTCAACAAC-3' and 5'GCAGATATCGGGCTTCTC GATAGACGGGG-3' (nucleotides 4331-5067) were used expecting a 760-bp product. For mouse γ2 cDNA, a paired primers of 5'-TATTAGCCAGAAGGTTGCGG-3' and 5'-TAGTCT CCAGCAGATGGAGG-3' (nucleotides 2973-3520) were used expecting a 549-bp product.

3. RESULTS

3.1 Laminin chain genes expressed in human keratinocytes

mRNAs extracted from human keratinocytes were amplified by RT-PCR using nine pairs of primers designed based on reported cDNA sequences of human laminin chains. After 20 cycles of PCR, the primers for α3 and γ2 mRNAs showed products with the expected sizes of 619 and 672 bp, respectively (not shown). The primers for β1, β3 and γ1 mRNAs showed the products of 690, 687 and 675 bp, respectively, after 25 cycles (Fig.1). Other primers did not show the product even after 30 cycle of PCR except for extremely weak signal of α1 (not shown), suggesting that the corresponding mRNAs are 32-fold less than β1, β3 and γ1 mRNAs. The results suggested that the amount of laminin mRNAs is in the order of γ2, α3, γ1, β1 and β3.

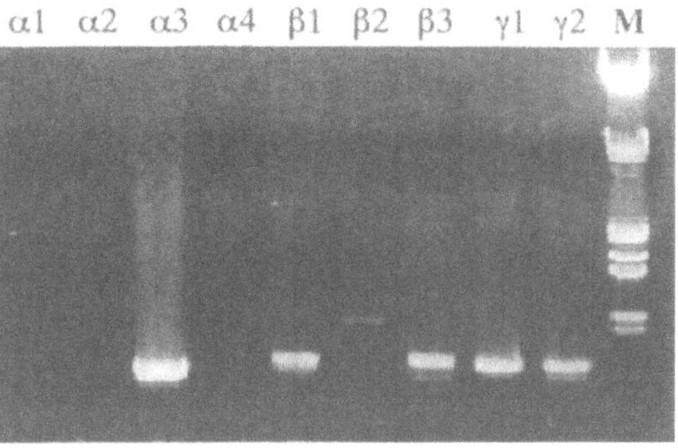

Fig.1 RT-PCR on mRNA extracted from human keratinocytes

3.2 Laminin chain genes expressed in mouse 3T3-L1 adipocytes

To explore laminin genes expressed in 3T3-L1 adipocytes, RT-PCR on mRNA extracted from the cell was carried out using ten pairs of primers designed based on reported cDNA sequences of mouse laminin chains. After 30 cycles of PCR, only the primers for α4, β1 and γ1 mRNAs gave amplified products with the expected sizes of 664, 702 and 760-bp, respectively, while other primers did not (Fig.2). The RT-PCR products from α4, β1 and γ1 mRNAs were undetectable at 20 cycles but became detectable at 25 cycles (not shown). These suggest that 3T3-L1 cells express α4, β1 and γ1 mRNAs in more than 32-fold excess of other laminin chain mRNAs. Northern blot analysis shoed that the quantity of these mRNA increase 2-3 fold by induction of the adipocyte differentiation (result not shown).

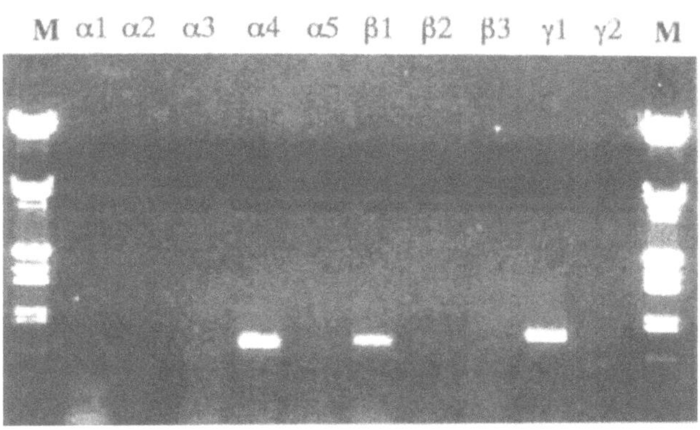

Fig.2 RT-PCR on mRNA extracted from mouse 3T3-Li adipocytes

3.3 Laminin chain gene expression in embryonal carcinoma F9

Embryonal carcinoma F9 cells can differentiate into primitive endoderm-like cells by treatment with retinoic acid and these cells can differentiate into either parietal endoderm-like cells of visceral endoderm-like cells by further treatment with dibutyryl cAMP or by culturing in suspension, respectively. Despite such differentiation of F9 cells, the type of expressed laminin chain mRNAs remained unchanged. The result in Fig.3 shows the result of RT-PCR on mRNA extracted from visceral endoderm-like F9 cells. RT-PCR using ten pairs of primers for mouse laminin chains showed the fragments for α1, α4, α5, β1 and γ1 chains.

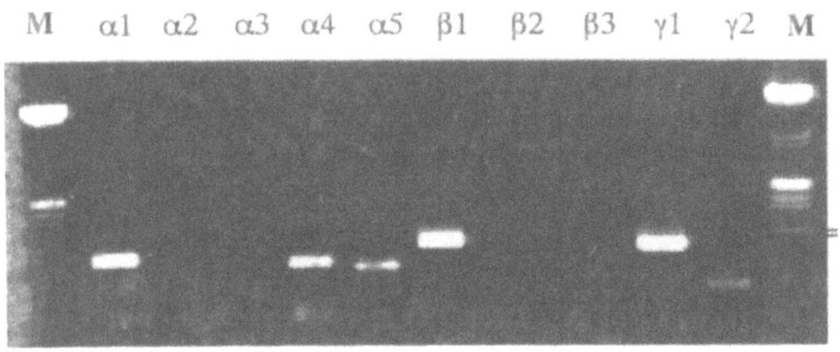

Fig.3 RT-PCR on mRNA extracted from embryonal carcinoma F9 cells

4. DISCUSSION

This study showed that variety of laminin chains are expressed in association with differentiated phenotype of cell lines. Expression of γ2, α3, γ1, β1 and β3 mRNA in human keratinocytes suggested the assembly of laminin-5 (α3β3γ2) and laminin-6 (α3β1γ1)which are a specific laminins of skin. Mouse 3T3-L1 adipocytes expressed α4, β1 and γ1 mRNAs suggesting the assembly of only laminin-8 (α4β1γ1). RT-PCR on mRNA extracted from F9 cells gave the fragments for α1, α4, α5, β1 and γ1, and the result remained unchanged by the differentiation of F9 cells into embryonic endoderm-like cells. Since the differentiation of F9 cells mimics the process of the early blastocyst development, laminin-1 (α1β1γ1) , laminin-8 (α4β1γ1) and laminin-11 (α5β1γ1) appear to be the variants expressed at very early stage of the embryogenesis.

REFERENCES

1) Niimi T, Kumagai C, Okano M, Kitagawa Y (1997) Differentiatio-dependent expression of laminin-8 (a4b1g1) mRNAs in mouse 3T3-L1 adipocytes. Mtrix Biol. 16: 223-230

Author index

Subject index

FSC
www.fsc.org
MIX
Papier | Fördert
gute Waldnutzung
FSC® C083411

Kolibri 360 GmbH
Ferdinand-Jühlke-Straße 7
99095 Erfurt, Deutschland
produktsicherheit@kolibri360.de